26th Symposium on Plasma Sciences for Materials

(SPSM26)

Journal of Physics: Conference Series Volume 518

Fukuoka, Japan
23 – 24 September 2013

Editors:

T. Watanabe
M. Sekine
T. Ichiki
M. Shiratani
A. Hatta

ISBN: 978-1-63439-299-0
ISSN: 1742-6588

Printed from e-media with permission by:

Curran Associates, Inc.
57 Morehouse Lane
Red Hook, NY 12571

CURRAN ASSOCIATES INC. proceedings.com

Some format issues inherent in the e-media version may also appear in this print version.

Printed by Curran Associates, Inc. (2014)

Institute of Physics
Dirac House, Temple Back
Bristol BS1 6BE UK

Phone: 44 1 17 929 7481
Fax: 44 1 17 920 0979

techtracking@iop.org

Additional copies of this publication are available from:

Curran Associates, Inc.
57 Morehouse Lane
Red Hook, NY 12571 USA
Phone: 845-758-0400
Fax: 845-758-2634
Email: curran@proceedings.com
Web: www.proceedings.com

TABLE OF CONTENTS

Journal of Physics: Conference Series **518** (2014) 011001 doi:10.1088/1742-6596/518/1/011001

Preface

26th Symposium on Plasma Science for Materials (SPSM-26)

Takayuki Watanabe

The 26th Symposium on Plasma Science for Materials (SPSM-26) was held in Fukuoka, Japan on September 23-24, 2013. SPSM has been held annually since 1988 under the sponsorship of The 153rd Committee on Plasma Materials Science, Japan Society for the Promotion of Science (JSPS). This symposium is one of the major activities of the Committee, which is organized by researchers in academia and industry for the purpose of advancing intersectional scientific information exchange and discussion of science and technology of plasma materials processing.

Plasma processing have attracted extensive attention due to their unique advantages, and it is expected to be utilized for a number of innovative industrial applications such as synthesis of high-quality and high-performance nanomaterials. The advantages of plasmas including high chemical reactivity in accordance with required chemical reactions are beneficial for innovative processing. In recent years, plasma materials processing with reactive plasmas has been extensively employed in the fields of environmental issues and biotechnology.

This conference seeks to bring different scientific communities together to create a forum for discussing the latest developments and issues. The conference provides a platform for the exploration of both fundamental topics and new applications of plasmas by the contacts between science, technology, and industry. The conference was organized in plenary lectures, invited, contributed oral presentations, and poster sessions. At this meeting, we had 142 participants from 10 countries and 104 presentations, including 11 invited presentations. This year, we arranged special topical sessions that cover Plasma Medicine and Biotechnologies, Business and Academia Cooperation, Plasma with Liquids, Plasma Processes for Nanomaterials, together with Basic, Electronics, and Thermal Plasma sessions.

This special issue presents 28 papers that are selected via strict peer-review process from full papers submitted for the proceedings of the conference. The topics range from basic physics and chemistry of plasma processing to a broad variety of materials processing and environmental applications. This proceeding offers an overview on the recent advances in thermal and non-equilibrium plasmas as well as the challenges ahead in the field of plasma research and applications among engineers and scientists.

It is an honor to present this volume of Journal of Physics: Conference Series and we deeply thank the authors for their enthusiastic and high-grade contribution. The editors hope that this proceeding will be useful and helpful for deepening our understanding of science and technology of plasma materials processing and also for stimulating further development of the plasma technology.

Finally, I would like to thank the organizing committee and organizing secretariat of SPSM-26, and the participants of the conference for contribution to a successful and exciting meeting. The conference was chaired by Prof. Masaharu Shiratani, Kyushu University. I would also like to thank the financial support from The 153rd Committee on Plasma Materials Science.

Journal of Physics: Conference Series **518** (2014) 011001 doi:10.1088/1742-6596/518/1/011001

Editors of SPMS-26

Prof Takayuki Watanabe, Kyushu University, Japan
Prof Makoto Sekine, Nagoya University, Japan
Prof Takanori Ichiki, The University of Tokyo, Japan
Prof Masaharu Shiratani, Kyushu University, Japan
Prof Akimitsu Hatta, Kochi University of Technology, Japan

Sponsors and Supporting Organization: The 153rd Committee on Plasma Materials Science, Japan Society for the Promotion of Science

Journal of Physics: Conference Series **518** (2014) 011002 doi:10.1088/1742-6596/518/1/011002

Peer review statement

All papers published in this volume of *Journal of Physics: Conference Series* have been peer reviewed through processes administered by the proceedings Editors. Reviews were conducted by expert referees to the professional and scientific standards expected of a proceedings journal published by IOP Publishing.

Microscopic structure and electrical transport property of sputter-deposited amorphous indium-gallium-zinc oxide semiconductor films

H Yabuta[1], N Kaji[1], M Shimada[1], T Aiba[1], K Takada[1], H Omura[1], T Mukaide[1], I Hirosawa[2], T Koganezawa[2] and H Kumomi[1]

[1] Corporate R&D Headquarters, Canon Inc., 3-30-2 Shimomaruko, Ohta-ku, Tokyo 146-8501, Japan
[2] SPring-8, Japan Synchrotron Radiation Research Institute, 1-1-1 Kouto, Mikazuki-cho, Hyogo 679-5198, Japan

E-mail: yabuta.hisato@canon.co.jp

Abstract. We report on microscopic structures and electrical and optical properties of sputter-deposited amorphous indium-gallium-zinc oxide (a-IGZO) films. From electron microscopy observations and an x-ray small angle scattering analysis, it has been confirmed that the sputtered a-IGZO films consist of a columnar structure. However, krypton gas adsorption measurement revealed that boundaries of the columnar grains are not open-pores. The conductivity of the sputter-deposited a-IGZO films shows a change as large as seven orders of magnitude depending on post-annealing atmosphere; it is increased by N_2-annealing and decreased by O_2-annealing reversibly, at a temperature as low as 300°C. This large variation in conductivity is attributed to thermionic emission of carrier electrons through potential barriers at the grain boundaries, because temperature dependences of the carrier density and the Hall mobility exhibit thermal activation behaviours. The optical band-gap energy of the a-IGZO films changes between before and after annealing, but is independent of the annealing atmosphere, in contrast to the noticeable dependence of conductivity described above. For exploring other possibilities of a-IGZO, we formed multilayer films with an artificial periodic lattice structure consisting of amorphous InO, GaO, and ZnO layers, as an imitation of the layer-structured $InGaZnO_4$ homologous phase. The hall mobility of the multilayer films was almost constant for thicknesses of the constituent layer between 1 and 6 Å, suggesting rather small contribution of lateral two-dimensional conduction It increased with increasing the thickness in the range from 6 to 15 Å, perhaps owing to an enhancement of two-dimensional conduction in InO layers.

1. Introduction

Amorphous In-Ga-Zn-O (a-IGZO) films have attracted much attention since the first report on a thin-film transistors (TFTs) with an a-IGZO channel [1] and a subsequent report on a TFT using a sputter-deposited a-IGZO thin film [2]. As a result, many researchers have improved characteristics of a-IGZO TFTs and consequently demonstrate a high field-effect mobility of about 10 $cm^2V^{-1}s^{-1}$, a small subthreshold swing down to 0.063 $Vdec^{-1}$ [3], a large on-off current ratio greater than 10^{10} with a very low off-current lower than 10^{-14} A [4], excellent short-range uniformity [5], and high frequency operation in ring oscillator circuits [6]. Lately, a lot of electronics companies have launched research and development on amorphous oxide semiconductors and their application to TFTs using them as channel materials, in order to replace amorphous silicon (a-Si) TFTs and low temperature polycrystalline silicon (LTPS) TFTs, which have limitations inhibiting realization of future flat panel displays (e.g., low carrier mobility of a-Si and difficulty in large area formation of LTPS). The companies have eagerly demonstrated prototype electronic devices using a-IGZO TFTs as switching and driving elements for flat-panel displays, including organic light-emitting diode (OLED) displays, liquid crystal displays (LEDs), and flexible displays, which exhibit attractive features such as a high speed, high resolution, and low power-consumption. Nowadays, at an exhibition of home-electronics or a conference on displays, we can witness prototypes of attractive displays which are announced to be released in the near future, such as high-resolution (4k2k) large-sized (over 50 inches) OLED/LCD TVs and full-colour flexible OLED displays using a-IGZO TFTs. More recently, attempts to apply amorphous oxide semiconductors also to other electronic devices have been started; e.g. combining sputter-deposited n-type IGZO TFT technology with sputter-deposited p-type tin oxide TFT technology [7] and exploiting their characters of low-temperature manufacturing, a high-voltage complementary inverter on Si-LSI Cu-interconnects with n-type IGZO and p-type SnO channels has recently been demonstrated [8].

As mentioned above, research on a-IGZO TFTs and their application to electronic devices have been eagerly conducted, but fundamental properties of a-IGZO films have not been so much reported, comparably. For further development of a-IGZO TFTs and their application, it should be important to understand the mechanism of electron conduction in a-IGZO, that is, the reason why this material has larger electron mobility than other amorphous semiconductors. In order to obtain information on the conduction mechanism, dependence of electric properties on thermal treatment conditions should be investigated, because thermal treatment can change the electric conduction properties, as deposition conditions such as sputtering gas composition do [2]. Understanding the effect of post-deposition thermal treatment is also important, in view of fabrication process optimization, because a-IGZO films should experience a lot of thermal processes in a flow of manufacturing a device incorporating them. Here, we present our experimental results on microscopic structures, electrical transport properties and optical properties of sputter deposited a-IGZO films having received post-deposition annealing, which would give a clear prospect of obtaining an insight into the mechanism of electric conduction in a-IGZO.

Further, we introduce an additional topic of a-IGZO: a multilayer artificial lattice fabricated by depositing constituent layers of amorphous InO_x, GaO_x and ZnO_x to imitate the two-dimensional layered structure of crystalline $InGaZnO_4$. Although investigation on artificial lattice films with layers of a conventional amorphous semiconductor material such as a-Si was reported a lot [9,10], there are only a few reports on artificial lattice films consisting of amorphous oxide semiconductor layers. This study on such an artificial lattice was carried out as a challenge of expanding possibility of amorphous oxide semiconductors.

2. Experimental details

The a-IGZO films were deposited on 20×20 mm^2 or 3 inch-diameter substrates by radio frequency (rf) magnetron sputtering using a polycrystalline $InGaZnO_4$ (or other compositions as needed) ceramic target of 3 inch in diameter (Toshima manufacturing Ltd.) in mixed gases of Ar and O_2. We employed two types of sputtering geometries in terms of the substrate to target positional relationship, that is, an off-axis (diagonal) one and an on-axis (face-to-face) one. Films prepared by the on-axis geometry deposition were used in room-temperature Hall effect measurements on 30 nm- and 300 nm-thick films and for optical measurements. Films prepared by the off-axis geometry deposition was used in the other experiments. In the off-axis geometry, a 3-inch diameter substrate holder was positioned to be diagonally opposite to the sputtering target with an off-axis angle of about 35 degrees, where the surface of a substrate placed on the holder was parallel to the target surface and the distance between the target-centre and the substrate-holder-centre was 130 mm. In the on-axis geometry, a substrate was placed in front of the target to face it at a distance of 125 mm. Standard deposition conditions were as follows: an input rf power of 200 W, a sputtering gas pressure of 0.50 Pa, no intentional substrate heating for the both sputtering geometries, and $O_2/(Ar+O_2)$ molar ratio of 5 and 1 mol% for the off-axis geometry and the on-axis geometry, respectively. It has been confirmed that there was no significant difference in electrical properties (conductivity, mobility, etc.) between the a-IGZO films prepared by the off-axis geometry deposition and those by the on-axis geometry deposition when deposited under the standard conditions.

Film compositions were analyzed using an x-ray fluorescence (XRF) apparatus (Rigaku ZSX 100e). In-*Lα*, Ga-*Kα* and Zn-*Kα* emissions exited by x-ray from a rhodium tube were detected, and In:Ga:Zn composition ratios were calculated by the fundamental parameter method. Typical composition ratios of a-IGZO films deposited with the off-axis geometry and the on-axis geometry were In:Ga:Zn of 1:0.9:0.6 and 1:1.1:0.8, respectively. We already confirmed before that the electrical properties of a-IGZO are hardly affected by this extent of fluctuation in composition around the nominal composition 1:1:1, in particular in Zn content [2]. For basic characterization, x-ray diffraction (XRD) and x-ray reflectivity (XRR) were carried out using a parallel beam x-ray diffractometer (Philips X'pert MRD) with a multilayer x-ray mirror and a copper tube. A typical value of the mass density of the sputter-deposited a-IGZO films evaluated by XRR was approximately 6.0 gcm^{-3}.

The deposited films were annealed in a tube-furnace with a fused-silica core-tube (1400 mm in length, 150 mm in diameter) under an N_2 or O_2 gas flow of a few litres per minute. The standard time duration of the annealing in N_2 (N_2-annealing) or O_2 (O_2-annealing) was 1 hour for 30 nm- and 40 nm-thick films, and 10 hours for 300 nm-thick films. Both the N_2- and O_2-annealings were applied to some films in a sequential combination (hereafter, referred to as "N_2-annealing+O_2-annealing"). A sealed-vessel-type tube-furnace was used for H_2 annealing treatment in a mixture of 2% H_2 and 98% He gases after gas-exchange by evacuating and flushing. Some of the a-IGZO films were crystallized by high-temperature annealing at above 700°C in air in a box furnace with molybdenum silicide heaters.

Most of the a-IGZO films used for electrical measurements were deposited on a thermally-grown SiO_2 (100 nm in thickness) layer over a Si substrate or on a glass (Corning #1737) substrate. Highly conductive a-IGZO films were formed on a Si substrate coated with a silicon nitride (SiN_x) layer deposited by plasma-enhanced chemical vapour deposition (PECVD), using an effective diffusion of hydrogen from the PECVD-SiN_x layer into the a-IGZO film.

Cross-sectional micro-structures of the a-IGZO films were observed by transmission electron microscopy (TEM) and scanning transmission electron microscopy (STEM) using a Tecnai F30 (FEI Co.) operating at 300 keV. The TEM specimens were prepared by focused ion-beam (FIB) irradiation followed by low-energy argon ion-milling (1 keV and 300 eV) using a Gentle Mill (Technoorg Linda Ltd. Co.) to remove a damage layer created by the FIB irradiation. Bright-field images were acquired under significantly underfocused conditions to emphasize the Fresnel contrast of grain boundaries in the films. Plan-view images of the film surfaces were observed by scanning electron microscopy (SEM) using an S-5500 (Hitachi High-Technologies Corp.) operating at 30 kV.

Grazing incidence small-angle x-ray scattering (GISAXS) was carried out at the BL46XU beamline of SPring-8. Incident x-rays were monochromatized to 12 keV and set at an angle of 0.15° with respect to the film surface, which was slightly lower than the critical angle of total reflection for the a-IGZO films. Scattered x-ray beams were detected using a scintillation counter scanned in an in-plane angle ($2\theta_\chi$) range from 0.05° to 3.50°. The specimens were held in a helium gas chamber to reduce a background signal generated by air-scattering.

To investigate the local coordination structure in the a-IGZO films, In, Ga, and Zn *K*-edge x-ray absorption fine structure (XAFS) spectra were collected at the BL19B2 beamline of SPring-8. We employed a transmission geometry using specimens of a-IGZO films removed from the substrates and then mixed with boron nitride powders in order to obtain more accurate information than that by the fluorescence method, because a background signal can be readily subtracted from thus obtained XAFS spectra . Polycrystalline $InGaZnO_4$ powders were also measured as a reference.

In order to investigate mesoscopic film structures, the effective surface area was estimated by gas-adsorption surface-area analysis based on the Brunauer-Emmett-Teller (BET) method using krypton (Kr) gas. A sample of a-IGZO film on a thin Si substrate (100 μm in thickness) whose surfaces are both mirror-polished and covered with a thermal SiO_2 layer was cut into small chips and then packed in a sample cell as dense as possible. Each sample cell thus containing approximately 2.5 g of a-IGZO-coated Si substrate chips packed in the cell was baked at 300°C for about 50 hours in vacuum, before introducing Kr gas into the cell at 77 K. Two sets of measurements using different cells were executed for each sample on a-IGZO film to confirm the reproducibility.

The electrical conductivity and the Hall coefficient were measured at room temperature by the van der Pauw method using a Hall effect measurement system (Accent HL5500PC) with a high-resistivity buffer amplifier. The applied voltage between the electrodes was 20 mV and the magnetic field was 0.5 Tesla. Temperature dependences of the electrical conductivity, Hall effect, and magneto-resistance were measured using a physical property measurement system (PPMS, Quantum Design) with a semiconductor characterization system (Keithley 4200-SCS) in a temperature range from 2 to 370 K. The magnetic field for the Hall measurement was 1 Tesla. The magnetic field was swept up to 9 Tesla in the magneto-resistance measurement. For these electrical measurements, stacked electrodes of a 100 nm-thick gold layer on a 5 nm-thick titanium layer were formed on the a-IGZO films by electron-beam evaporation or sputtering thorough a metallic shadow mask.

If either of n- and p-type carriers is dominant in electrical conduction, the carrier density N can be calculated from the Hall coefficient, $R_H = r_H/q_cN$, where q_c is the elementary charge of a carrier and r_H is the Hall factor which is equivalent to unity for degenerate semiconductors and slightly deviates from unity for non-degenerate semiconductors, depending on the carrier energy (type of scattering) [11]. We assume that there is no contribution of holes to electrical conduction in a-IGZO and $r_H = 1$ can be used even for non-degenerate semiconductor a-IGZO films.

To evaluate the optical coefficients of a-IGZO, optical transmission and reflection spectra were measured on a-IGZO films on fused-silica substrate at a wavelength range of 240-2400 nm covering ultraviolet (UV), visible (VIS) and near-infrared (NIR) regions, using a Hitachi U-4100 spectrometer. The optical absorption coefficient, α, was determined by $T_r = \exp(-\alpha d)$, where T_r is the transmittance and d is the film thickness. The transmittance was corrected as $T_r = T_m/(1-R_m)$, where T_m and R_m are the measured transmittance and reflectance of the films, respectively, to eliminate the interface-interference effect which is often observed as fringes in the T_m and R_m spectra in transparent regions [12].

Spectroscopic ellipsometry measurement was carried out on a-IGZO films on heavily-doped silicon substrate coated with a thermal SiO_2 layer (100 nm in thickness), using an ellipsometer with a rotating-compensator instrument (M-2000, J. A. Woollam Co., Inc.) at wavelengths ranging 190-1690 nm. The measured spectra were analysed by minimum mean-square fitting with appropriate optical models.

For fabricating amorphous multilayer thin films, constituent layers of amorphous InO, GaO, ZnO and MgO were deposited at room temperature by pulsed laser deposition (PLD) using a computer-controlled PLD system (Pascal Co. Ltd.). A KrF excimer laser source (Lambda physik) with a wavelength of 248 nm and laser pulse duration of 25 ns was employed as a light source of the PLD. The laser pulse energy and repetition frequency were 50 mJ and 2 Hz, respectively, and focused spot size of the laser onto a target was controlled to 1 mm^2 or less. Commercially available In_2O_3, Ga_2O_3, ZnO and MgO ceramic pellets were employed as the targets. The deposition atmosphere was pure oxygen kept at 1 Pa. Constituent layers were stacked one by one, using the computer-controlled PLD system, where the thickness of each constituent layer was automatically controlled to be an intended value by irradiating a corresponding number of laser pulses determined in advance.

3. Results

3.1. Microscopic structural analyses

3.1.1. Electron microscopy

A cross-sectional TEM image of an 100 nm-thick a-IGZO film on a Si substrate is shown in figure 1. The film looks uniform with no microstructure in the lower region extending from the a-IGZO/Si interface to 30 nm above it. In contrast, in the upper region, a number of bright lines are seen running perpendicularly to the film surface, which can be regarded as boundaries of a columnar structure with a column width of 20 nm. No electron-diffraction spot was observed with a plan-view TEM geometry, which implies that this IGZO film is in an amorphous state even though it has the columnar structure in the upper region. This columnar-structure in the a-IGZO films on Si was clearly observed also by cross-sectional scanning transmission electron microscopy (STEM), as shown in figure 2(a). Simultaneously, electron diffraction data obtained with a nm-size electron beam in the STEM system revealed that this IGZO film is wholly amorphous. As can be seen in figure 2(b), an a-IGZO film on SiO_2 shows an amorphous columnar structure, similar to that on Si.

Figure 1. Bright-field cross-sectional TEM image of a 100 nm-thick a-IGZO film on Si substrate. Carbon and Cr films were deposited on the a-IGZO film to protect and define the surface of the a-IGZO film.

Formation of the columnar structure observed in the upper region of the films is probably related to the film deposition conditions. It is known that sputter-deposited films show a variety of microstructures depending on sputtering-gas pressure, substrate temperature, input power, and so on [13]. It was reported that amorphous silicon films deposited by sputtering, for example, showed a columnar microstructure when deposited under a certain deposition condition, and other structures (uniform, granular, etc.) under different conditions [14]. Referring to these previous studies, it is considered that the deposition conditions employed in this study led to the formation of the uniform structure near the film/substrate interface and the micro-columnar structure in the upper region of the films.

Surface plan-view SEM images of 20 nm-, 40 nm-, and 100 nm-thick a-IGZO films are shown in figures 3(a), (b), and (c), respectively. The contrast seen in these images originates in not only the surface roughness but also inhomogeneity of mass density in the films, since the SEM apparatus employed in this study possesses a higher sensitivity to mass density than conventional SEM apparatuses because of the special geometrical alignment of the electron detector placed in front of the specimen. The surfaces of the a-IGZO films having thicknesses of 40 nm [figure 3(b)] and 100 nm [figure 3(c)] show granular-like-structures with grain sizes of about 20 and 30 nm, respectively, which are comparable to the column width observed in the cross-sectional TEM/STEM images [figures 1 and 2]. Also in the surface SEM image of the 20 nm-thick a-IGZO film [figure 3(a)], which exhibits a lower contrast than the others, we can see a similar granular-like structure on the surface. Although the 20 nm-thick region near the film/substrate interface looks uniform in the cross-sectional TEM/STEM images [figures 1 and 2], a columnar structure could exist even in the 20 nm-thick film as well as the thicker films. The diameter of the columns in the 20 nm-thick film is about 15 nm, which is slightly smaller than that of the 100 nm-thick films (~30 nm) and of the 40 nm-thick films (~20 nm). The grain size seems to increase with increasing film thickness.

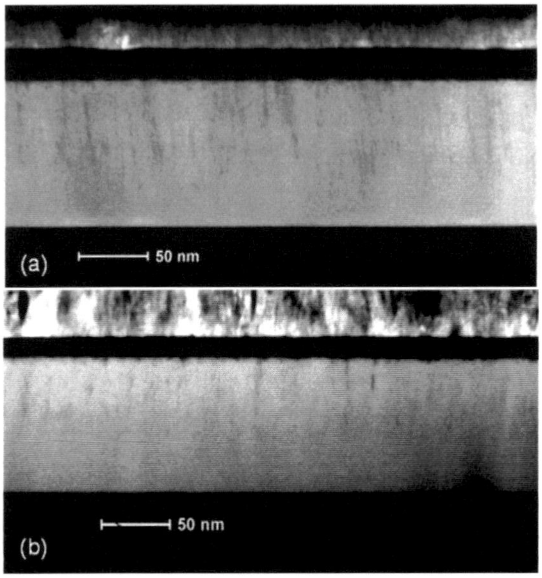

Figure 2. Cross-sectional STEM images of 100 nm-thick a-IGZO films on (a) Si substrate and (b) SiO_2/Si substrate.

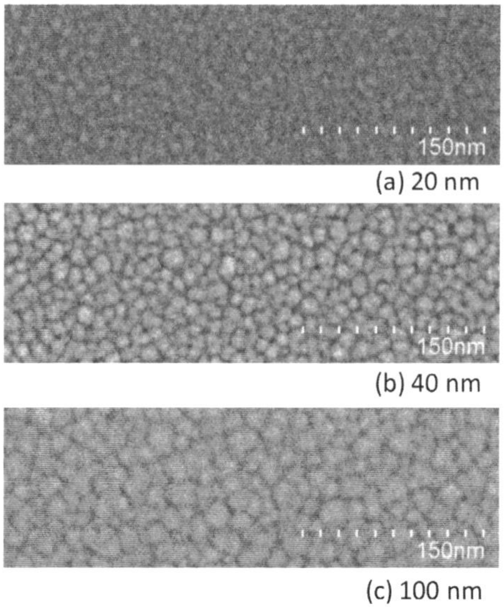

Figure 3. Plan-view SEM images of the surfaces of a-IGZO films with thicknesses of (a) 20 nm, (b) 40 nm and (c) 100 nm. The electron acceleration voltage was 30 kV, which elicited the granular-like structures more clearly than 1 kV. The contrast in the images reflects not only the surface roughness but also a difference in mass density.

3.1.2. In-plane GISAXS

Figure 4 shows in-plane GISAXS profiles of 20 nm-thick a-IGZO films deposited on (a) SiO_2/Si and (b) Si substrates. In the figures, GISAXS profiles of the respective substrates without an a-IGZO films are also shown as references. As can be seen there, the GISAXS profile of the a-IGZO films on SiO_2 has an obvious peak at a scattering vector value, $Q \approx 0.04$ Å^{-1}, and the scattering profile of the a-IGZO films on Si shows significant scattering intensity in the similar range of Q. These GISAXS profiles suggest a lateral correlation between grains in the a-IGZO films with an average in-plane distance of 10 nm between the centres of neighbouring grains. This result is consistent with that obtained from the plan-view SEM image of a 20 nm-thick a-IGZO film [figure 3(a)], which gave an average column width of 15 nm.

It is also found in comparison between the GISAXS profiles of a-IGZO on SiO_2/Si and Si that distribution of the grain distance is dependent on the substrate. The peak of a-IGZO on the Si substrate is broader than on the SiO_2/Si substrate, indicating broader distribution of the grain distance.

3.1.3. X-ray diffraction for high-temperature-annealed IGZO films

The GISAXS results suggest that the sputter-deposited a-IGZO films each contain nano-regions whose density is different from the neighbours. A probable constitution is that with high density grains each surrounded by low density grain-boundaries, which is suggested by the TEM/STEM and SEM observations [figures 1, 2 and 3]. Another possibility is a compositional phase separation into $InGaZnO_4$-like and $In_2Ga_2ZnO_7$-like amorphous phases, because as described in Section 2, the films have a composition of In:Ga:Zn $\approx$ 1:0.9:0.6, which is between $InGaZnO_4$ (In:Ga:Zn = 1:1:1) and $In_2Ga_2ZnO_7$ (1:1:0.5), and there is no solid solution between $InGaZnO_4$ and $In_2Ga_2ZnO_7$ in the crystalline In-Ga-Zn-O system. In order to confirm whether the compositional phase separation took place or not, transient states of the films from amorphous to crystalline ones were checked by means of x-ray diffraction.

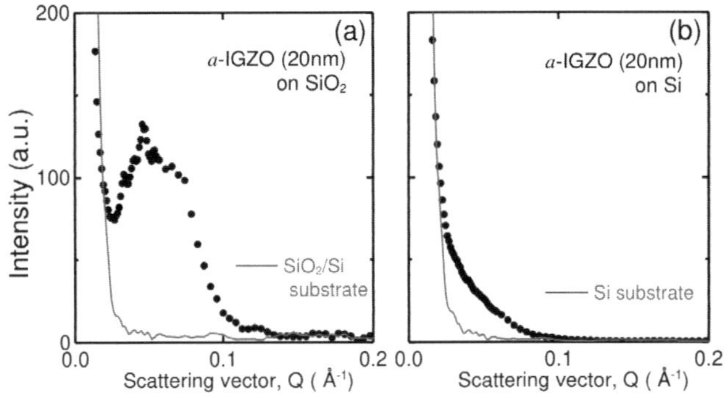

Figure 4. GISAXS profiles of 20 nm-thick a-IGZO films on (a) SiO_2/Si and (b) Si substrates. GISAXS profiles of the substrates only are also shown with solid lines.

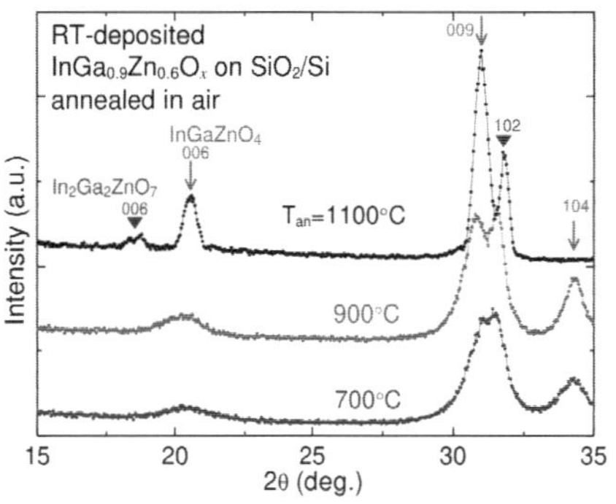

Figure 5. X-ray diffraction patterns of polycrystalline $InGa_{0.9}Zn_{0.6}O_x$ films on SiO_2/Si deposited at room temperature and annealed in air at (T_{an} =) 700, 900 and 1100°C. The diffraction patterns were measured by ω-2θ scan method. Arrows and inverse triangles indicate peaks of $InGaZnO_4$ and $In_2Ga_2ZnO_7$, respectively, with diffraction indices. For the 1100°C-annealed film, the peak at 34° is absent because the *c*-axis of $InGaZnO_4$ crystallites in the film are oriented normal to the substrate surface.

Figure 5 shows the x-ray diffraction patterns of IGZO films crystallized by post-deposition annealing in air at 700, 900 and 1100°C. The IGZO film annealed at 1100°C showed the phase separation into $InGaZnO_4$ and $In_2Ga_2ZnO_7$. According to the rocking curve (ω-scan) measurement, the $InGaZnO_4$ phase was *c*-axis oriented, while $In_2Ga_2ZnO_7$ phase was randomly oriented. On the other hand, the IGZO films annealed at 700°C and 900°C showed broad diffraction peaks at angles between the diffraction peaks of $InGaZnO_4$ and $In_2Ga_2ZnO_7$. The peak at around 31° for the 900°C annealed film indicated the onset of the phase separation. We suppose that the broad peaks for the 700°C-annealed IGZO film come from an intermediate phase between $InGaZnO_4$ and $In_2Ga_2ZnO_7$, which is probably a non-equilibrium phase having a layered structure similar to $InGaZnO_4$ or $In_2Ga_2ZnO_7$ with stacking faults, thus showing a lattice constant between those of $InGaZnO_4$ and $In_2Ga_2ZnO_7$. Since a temperature higher than 900°C is necessary to complete the phase separation into $InGaZnO_4$ and $In_2Ga_2ZnO_7$ phases, the GISAXS results are not due to such a kind of phase separation in the a-IGZO films with an intermediate composition.

3.1.4. Kr gas adsorption surface area analysis

O_2 or N_2 atmosphere affects the conductivity of a-IGZO films when they undergo annealing at a low temperature of 300°C or below, as will be described later. It was already reported that the carrier density of a-IGZO films and their TFT characteristics significantly change when exposed to an O_2 atmosphere [15]. Some polycrystalline oxide-semiconductor films used for gas sensors, such as SnO_2 films, change their conductivity by adsorbing gas molecules at the grain boundaries in the films [16,17], and accordingly the sidewall surface area of grains in such films can be evaluated by the Kr gas adsorption technique [18]. If the column (or grain) boundaries of a-IGZO films observed in the cross-sectional TEM/STEM images (figures 1 and 2) were fine pores piercing the film from the surface, gas molecules could easily penetrate the films through the pores and react to form or dissipate potential barriers at the boundaries, thus resulting in wide variation of conductivity, as observed on polycrystalline oxide-semiconductor gas-sensor films and in hydrogen passivation of grain boundaries in polycrystalline Si films. The gas penetration and reaction should be enhanced by thermal treatment.

To investigate whether the column boundaries are such pores or not, Kr-gas-adsorption surface-area analysis was carried out on 40 nm- and 400 nm-thick a-IGZO films. The results are summarized in table 1. No inner-surface area is detectable for the 40 nm-thick a-IGZO film, since the measured surface area of the film is exactly the same as its planar dimension. On the other hand, the measured surface area of the 400 nm-thick a-IGZO film is about 50% larger than its planar dimension. According to the following analysis, this difference between the surface area and the planar dimension is probably due to the surface roughness of the 400 nm-thick film, and there should exist no such pores at the column boundaries.

Table 1. Surface areas of a-IGZO films of 40 and 400 nm in thickness, estimated by Kr-gas-adsorption surface-area analysis with the BET method. Planar dimensions of the samples are also shown.

	surface area of film (m^2)*	planar dimension (m^2)
40 nm-thick a-IGZO film	0.011	0.011
400 nm-thick a-IGZO film	0.016	0.011

*Surface area of film was derived by subtracting the surface area of the sample cell from the measured gross surface area and then subtracting the area of back surface of SiO_2/Si substrate equal to the planar dimension of sample.

From the cross-sectional TEM image shown in figure 1, the surface roughness of the 100 nm-thick a-IGZO film is estimated to be 2-3 nm. The surface roughness could increase to 10 nm in the 400 nm-thick a-IGZO film, because the surface roughness usually increases monotonously with increasing film thickness. Assuming a simplified geometrical model for surface roughness estimation in which pyramids of 10 nm-height with a 20×20 nm^2 basal plane are laterally aligned, its total surface area is estimated to be approximately 1.4 times as large as its planar dimension. On the other hand, for the case where cylinders of 20 nm diameter are arranged into a square lattice, the total surface area of the columns' sidewalls should be much larger than the planar dimension (about 60 times as large as the planar dimension). Therefore, the adsorption on the inner-surfaces of the columns is not likely to be the cause of the observed difference of only 50%. Thus, the 50% increase of the measured surface area compared to the planar dimension should be explained as the effect of the surface roughness. We conclude, therefore, that the column boundaries observed in the cross-sectional TEM images are not pores through which gas molecules easily penetrate into the film. Variation in the conductivity of the a-IGZO films shown later in figure 6 is not attributed to the same mechanism as that dominating in the polycrystalline oxide semiconductor films used for O_2-gas sensors, such as SnO_2.

3.1.5. XAFS

In radial structural functions (RSFs) of a-IGZO films obtained from their XAFS spectra of In-*K*, Ga-*K* and Zn-*K* edges, obvious peaks of the nearest-neighbour M-O (M = In, Ga or Zn) pairs were observed, but no peak of the next-nearest-neighbour M-M pairs appeared, while both kinds of peaks were clearly seen in RSFs of crystalline $InGaZnO_4$ powders. These results confirm that the a-IGZO films have a disordered atomic arrangement unlike the crystalline IGZO.

The coordination numbers, each indicating the number of nearest-neighbour oxygen atoms around a metal atom, and the distance between nearest-neighbour M and O were estimated from each of the RSFs. Table 2 collects the first-shell fitting results on a sputter-deposited 20 nm-thick a-IGZO film using the ideal values calculated from the structural data on crystalline $InGaZnO_4$ [19]. Almost the same results as the above were obtained for 100 nm-thick a-IGZO films formed with various conditions (e.g., no annealing or 300°C-annealing after room temperature deposition, and no annealing after 240°C-deposition). Radial distribution functions of 20 nm- and 100 nm-thick a-IGZO films formed with the same conditions as the above were also estimated by the grazing incident x-ray scattering and conventional x-ray scattering, using film-samples removed from the substrates. The results with respect to short-range inter-atomic distances are consistent with the RSFs obtained by the XAFS measurement.

Table 2. First shell fitting results on In-*K*, Ga-*K* and Zn-*K* edges EXAFS spectra of a sputter-deposited 20 nm-thick a-IGZO film and of commercial crystalline $InGaZnO_4$ powders, along with the ideal values of crystalline $InGaZnO_4$. CN is the coordination number of nearest-neighbour oxygen atoms around In, Ga or Zn, and *r* is the distance between the nearest-neighbour oxygen atoms and In, Ga or Zn..

	a-IGZO		crystalline $InGaZnO_4$		ideal*	
	CN	*r*(Å)	CN	*r*(Å)	CN	*r*(Å)
In-O	4.6	2.14	6.0	2.18	6	2.18
Ga-O	4.3	1.87	4.5	1.88	(Ga,Zn)-O 3	 1.93
Zn-O	4.3	2.01	4.1	1.99	1 1	1.97 2.32

* Calculated from the crystal structure data of an $InGaZnO_4$ single crystal. Ga and Zn occupy the same site randomly, and this (Ga, Zn) site is coordinated by five oxygen atoms [19].

It should be noted that the coordination number (CN) of In-O in the a-IGZO films is remarkably smaller than that of the crystalline $InGaZnO_4$, differently from the result reported by Cho *et al* [20]. Furthermore, the difference in the inter-atomic distance (*r*) of In-O between the a-IGZO film and the crystalline IGZO is comparably larger than those of Ga-O and Zn-O. These suggests that InO_6 octahedrons in the a-IGZO film have an imperfect structure with a considerable number of oxygen vacancies.

3.2. Electrical transport properties

3.2.1. Conductivity variation with annealing conditions

The electrical conductivity of a-IGZO films is known to be able to be controlled by some deposition conditions such as the O_2 partial pressure in sputtering atmosphere [2] and the substrate temperature. It is noteworthy that the conductivity of a-IGZO films widely change when annealed **after** their deposition and can be controlled by conditions of the annealing temperature and atmosphere. Figures 6(a) and (b) show annealing temperature dependences of the carrier density (*N*) and those of the conductivity (σ), respectively. When the films were annealed in N_2 ambient, the carrier density drastically increased with increasing the annealing temperature up to nearly 300°C [closed circles in both figures 6(a) and (b)]. For example, an increase in the carrier density of five orders of magnitude was observed after N_2-annealing at 300°C on 30 nm-thick a-IGZO films on glass, and that of nearly seven orders of magnitude was observed on the 40 nm-thick a-IGZO/SiO_2/Si samples [open circles in figure 6(b)]. By contrast, the carrier density kept as low as 10^{13} cm^{-3} after annealing at 300°C in an O_2 ambient [open triangle in figure 6(a)]. Furthermore, the O_2-annealing after N_2-annealing reduced the carrier density of the N_2-annealed films [closed rhombus symbols in figure 6(a)] to almost the same value as that of the film O_2-annealed at 300°C or of the non-annealed (as-deposited) films, when the temperature of O_2-annealing was higher than 200°C. On the other hand, addition of N_2-annealing increased the carrier density of the O_2-annealed films to the state of having values of and temperature dependence of carrier density both similar to those measured after the N_2-annealing without pre-annealing in O_2 [inverse triangle symbols in figure 6(b)]. These results suggest that the carrier density of the a-IGZO films can reversibly change such that it is increased by N_2-annealing and decreased by

O_2-annealing at relatively low temperatures of ≤ 300°C. This is similar to an already reported reversible conductivity change of crystalline $Ti_{0.94}Nb_{0.06}O_{2+d}$ films by O_2- and H_2-annealing at 400°C [21]. The extent of the conductivity change of the a-IGZO films was remarkably wide (5-7 orders of magnitude). To our knowledge, there are few materials that show such a wide-range reversible change of carrier density by annealing at such low temperatures as below 300°C, while irreversible changes are often observed in crystallization, structural reorientation (relaxation), or termination of defects (e.g. oxygen vacancy) by low temperature annealing.

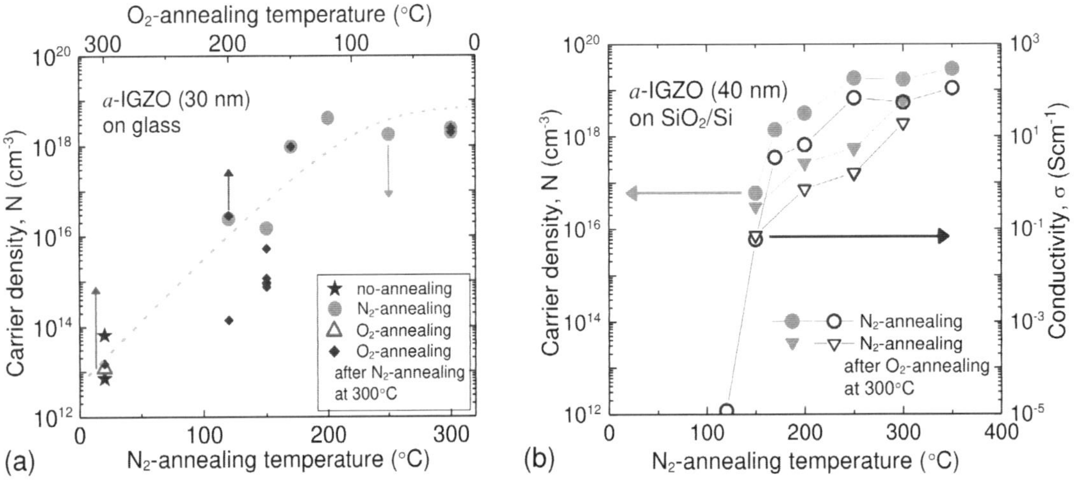

Figure 6. (a) Dependence of carrier density on annealing conditions measured on sputter-deposited a-IGZO films (30 nm in thickness, on glass substrate): [star] with no annealing treatment (as deposited), [circle] with N_2-annealing at various temperatures, [triangle] with O_2-annealing at 300°C, and [rhombus] with O_2-annealing at various temperatures after N_2-annealing at 300°C. A dotted line is indicated as a guide for eyes. (b) Dependences of carrier density [closed symbols] and conductivity [open symbols] on annealing conditions measured on sputter-deposited a-IGZO films (40 nm in thickness, on SiO_2/Si substrate): [circle] with N_2-annealing at various temperatures, [inverse triangle] with N_2-annealing at various temperatures after O_2-annealing at 300°C.

In order to investigate the dynamics of the above-described conductivity change by annealing, we studied thickness dependence of conductance on N_2-annealed a-IGZO films with various annealing time durations. Figure 7 shows thickness dependences of sheet conductance for the films annealed at 300°C in N_2 for 1 hour and 10 hours. For both annealing durations, the sheet conductance linearly increases with the film thickness in a small thickness region, but deviates from the linearity and turns into a state of decreasing with the thickness at 200 nm in the 1 hour annealing case and at 400 nm in the 10 hour annealing case. These observations suggest that some substances which generate conduction carriers like dopants could be created at the film surface by the N_2-annealing, and diffuse into the films with time. A generator of the conduction carriers would not be an injected nitrogen, because annealing in vacuum also increased the conductivity of a-IGZO films, and instead, it would something created by reduction of oxygen such as oxygen deficiency, which works as a donor, or that by removal of excess oxygen as a compensated acceptor.

It is also observed in figure 7 that linear extrapolation of the sheet conductance to zero thickness does not pass through the origin but does a definitely positive intercept (23 nm) at the zero conductance. This observation suggests the existence of a low conductivity layer of about 20 nm in thickness in each of the a-IGZO films. The supposed low-conductivity layer could be a depletion layer generated just above the a-IGZO/SiO_2 interface. Yamaguchi *et al* estimated such a depletion layer to be 65 nm in thickness [22]. However, it is not likely that the thickness of the depletion layer is more than 40 nm in the films of the present study, because the thinnest (40 nm-thick) films in figure 7 were still conductive after annealing.

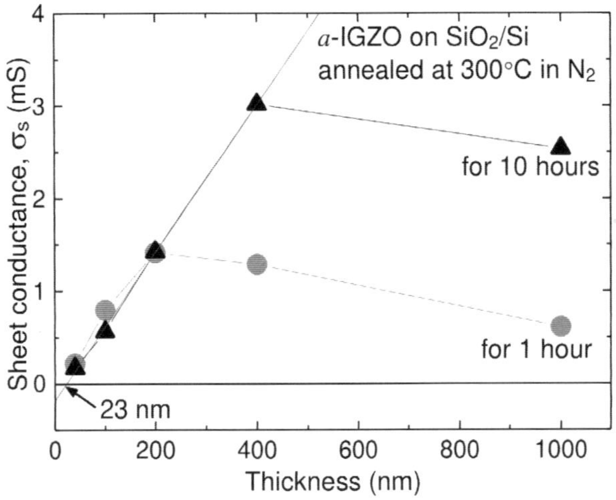

Figure 7. Thickness dependences of the sheet conductance of a-IGZO films on SiO_2/Si substrate annealed at 300°C in N_2 for 1 hour and 10 hours.

3.2.2. Relation between mobility and carrier density

A simple relation between the Hall mobility (μ) and the carrier density (N) with monotonous increase of μ with increasing N is often observed on amorphous oxide semiconductor films such as of In-Zn-O (a-IZO) with $N \leq 10^{20}$ cm^{-3} [23-25], PLD-deposited a-IGZO [26,27] and Zn-Sn-O (a-ZTO) [28]. The μ-N relations of these materials each can be approximated by a power law of $\mu \propto N^y$, which is observed as a linear relation between logμ and log$N$. Some models have been proposed for explaining the mechanism of the monotonous increase of μ with N. Nomura *et al* gave a basic view based on scattering of carriers by potential barriers attributed to randomness of atomic positions, in which the potential barrier height relatively lowers when the Fermi level rises with increase of the carrier density, resulting in an increase of the mobility [1,29]. Recently, Kamiya *et al* [30,31] formalized this view using an idea of a percolation model in disordered systems proposed by Adler *et al* [32]. On the other hand, Martins *et al* [23] interpreted this phenomenon with carrier scattering by defects, like scattering by dislocations in crystalline semiconductors [33].

Figure 8 shows μ-N relations of sputter-deposited a-IGZO films of various thicknesses and preparing conditions. Closed triangle symbols represent data on 300 nm-thick films formed with various O_2 contents in the sputtering gas and are replotted in terms of O_2 content in the inset. While most of the films were deposited on glass substrates, some films deposited in 1% O_2/99% Ar gas (hereafter referred to as "1% O_2 gas") were deposited on Si substrates coated with a PECVD-SiN_x film (200-300 nm in thickness), to obtain highly conductive a-IGZO films. Hydrogen molecules or atoms incorporated in the PECVD-SiN_x film diffuse into the a-IGZO film on the SiN_x, particularly during annealing even at low temperatures such as 300°C, and thus drastically increase the conductivity of the a-IGZO film. It is predicted that the increase of conductivity by hydrogen incorporation is caused by not only reduction of oxygen but also a chemical doping effect of hydrogen atoms placed at vacant sites [34] or intersites [35,36] in a-IGZO. These a-IGZO films on SiN_x/Si substrates were annealed at 300°C in N_2, while those on glass substrate were annealed at 300°C in N_2 (N_2-annealing), in O_2 (O_2-annealing), or in O_2 after N_2-annealing (N_2-annealing+O_2-annealing). Similarly to the amorphous oxide semiconductors described in the above-mentioned previous reports, μ of the 300 nm-thick a-IGZO films monotonously increase with N, and seems to obey a power law of $\mu \propto N^y$. The index y was estimated to be 0.045, which is smaller than the values 0.2-0.5 estimated for other amorphous oxide semiconductors [23,25,28].

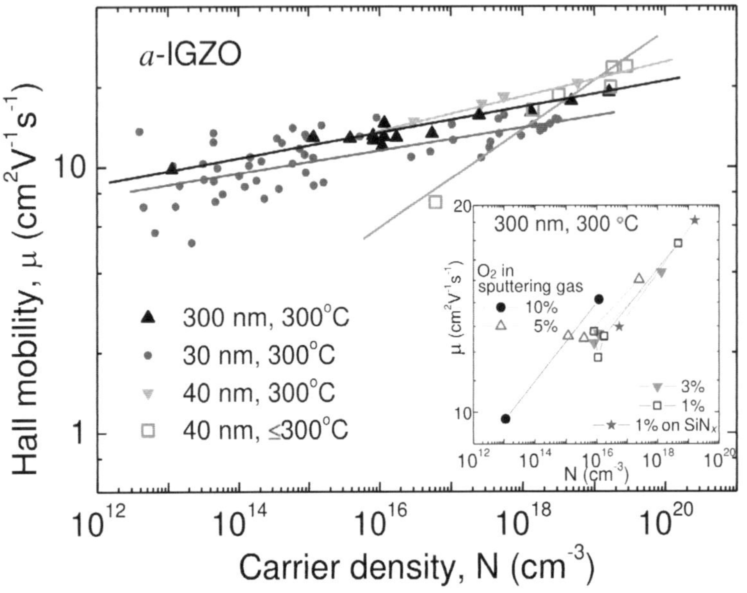

Figure 8. Relation between the carrier density N and Hall mobility μ of sputter-deposited a-IGZO films with various thicknesses, substrates, and annealing conditions: [triangle] 300 nm-thick a-IGZO films formed in a sputtering gas containing 1, 3, 5, or 10 % O_2 on glass substrate or PECVD-SiN_x/Si substrate, and treated with O_2- or N_2-annealing or N_2-annealing+O_2-annealing at 300°C. The data are also plotted in the inset in terms of O_2 content. Among the films deposited with the same sputtering condition, the carrier density of films with N_2-annealing is larger than the others, and those with O_2-annealing and N_2-annealing+O_2-annealing are comparable, as notices in figure 6(a). [dot] 30 nm-thick a-IGZO films on glass substrate with various sputtering powers (50-400 W), sputtering pressures (0.1-3.0 Pa) and O_2 contents in sputtering gas (1-10%). These films were treated with O_2- or N_2-annealing at 300°C or O_2-annealing at various temperatures after N_2-annealing at 300°C, [inverse triangle] 40 nm-thick a-IGZO films on SiO_2/Si substrate annealed in N_2 at various temperatures after O_2-annealing at 300°C, [open square] 40 nm-thick a-IGZO films on SiO_2/Si substrate annealed once in N_2 at various temperatures.

Closed-dot symbols in the main graph of figure 8 represent data on sputter-deposited 30 nm-thick films on glass substrate. The carrier density of these films were controlled by adjusting the sputtering power (50-400 W), total sputtering pressure (0.1-3.0 Pa), O_2 content in the sputtering gas (1-10%) at their depositions. All the films were annealed at 300°C in O_2 or N_2, and some of them were annealed in O_2 at various temperatures after the N_2-annealing at 300°C, also to change the carrier density. As seen in figure 8, no apparent dependence of the logμ-log$N$ slopes on the deposition conditions (power, pressure and O_2 content) was observed. All the data, regardless of the sputtering conditions, seem to obey a single power-law relation with some degree of fluctuation, and its index ($y \approx 0.042$) is almost the same as that of the data on the 300 nm-thick films. Also 40 nm-thick films on SiO_2/Si annealed in N_2 at various temperatures after O_2-annealing at 300°C show a similar μ-N relation [inverse triangles]. However, 40 nm-thick films on SiO_2/Si substrate annealed only once in N_2 at 300°C or various temperatures lower than 300°C [open squares] exhibit a steeper slope of logμ-log$N$ ($y \approx 0.18$) than the other films. We suppose that the index in the power law μ-N relation depends on the annealing temperature, where lower annealing temperature make Hall mobility of a-IGZO lower than that expected from its carrier density.

3.2.3. Temperature dependences of electrical conductivity, carrier density and Hall mobility

Temperature dependences of electric transport properties of a-IGZO films were investigated to achieve a clues to the conduction mechanism. Figure 9 shows the conductivity (σ) of 40 nm-thick films with various carrier densities in a range from 1×10^{16} to 3×10^{19} cm^{-3} at room temperature, as a function of inverse temperature (T^{-1}). Table 3 shows a list of substrates, annealing conditions, and carrier density values at room temperature for the samples labelled as A to F in figure 9. The conductivity of every a-IGZO film in figure 9 increases with increasing temperature, and temperature dependence of the conductivity becomes weaker as the carrier density increases.

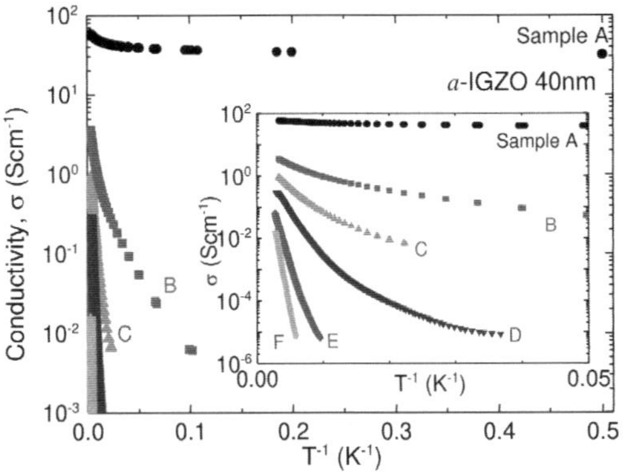

Figure 9. Dependence of the conductivity σ on inverse temperature T^{-1} for 40 nm-thick a-IGZO films (sample details are shown in table 3). The inset shows a detailed view above 20 K.

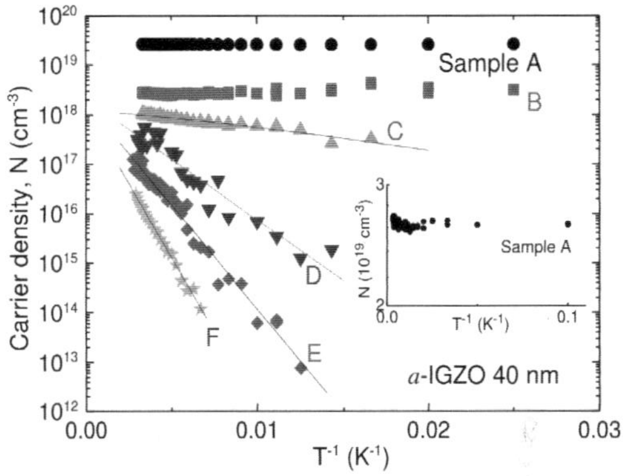

Figure 10. Dependence of the carrier density N on T^{-1} for 40 nm-thick a-IGZO films (sample details are shown in table 3). The solid lines are fitted lines according to the Arrhenius-type thermal activation. The inset shows a view in a wider temperature range for the data of Sample A.

In general, temperature dependence of conductivity in disordered semiconductors exhibits various behaviours originating from different conduction mechanisms according to temperature range. As temperature decreases, the band conduction at high temperature ($\sigma \propto \exp(-T^{-1})$) is followed by the nearest-neighbour hopping (impurity-band) conduction ($\sigma \propto \exp(-T^{-1})$ with a gentler slope than the band conduction), and finally by variable range hopping (VRH) conduction at low temperature ($\sigma \propto \exp(-T^{1/q})$ where q = 2, 3 and 4 for one, two and three dimensional (1D, 2D and 3D) systems, respectively [37]. In figure 9, the conductivity looks proportional to $\exp(-T^{-1})$ at high temperatures of hundreds Kelvin, and deviates from the relation at low temperatures. However, it seems that the conduction of these samples at low temperatures cannot always be attributed to 3D-VRH. Probably, the VRH conduction of Coulomb-gap system or/and other conduction mechanisms contribute to this behaviour.

Table 3. Substrates, annealing conditions, and values of the carrier density N at room temperature of the respective a-IGZO samples used in figures 9, 10, and 12.

Sample	Substrate	Annealing conditions	N at 300 K (cm^{-3})
A	SiN_x/Si	300°C in N_2	2.7 x 10^{19}
B	Glass	300°C in 2% D_2/98% Ar	2.8 x 10^{18}
C	Glass	300°C in N_2	1.1 x 10^{18}
D	Glass	300°C in N_2 + 150°C in O_2	2.3 x 10^{17}
E	Glass	300°C in N_2 + 170°C in O_2	5.6 x 10^{16}
F	Glass	300°C in N_2 + 200°C in O_2	1.3 x 10^{16}

Figure 10 shows dependence of the carrier density (N) on inverse temperature (T^{-1}). While the carrier densities of Samples A and B are independent of temperature, indicating that the samples are degenerated, those of Samples C, D, E, and F (having carrier density values lower than Samples A and B) show an Arrhenius-type thermal activation being proportional to $\exp(-T^{-1})$. Temperature dependence of carrier density is expressed with the donor density N_d by

$$\frac{N^2}{N_d - N} = \frac{N_c}{2}\exp\left(-\frac{E_d}{k_B T}\right) \tag{1}$$

where k_B is the Boltzmann constant, E_d the the ionization energy of the donor (i.e., energy difference between the donor state and the conduction band minimum) and N_c the effective density-of-state of conduction band. If $N_d >> N$, equation (1) can be approximated as

$$N \approx \sqrt{\frac{N_c N_d}{2}}\exp\left(-\frac{E_d}{2k_B T}\right) \tag{2}$$

E_d and N_d of Sample C are estimated by using equation (1) because N_d could not be so much larger than N, and those of Sample D, E, and F are by equation (2). It is assumed here that $N_c = 2(m^* k_B T/2\pi\hbar^2)^{3/2}$ (where m^* is the effective mass of electron and $\hbar$ is equal to Planck's constant h divided by 2π) does not depend on temperature because the temperature dependence of N_c should be weaker than $\exp(-E_d/k_B T)$ at high temperatures. Thus, we employed a value of N_c at 300 K along with $m^* = 0.4m_0$ (m_0 : the free electron mass) to calculate E_d and N_d from the fitted lines drawn in figure 10. The relation between thus estimated E_d and N_d values are shown in figure 11(a), where $N = N_d$ is assumed for the degenerate samples (Samples A and B) as often used for transparent conductive oxide (TCO) films such as ITO [38-40]. This figure indicates that Samples D, E, and F have somewhat deep donors, and E_d of these decreases with increasing N_d. This lowering of E_d is probably attributed to an electrostatic shielding effect of electrons loosely bound by impurity ions.

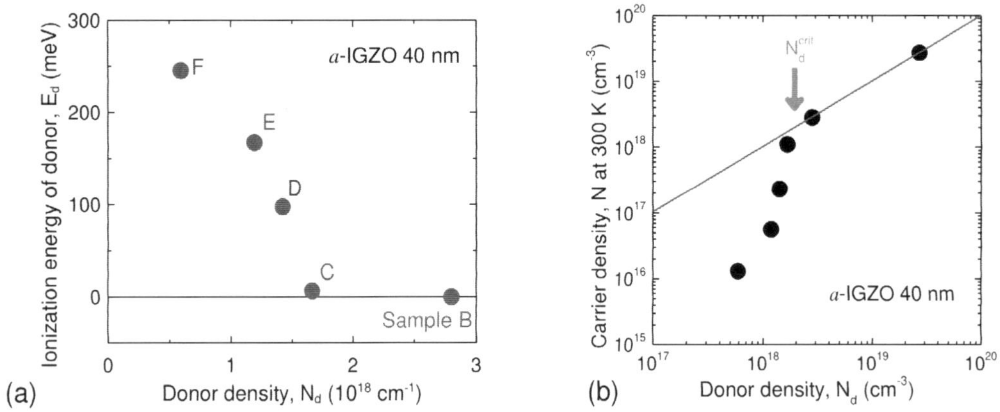

Figure 11. (a) Dependence of the donor ionization energy E_d on the donor density N_d, both calculated from the data shown in figure 10. (b) Relation between carrier density N at 300 K and donor density N_d obtained from the data in figure 10. A solid line indicates the relation of $N = N_d$. An arrow indicates the point of $N_d = N_d^{crit}$ above which a-IGZO is degenerate.

Figure 11(b) shows a relation between the carrier density (N) measured at 300 K and the estimated donor density (N_d). The solid straight line in the figure indicates the relation of $N = N_d$ for the degenerate samples. For the non-degenerate samples having $N_d < 2\times10^{18}$ cm^{-3} ($\approx N_d^{crit}$), deviation from the line of $N = N_d$ increases as N_d decreases. It is worth noting that the decrease of N_d from 2×10^{18} cm^{-3} to 6×10^{17} cm^{-3} (i.e., N_d changes only by 1/4 times) results in a large reduction of N by almost two orders of magnitude. This is probably the reason why the conductivity of a-IGZO films is changed so easily and widely by low-temperature annealing, as shown in figure 6.

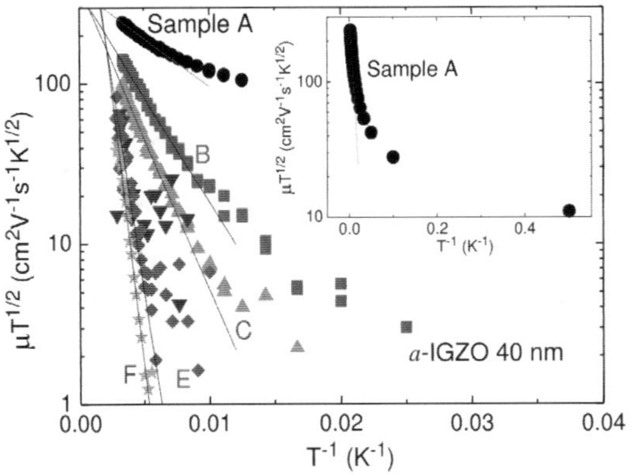

Figure 12. Temperature dependence of the Hall mobility μ, represented by a relation between $\mu T^{1/2}$ and T^{1}. The solid lines are fitted lines using equation (3). Data on Sample D was excluded from the thermal activation analysis because it is too much scattered to fit. The inset shows a wide temperature-range view for the data on Sample A.

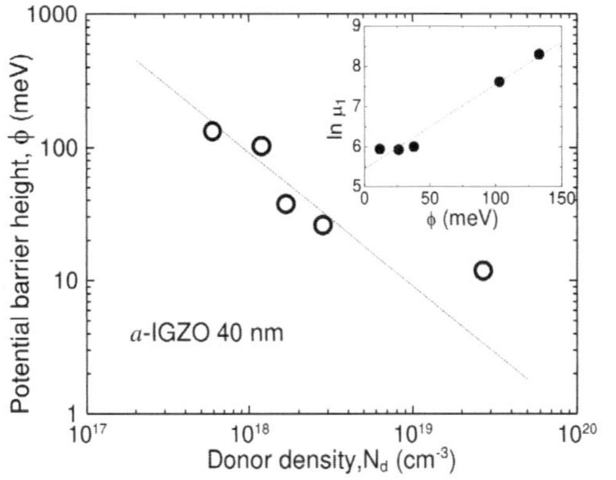

Figure 13. Potential barrier height ϕ calculated from the temperature dependence of Hall mobility shown in figure 12 using equation (3), as a function of donor density N_d estimated from the data of figure 10. The inset shows dependence of the prefactor μ_1 of equation (3) on ϕ, to verify the relation expressed by equation (5).

Since a granular-like structure has been observed by TEM/STEM and SEM in this study, we analyzed the temperature dependence of Hall mobility, based on a hypothesis that the carrier electrons pass through the potential barriers at grain-boundaries by thermionic emission (grain-boundary scattering model) in the a-IGZO films. It has been reported that conductors dominated by the grain-boundary scattering usually show thermal activation of carrier mobility similar to the Arrhenius-type [41,42], which is expressed as

$$\mu = \mu_0 \exp(-\phi/k_B T) = \mu_1 T^{-1/2} \exp(-\phi/k_B T), \quad (3)$$

where ϕ is the potential barrier height (in the dimension of energy) at grain boundaries and μ_1 a pre-factor. Figure 12 shows application of this model to the Hall mobility data, in which the relation between $\log \mu T^{1/2}$ and T^{1} can be fitted with straight lines at high temperatures. The deviation from the straight lines at low temperatures could be due to tunneling conduction through the potential barriers at the grain boundaries [42], and/or distribution of the potential barrier height ϕ at the grain boundaries [43]. In figure 13, values of the potential barrier height ϕ calculated from the fitting result in figure 12 using equation (3) are plotted as a function of values of the donor density N_d estimated from the result in figure 10 and shown in figure 11. According to Seto [41], the potential barrier height at grain boundaries can be expressed with the donor density N_d and the grain-boundary trap density Q_t as

$$\phi = e^2 Q_t^2/(8\varepsilon_r \varepsilon_0 N_d), \quad (4)$$

under partially-depleted conditions (i.e., the grain size is larger than Q_t/N_d). The solid line in figure 13 is a theoretical curve calculated by equation (4) with $\varepsilon_r = 12$ [44] and $Q_t = 2.2\times10^{12}$ cm^{-2}, which seems to represent well the estimated dependence using figure 12 and equation (3). A possible reason of the somewhat large deviation at $N_d = 3\times10^{19}$ cm^{-3} for Sample A will be discussed in the next subsection. The value of $Q_t = 2.2\times10^{12}$ cm^{-2} is comparable to those reported on other polycrystalline semiconductor films, e.g., 3.0-3.6$\times10^{12}$ cm^{-2} for silicon [41], 0.5-3.3$\times10^{12}$ cm^{-2} for InP [45], and 6.7$\times10^{12}$ cm^{-2} for epitaxial ZnO [46]. For semiconductors with grain-boundary scattering, the pre-factor μ_1 in equation (3) is often considered to be dependent on ϕ as

$$\mu_1 = \mu_2 \exp(\gamma\phi), \quad (5)$$

where the pre-factor μ_2 is a constant and γ is a coefficient. Orton *et al.* [47] derived equation (5) by assuming a linear temperature dependence of actual potential barrier height, $\phi_b = (1-\gamma k_B T)\phi$, as in Si-bycrystals [48], and by substituting ϕ_b for ϕ in equation (3). A dependence of μ_1 on ϕ, which was evaluated from the fitting results on Hall mobility in figure 12 using equation (3), is shown in the inset of figure 13. It seems to obey equation (5) indicated by the straight line in the inset. Values of γ and μ_2 are estimated to be 21 eV^{-1} and 230 cm^{2}V^{-1}s^{-1}, respectively. This γ value is close to 25 eV^{-1} for polycrystalline CdS films [47].

From the above-described results of the analyses based on the hypothesis of a granular-like structure found in TEM/STEM, SEM, and GISAXS observations, the behaviour of Hall mobility in sputter-deposited a-IGZO films can be explained by the grain boundary scattering model, as well as other proposed models.

3.2.4. Anderson localization

Weak localization behaviour of amorphous oxide semiconductors was discussed by Shimakawa *et al.* in their study on electronic transport property of amorphous 2CdO·1GeO$_2$ and 2CdO·PbO$_2$ films [49]. Recently, it has been reported that a-IZO films with a large carrier density behave as metallic films and show the weak localization behaviour [50]. We also analysed the electrical conduction of the sputter-deposited a-IGZO films from the viewpoint of the Anderson localization. For Sample A having the largest carrier density in this study, the temperature dependence of conductivity shown in figure 9 is transformed into figure 14, where the conductivity σ are plotted as a function of $T^{1/2}$. According to the weak localization theory, a three-dimensional (3D) weak localization system shows $\sigma \propto T^{1/2}$, and the conductivity of Sample A obeys this law in the range of 10 to 140 K. This suggests that 3D weak localization occurs in this temperature range. Below 10 K, σ deviates from the $T^{1/2}$ law, and instead, as shown in the inset, the sheet conductance σ_s obeys the relation concerning 2D weak localization expressed by $\sigma_s = \sigma_0 + (e^2/2\pi^2\hbar)p_T\ln T$ with a universal parameter $p_T \sim 1$ [51]. This crossover between 3D and 2D states in a weak localization system could occur when the inelastic scattering length (L_{in}) increases at low temperatures to be comparable to the film thickness.

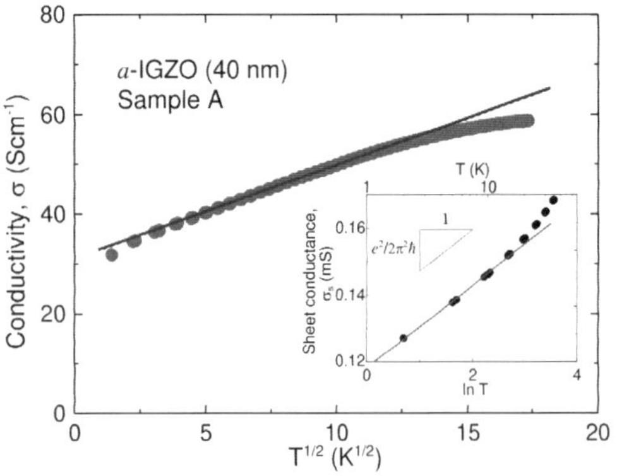

Figure 14. Plot of conductivity σ as a function of $T^{1/2}$. The fitted line indicates a linear relation between σ and $T^{1/2}$ (3D weak localization behaviour) in the range of 10 to 140 K. The inset shows sheet conductance σ_s as a function of ln$T$. The gradient of the fitted line at low temperatures is $e^2/2\pi^2\hbar$, indicating 2D weak localization behaviour.

One of the characteristics of weak localization systems is to exhibit negative magneto-resistance even without any magnetic ion, in contrast to that common non-magnetic systems usually show positive magneto-resistance. Figure 15(a) shows magnetic-field (H) dependence of the magneto-conductance (inverse of magneto-resistance) $\sigma_s(H)$-$\sigma_s(0)$ measured at various temperatures on the a-IGZO film (Sample A). The a-IGZO film exhibits obvious negative magneto-resistance (positive magneto-conductance) up to 150 K. The decrease above 4 Tesla in the $\sigma_s(H)$-$\sigma_s(0)$ curve measured at 2 K is attributed to the spin-orbit scattering, which contributes to a positive component in magneto-resistance [52]. Figure 15(b) shows relations between σ_s and H^2 for the respective measurement temperatures. According to the theory, σ_s is proportional to H^2 at low fields ($H << \hbar c/4eL_{in}^2$) in both 2D and 3D weak localization systems. It is apparent that, for each of the measurement temperatures, σ_s is proportional to H^2 under low fields, and the field range showing the $\sigma_s \propto H^2$ relation expands as temperature increases, which corresponds to decrease of the inelastic scattering time with temperature expressed as $\tau_{in} = L_{in}^2/D \propto T^p$, where D and p are the electron diffusion coefficient and a universal parameter, respectively.

According to the above analyses of temperature and magnetic-field dependences of sheet conductance, the sputter-deposited 40 nm-thick a-IGZO film with a relatively large carrier density (10^{19} cm^{-3}) shows weak localization like a-IZO films [50]. This suggests that this a-IGZO film is metallic but weakly acts as insulator because of the weak localization effect. This weak localization effect probably leads to overestimation of the potential barrier height at grain boundaries for the sample shown in figure 13, because the dominant scattering mechanism in such a-IGZO with a large carrier density as Sample A is no longer the grain boundary scattering but is the weak localization.

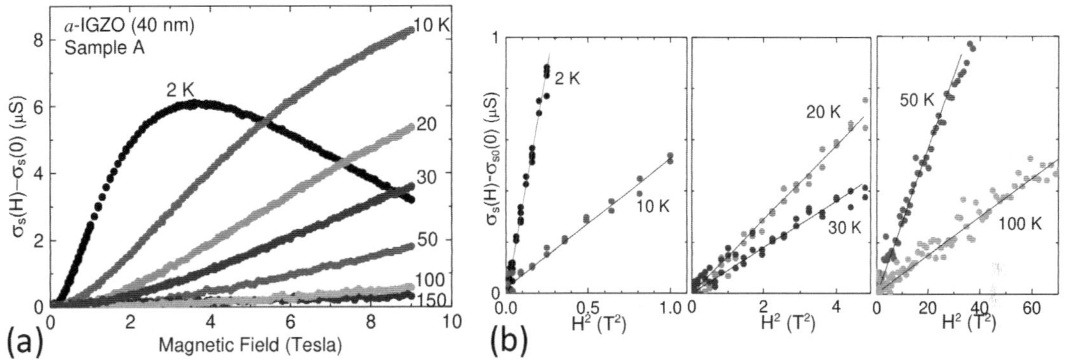

Figure 15. (a) Magnetic-field dependence of the magneto-conductance $\sigma_s(H)$-$\sigma_s(0)$ for a 40 nm-thick a-IGZO film (Sample A) at various measurement temperatures. (b) Replot of the magneto-conductance values as a function of the square of the magnetic field.

3.3. Optical analyses

3.3.1. UV-VIS-NIR absorption

The optical absorption coefficient, α, of amorphous semiconductors can usually be expressed by

$$\alpha h\nu = B(h\nu - E_g^{opt})^2, \tag{6}$$

where B is a coefficient, h Planck's constant, ν the photon frequency, and E_g^{opt} the optical band-gap (Tauc' gap) energy [53,54]. Figure 16 shows for 300 nm-thick a-IGZO films deposited in 1% O_2 gas with various annealing conditions (on silica substrates; 300°C N_2-annealing, O_2-annealing, N_2-annealing+O_2-annealing and no annealing (as-deposited), and on a SiN_x/silica substrate; 300°C N_2-annealing), values of E_g^{opt} and B in equation (6) calculated from $(\alpha h\nu)^{1/2}$-$h\nu$ plots of measured spectra, two of which are shown in the inset. E_g^{opt} shows almost the same values of around 3.1 eV for all the annealed films, but that of the as-deposited film shows a slightly smaller value of 3.0 eV. This shift of E_g^{opt} is not due to the Burstein-Moss effect, because these E_g^{opt} values are independent of carrier density, referring to figure 6 and table 3. This E_g^{opt} shift is probably due to intrinsic change of band-gap through structural relaxation caused by annealing. The coefficient B shows the same tendency as E_g^{opt}: almost the same values for the annealed films but a smaller value for the as-deposited one.

3.3.2. Spectroscopic ellipsometry

We employed the Tauc-Lorentz (TL) model [55,56] to analyze measured ellipsometry spectra of a-IGZO films. The TL model was originally developed to parameterize the optical functions of amorphous materials, and has given excellent fitting results also on crystalline transparent conductive oxide (TCO) films [57,58] and a-IGZO films [15,59]. The present analyses using only the TL model were also successful for most of the a-IGZO films except extremely conductive films with a carrier density of 10^{20} cm^{-3} or more. Figure 17 shows $\Psi(h\nu)$ and $\Delta(h\nu)$ of the ellipsometry spectra (symbols) measured on an a-IGZO film deposited in 1% O_2 gas and then received 300°C N_2-annealing, along with fitting results using the TL model (solid lines), which represent the measured spectra very well.

Values of the optical bandgap energy (E_g^{TL}), refractive index (n_f) and extinction coefficient (k_f) at 1 eV, obtained by the TL model fitting, are listed in table 4. Each E_g^{TL} value is in good agreement with the corresponding optical bandgap energy E_g^{opt} value shown in figure 16, which was estimated from the optical absorption using equation (6) as already described. Values of k_f must be zero at $h\nu \leq E_g^{TL}$ if an ellipsometry spectrum can be expressed by solely the TL model, because the TL model relies on the assumption that the imaginary part of dielectric constant (corresponding to optical absorption) is extinguished below its bandgap energy. Actually, the measured spectra of the a-IGZO films in table 4 except the H_2-annealed film (listed at the bottom of the table) were well expressed by solely the TL model, especially at $h\nu \leq E_g^{TL}$. In other words, absorption below E_g^{TL} indicating a sub-band-gap transition or free carrier absorption was not observed by ellipsometry. The refractive indices of the a-IGZO films except that of the H_2-annealed one were $n_f \approx 2.0$ independent of the conditions for annealing and sputtering deposition.

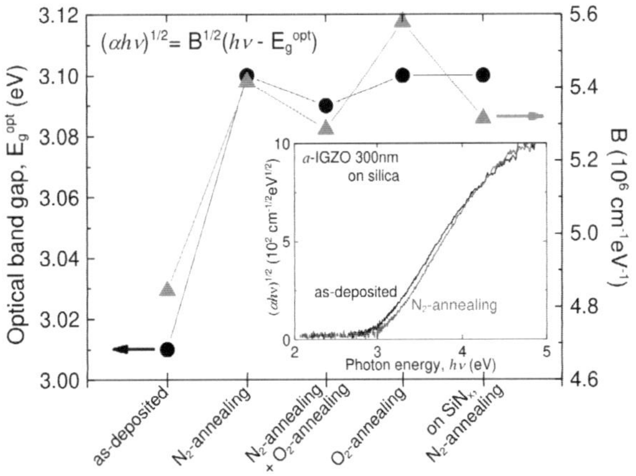

Figure 16. Variation of the optical band gap E_g^{opt} and the coefficient B in equation (6) for 300 nm-thick a-IGZO films with changing annealing treatments. Optical absorption spectra in a $(\alpha h\nu)^{1/2}$ vs. $h\nu$ plot are shown in the inset, for the films with N_2-annealing and no annealing treatment (as-deposited).

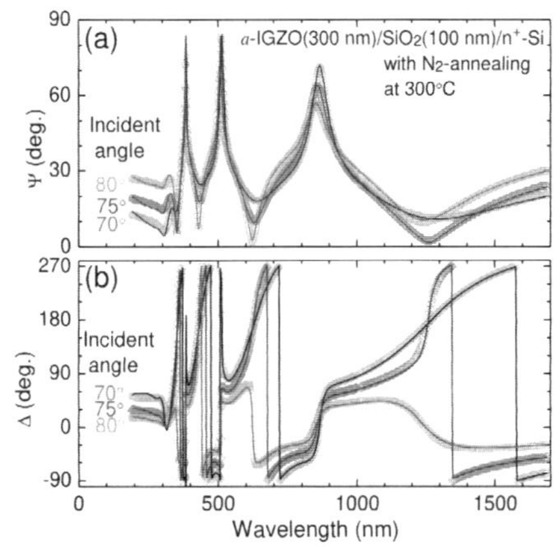

Figure 17. Ellipsometry spectra of (a) $\Psi(h\nu)$ and (b) $\Delta(h\nu)$ measured with incident angles of 70, 75 and 80 degrees for a 300 nm-thick a-IGZO film deposited in 1% O_2 gas on SiO_2(100 nm)/n$^+$-Si substrate and then N_2-annealed at 300°C. Solid lines indicate fitting results using the Tauc-Lorentz model.

The anomaly described above in the optical constants of the 400°C H_2-annealed a-IGZO film having a high carrier density of 1.2×10^{20} cm^{-3} is due to free carrier absorption. The fitting by solely the TL model ended up with large deviations from the measured $\Psi(h\nu)$ and $\Delta(h\nu)$ values at wavelengths ≥ 0.9 μm. Thus we analyzed the ellipsometry spectra of high carrier-density films by a linear combination of the TL model and the Drude model expressed as $\varepsilon_C(h\nu) = \varepsilon_{TL}(h\nu) + \varepsilon_D(h\nu)$, where ε_{TL} and ε_D are the dielectric functions of the TL model and the Drude model, respectively. According to the Drude theory, the complex dielectric function $\varepsilon_D(h\nu)$ can be expressed with the relaxation time τ, carrier density N, and mobility $\mu = e\tau/m^*$ as

$$\varepsilon_D = \frac{-\hbar^2 e\mu N}{\varepsilon_0 \{\tau \cdot (h\nu)^2 + i\hbar \cdot (h\nu)\}} \quad (7)$$

for convenience of optical analyses [60]. Consequently, we found that a parameter set of $m^* = 0.40m_0$, $N = 1.5 \times 10^{20}$ cm^{-3}, and $\mu = 20$ cm^{2}V^{-1}s^{-1} gives the best fit to the measured spectra. The value of $m^* = 0.40m_0$ is close to $0.35m_0$ reported on PLD-deposited a-IGZO films by optical model analysis [61]. The optical mobility is close to but slightly larger than the Hall mobility of 18 cm^{2}V^{-1}s^{-1}. This is consistent with the small potential barrier height at grain boundaries of the a-IGZO films with a large carrier density, shown in figure 13.

Table 4. Optical bandgap E_g^{TL}, and refractive index n_f and extinction coefficient k_f at $h\nu = 1$ eV estimated from spectroscopic ellipsometry spectra by applying the Tauc-Lorentz model, for 300 nm-thick a-IGZO films deposited in sputtering gas with various O_2 contents and then annealed under various annealing conditions. Exceptionally, a linear combination of the Tauc-Lorentz and the Drude models was used for the fitting of the spectra of the film with H_2-annealing treatment. The annealing temperature was 300°C for all the samples except those with H_2-annealing (400°C) and no annealing.

Sputtering gas	Annealing condition	E_g^{TL} (eV)	n_f at 1 eV	k_f at 1 eV
1% O_2	no annealing treatment	3.06	1.98	0
	N_2-annealing	3.10	1.97	0
	N_2-annealing + O_2-annealing	3.10	1.97	0
	O_2-annealing	3.09	1.98	0
3% O_2	N_2-annealing	3.11	1.98	0
5% O_2	N_2-annealing	3.12	1.98	0
10% O_2	N_2-annealing	3.13	1.99	0
1% O_2	H_2-annealing* (at 400°C for 40 hours)	3.10	1.83	0.02

*Annealing treatment in 2% H_2/98% He mixture gas.

4. Discussion

As described in Sec. 3.1, the TEM/STEM, SEM observations and the GISAXS analysis revealed that the sputter-deposited a-IGZO films have columnar-like structures. This suggests the existence of grain boundaries even though the films are amorphous. This is consistent with the fact that the thermal activation behaviour of the Hall mobility observed on the sputter-deposited a-IGZO films [figure 12] can be explained by a model of thermionic emission through potential barriers at grain boundaries. According to the model, the a-IGZO films with a smaller carrier density N (donor density N_d, to be exact) have higher potential barriers as shown in figure 13, resulting in a smaller carrier mobility μ. This relation leads to positive correlation between μ and N. In fact, such μ-N relations have already been observed on some crystalline semiconductor films, e.g., polycrystalline Si [41], S-doped polycrystalline InP [45], Al-/Li-doped epitaxial ZnO [46] and polycrystalline and amorphous In_2O_3 [62] with low carrier densities compared with those for electrode applications. The mobility of these semiconductor films were well explained by the grain-boundary scattering model. However, theoretically, the mobility of these films should drastically decrease with decreasing carrier density if the donor ionization energy E_d is almost constant, and the power-law-like μ-N relation can be observed in only a narrow range of N.

In the case of the present sputter-deposited a-IGZO films, E_d increases with decreasing N_d, as shown in figure 11(a), resulting in a significant decrease of N by a slight decrease of N_d. This causes a moderate decrease of μ with decreasing N, thus expanding the range of N in which the μ-N relation obeys the power-law. This may be one possible reason of the characteristic μ-N relation of the sputter-deposited a-IGZO films.

5. Amorphous InO-GaO-ZnO multilayered films

The family of crystalline $InGaO_3(ZnO)_m$ each has a layered crystal structures consisting of stacked InO layers and (Ga,Zn)O layers with random distribution of Ga and Zn. For example, each unit cell of $InGaZnO_4$ (m = 1) has a nine-layer stacks along the c-axis, as shown in figure 18(a). Since the lattice constant c of $InGaZnO_4$ is about 27 Å, the thickness of one constituent layer is estimated to be about 3 Å on average [63]. As shown in figure 18(b), we designed an artificial lattice of amorphous $InGaZnO_4$ with InO and 2(Ga,Zn)O layers stacked alternately, imitating the crystalline $InGaZnO_4$ layer structure. For convenience of the deposition, a stacked of one GaO layer and one ZnO layer were employed in place of the 2(Ga,Zn)O layer. We chose not a simple stacking pattern with a repetition of InO-GaO-ZnO but a pattern with a repetition of InO-GaO-ZnO-InO-ZnO-GaO, because an artificial lattice with the simplest stacking pattern possibly has a polarization in the stacking direction due to a lack of mirror inversion which may affect electrical properties of the multilayer-films. When the thickness of each constituent layer (L) is smaller than 3 Å, which is the above-mentioned average thickness of monolayers in crystalline $InGaZnO_4$, the film becomes in an usual isotropic amorphous state as shown in figure 18(c), and multilayered structures could be realized when L is larger than 3 Å.

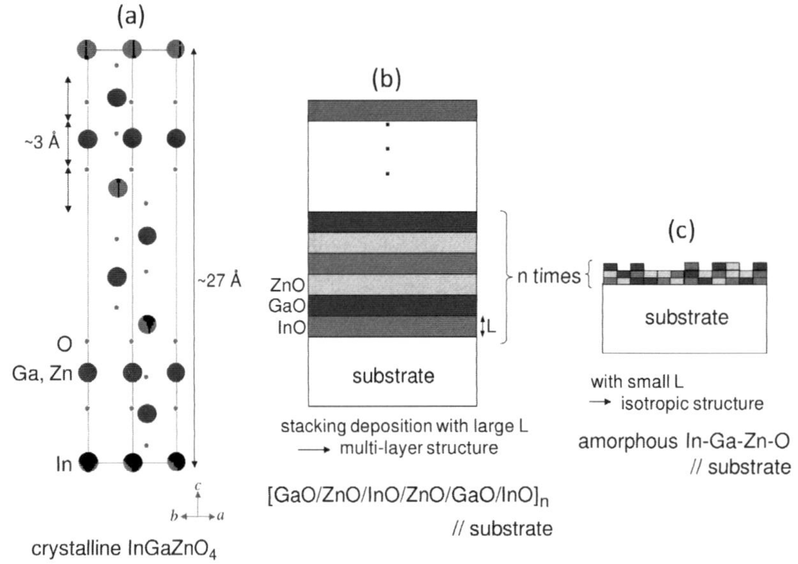

Figure 18. (a) Crystal structure of $InGaZnO_4$. (b) Designed structure of multilayered $[GaO/ZnO/InO/ZnO/GaO/InO]_n$ film (n: repetition number) when the constituent layer thickness L is large enough (>3 Å). (c) Schematic drawing of an isotropic amorphous In-Ga-Zn-O film to be formed by the same method as that for the multilayered film shown in (b) when L is too small to form a continuous monolayers.

The deposition rate for each constituent layer was estimated from film thicknesses measured as a function of the number of input laser pulses on test samples deposited on a thermally-grown SiO_2 (100 nm)/Si substrate. Figure 19 shows XRR patterns for GaO thin layers deposited with various number of input laser pulses (60-840 pulses). The number of fringes on the XRR profiles increases with increasing the number of input pulses. As the number of fringes corresponds to the thickness, the film thickness for each of the test samples was estimated from the corresponding XRR pattern (fitting results for GaO layers are shown in the figure). Thus obtained film thicknesses are plotted against the number of input laser pulses in figure 20, along with the results on InO and ZnO. For InO and GaO, in the thickness range up to 80 Å, the relations between thickness and the number of input pulses can be fitted almost with the same linear function. A thickness offset of the fitted line at zero pulse would correspond to a thin layer deposition during the pre-deposition step for target cleaning, due to the

Journal of Physics: Conference Series **518** (2014) 012001 doi:10.1088/1742-6596/518/1/012001

existence of a slight number of particles leaking from around the shutter in the PLD chamber. We estimated the deposition rate common to InO and GaO from the gradient of the fitted line. On the other hand, the deposition rate of ZnO shows two different values depending on the thickness; the lower value for thicknesses above 20 Å. This could arise from crystallization of ZnO, because it is well-known that ZnO films deposited by sputtering or evaporation are easily crystallized even at room temperature [64].

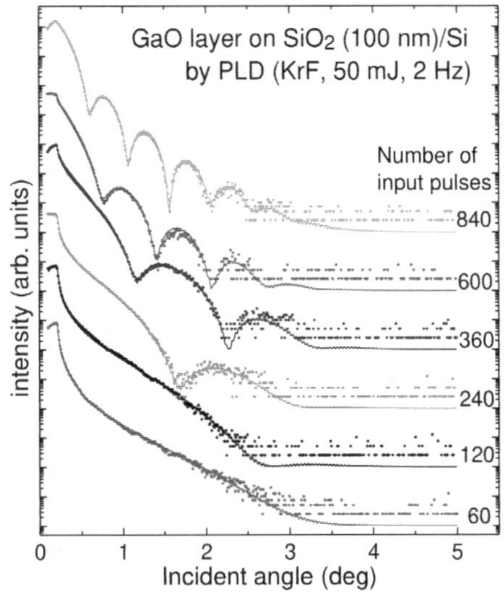

Figure 19. X-ray reflectivity (XRR) profiles of GaO layers deposited with various numbers of input laser pulses on SiO_2(100 nm)/Si substrates. Solid lines indicate fitting.

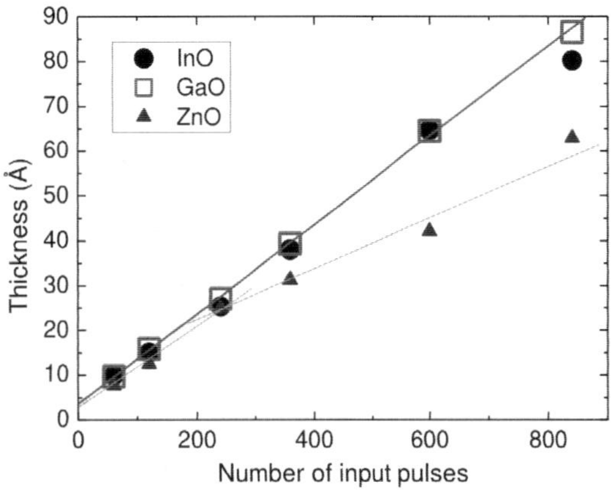

Figure 20. Dependences of film thickness on the number of input laser pulses for InO, GaO and ZnO layers.

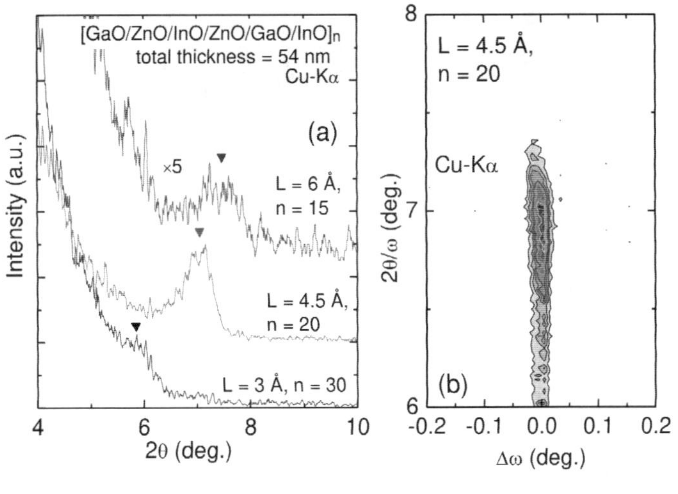

Figure 21.
(a) X-ray diffraction patterns taken by 2θ-ω method, in the range of 2θ = 4° to 10°, on multilayered $[GaO/ZnO/InO/ZnO/GaO/InO]_n$ films with the constituent-layer thickness L of 3, 4.5 and 6 Å.
(b) Reciprocal space map of the intensity of 200 diffraction of a multilayered $[GaO/ZnO/InO/ZnO/GaO/InO]_{20}$ film with L = 4.5 Å. The vertical-axis ($2\theta/\omega$) denotes 2θ in 2θ-ω scans, and the horizontal-axis ($\Delta\omega$) represents the difference of ω from a half of 2θ.

First, $[GaO/ZnO/InO/ZnO/GaO/InO]_n$ multilayered films with a total thickness of 54 nm were formed by n-times repetition of a deposition of the InO-GaO-ZnO-InO-ZnO-GaO layer on glass substrates with L = 3, 4.5 and 6 Å. In XRD patterns of these samples, no peak from crystalline InO, GaO or ZnO was found, which indicates that each constituent layer remains amorphous. Instead, as shown in figure 21(a), broad peaks were observed at very low angles of 5.8°, 7.1°, and 7.5° for the multilayered films with L = 3, 4.5 and 6 Å, respectively. These peaks each can be assigned to a Bragg peaks of the respective artificial lattices. For instance, the unit lattice parameter of the multilayer with L = 4.5 Å is calculated as 6 × L = 27 Å, while the peak observed at 7.1° indicates a spacing of about a half of 25 Å. Thus, this XRD peak is in agreement with the expected unit lattice parameter. Similarly, the peak at 5.8° for the multilayered film with L = 3 Å (unit lattice parameter: 18 Å) corresponds to a spacing of about 15 Å, and the peak at 7.5° for the multilayered film with L = 6 Å (unit lattice parameter: 36 Å) corresponds to a spacing of about 12 Å, which is exactly one-third of the unit lattice parameter.

Figure 21(b) shows a reciprocal space map of the diffraction at 7.1° of the multilayered film with L = 4.5 Å. The spreads of the diffraction intensity along the vertical-axis ($2\theta/\omega$) and that along the horizontal-axis ($\Delta\omega$) indicate fluctuations in periodicity and that in orientation of the lattice, respectively. As can be seen in the map, the fluctuation in lattice periodicity is fairly large, while that in lattice orientation is very small. This observation suggests that highly parallel stacking of the constituent layers were realized, even though the actual thickness of the constituent layers fluctuated. It is thus confirmed that amorphous artificial lattices composed of InO, GaO and ZnO constituent layers were successfully formed.

Figure 22. (a) Cross-sectional TEM image of a multilayered $[GaO/ZnO/InO/ZnO/GaO/InO]_6$ film with L = 15 Å on SiO_2/Si substrate. Carbon and Metal (Cr) films were deposited on the film to define the film surface. (b) Expanded view of the cross-sectional TEM image in (a).

Cross-sectional TEM observation of this multilayered film (L = 4.5 Å) was carried out. As shown in figure 22, an obvious layered structure was confirmed. The dark and bright layers correspond to InO and 2(Ga,Zn)O layers, respectively, and the thickness of each layer is close to the intended thickness. This observation is consistent with the XRD results described above. Since the tolerance of the observation conditions was narrow for such thin layers and the conditions were optimised for an area around the film/substrate interface, the layered structure looks blurred at the upper area of the film. Additionally, surface roughness of the films slightly increases with increasing film thickness, resulting in an unclear layered structure at the upper area of the film. For a multilayered film with thicker composition layers of L = 15 Å, each constituent layer can be observed clearly even in the upper area of the film, as shown in figure 23(a). In this film, a definite 2D structure was attained, as can be seen in figure 23(b).

Subsequently, Hall measurement samples were prepared from 50-60 nm-thick $[GaO/ZnO/InO/ZnO/GaO/InO]_n$ multilayered films having various L values. Since the actual constituent layer thickness was found to become slightly smaller than the nominal thickness as mentioned above, $L = 3.5$ Å was employed instead of $L = 3$ Å. Figure 24 shows dependences of the Hall mobility and the carrier density on L. The dependence of the Hall mobility on L exhibits curious behaviour, while the carrier density shows simple and monotonous increase with L. We assume that the multilayered film with $L = 1$ Å possesses an isotropic amorphous structure like that shown in figure 18(c), while the films with $L > 3$ Å have a 2D multilayer structures forming an artificial lattices as seen in figure 22. Thus expected is a difference in the Hall mobility between them. However, the Hall mobility was almost independent of L up to 6 Å, i.e., regardless of the layer structure, with the values around 4 $cm^2V^{-1}s^{-1}$. This Hall mobility value seems reasonable for uniform a-IGZO films under this PLD condition (KrF excimer laser power density of approximately 5 Jcm^{-2}), compared to the reported values of about 2 and 10 $cm^2V^{-1}s^{-1}$ for a-IGZO films by PLD with 2 and 9 Jcm^{-2}, respectively [27]. This observation suggests that, at least in this small thickness range, the carrier transport in the lateral direction of the constituent layers is almost comparable with that in the vertical direction. Nearly isotropic conduction of $InGaZnO_4$ despite its highly anisotropic 2D layer-structure has been theoretically predicted with first-principle calculation [65], which is consistent with the presently observed behaviour of the Hall mobility for $L \leq 6$ Å. Furthermore, the imperfection of 2D structure observed in the cross-sectional TEM observation (figure 22) may have affected the Hall mobility in some degree.

(a)

(b)

Figure 23. (a) Cross-sectional TEM image of a multilayered $[GaO/ZnO/InO/ZnO/GaO/InO]_6$ film with $L = 15$ Å on SiO_2/Si substrate. Carbon and Metal (Cr) films were deposited on the film to define the film surface. (b) Expanded view of the cross-sectional TEM image in (a).

As L increases beyond 6 Å, the Hall mobility shows a drastic increase in the range of $L \leq 15$ Å. Such an increase in the Hall mobility could be attributed to enhancement of the character of InO of having a higher carrier mobility than ZnO, GaO and $InGaZnO_4$, with the increase in the layer thickness. It is also noteworthy that the peak Hall mobility of 23 $cm^2V^{-1}s^{-1}$ is comparable to or even slightly higher than those reported on epitaxial $InGaZnO_4$ films [66]. Also reduction of the imperfection of 2D structure shown in the TEM observation (figure 23) may have contributed to this rise of the Hall mobility.

Above L = 15 Å, the Hall mobility abruptly decreases with increasing L, which is presumably due to the increase of the roughness of each constituent layer with increasing L, caused by crystallization of ZnO layers accompanied by the ZnO deposition rate change shown in figure 20. The crystallization of ZnO layers would lead to facet formation and enlarge the roughness at the interfaces between layers, resulting in deterioration of the periodicity of the artificial lattice. The interface roughness and the deterioration of periodicity should enhance carrier scattering.

The monotonous increase of the carrier density in the whole L range [figure 24] could be attributed to enhancement of the above-mentioned indium oxide's character in InO layers, because amorphous In_2O_3 films usually show a carrier density value of the order of 10^{20} cm^{-3}. Consequently, L of 10-15 Å is an optimum condition to realize a high conductivity in the plane of the multilayered film.

Multilayered films of InO-GaO-MgO were also formed by substituting MgO for ZnO, and their carrier density and Hall mobility were measured. As shown together in figure 24, the carrier density of the InO-GaO-MgO multilayered films increases with L, and the Hall mobility is almost independent of L in the range of L = 1 to 6 Å, similarly to those of InO-GaO-ZnO multilayered films. However, the carrier density are remarkably small and the Hall mobility are rather small, compared to those of InO-GaO-ZnO multilayered films. These observations are consistent with the difference in the carrier density between $InGaMgO_4$ and $InGaZnO_4$ ceramics [67] and with the larger effective electron mass of $InGaMgO_4$ than $InGaZnO_4$ [65] This suggests that ZnO layers play a role of a carrier supplier and provide a large carrier mobility in $InGaZnO_4$, compared to MgO layers in $InGaMgO_4$. Furthermore, we formed 50 nm-thick YO-GaO-ZnO multilayer films, using yttrium oxide (YO) layers in place of InO layers, with L of 3.5-6 Å. The samples had a resistivity too high to measure the Hall effect. This result confirms that InO layers are further important in carrier supply and/or provision of a large carrier mobility in $InGaZnO_4$.

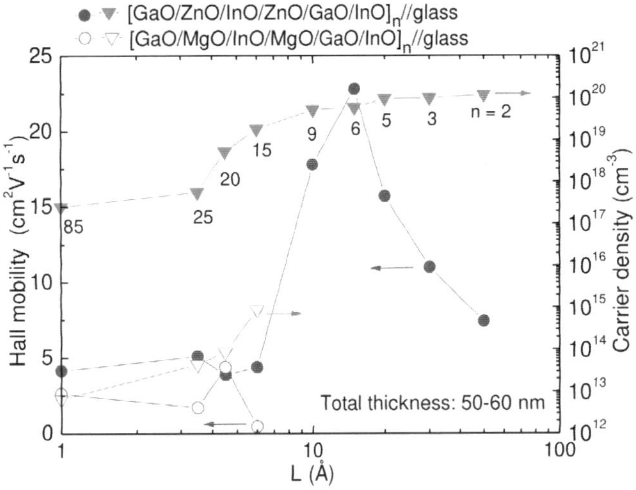

Figure 24. Dependences of the Hall mobility and the carrier density on L for multilayered $[GaO/ZnO/InO/ZnO/GaO/InO]_n$ and $[GaO/MgO/InO/MgO/GaO/InO]_n$ films.

6. Summary

We investigated microscopic structures, electrical and optical properties of sputter-deposited a-IGZO films, and discussed correlations among them.

TEM/STEM and SEM observations revealed that the sputter-deposited a-IGZO films consist of a columnar structure, which was confirmed by GISAXS analysis. Gas-adsorption analysis revealed, however, that the sidewall surfaces of the columnar grains are little active in adsorbing ambient gas molecules and thus, the grain boundaries observed by cross-sectional TEM/STEM are not open-pores. XAFS analysis indicated imperfection in In-O coordination in the a-IGZO films.

The conductivity of the sputter-deposited a-IGZO films exhibits a change as large as seven orders of magnitude almost reversibly between N_2- and O_2-annealings even at a temperature as low as 300°C, such that the high conductivity state obtained by N_2-annealing turned into the low conductivity state by O_2-annealing and vice versa. The sputter-deposited a-IGZO films exhibit a power-law relation between the Hall mobility μ and the carrier density N. Temperature dependence of N in the sputter-deposited a-IGZO films with $N \leq 1 \times 10^{18}$ cm^{-3} shows the Arrhenius-type thermal activation at temperatures around and below room temperature. The ionization energy of donor E_d increases with decreasing the donor density N_d and reaches 250 meV at $N_d \approx 6 \times 10^{17}$ cm^{-3}, resulting in a small N value of 1×10^{15} cm^{-3} at room temperature. For a-IGZO films having widely different conductivities, temperature dependence of μ suggests Arrhenius-type-like thermal activation above 150-200 K. This behaviour can be explained by a model of thermionic emission through potential barriers at grain boundaries (grain-boundary scattering model), which is consistent with the TEM/STEM and SEM observations.

The optical band-gap energy of the sputter-deposited a-IGZO films showed a difference between before and after annealing but did not change with annealing conditions, in contrast to a significant dependence of the carrier density on the annealing conditions. By spectroscopic ellipsometry analysis, electron effective mass m^* of the sputter-deposited a-IGZO films was estimated to be $0.4m_0$, which is close to the reported value on PLD-deposited a-IGZO films ($0.35m_0$).

Further, multilayered films composed of amorphous InO, GaO and ZnO (or MgO) constituent layers were formed using a computer-controlled PLD system, to imitate the layered structure of the $InGaZnO_4$ homologous phase. An artificial periodic lattice structure was realized in each of 54 nm-thick $[GaO/ZnO/InO/ZnO/GaO/InO]_n$ multilayer films with the single constituent layer thickness $L = 3$, 4.5 and 6 Å, which was confirmed by superlattice peaks in XRD profiles and cross-sectional TEM images. The Hall mobility of $[GaO/ZnO/InO/ZnO/GaO/InO]_n$ was almost constant for $L \leq 6$ Å, and increased for 6 Å $\leq L \leq$ 15 Å, suggesting that the conduction is rather isotropic even in the multilayer structures when L is small, and 2-dimensional conduction in InO layers is enhanced when L is larger. Substitution of MgO and YO layers for ZnO and InO, respectively, resulted in drastic increase of resistivity, suggesting that ZnO and InO layers each play an important role in the electrical conduction in $InGaZnO_4$.

Acknowledgments
We would like to thank H Hosono, T Kamiya, K Nomura, B Shinozaki, K Shimakawa, R Hayashi, K Abe for valuable discussion.

References

[1] Nomura K, Ohta H, Takagi A, Kamiya T, Hirano M and Hosono H 2004 *Nature (London)* **432** 488

[2] Yabuta H, Sano M, Abe K, Aiba T, Den T, Kumomi H, Nomura K, Kamiya T and Hosono H 2006 *Appl. Phys. Lett.* **89** 112123

[3] Kawamura T, Uchiyama H, Saito S, Wakana H, Mine T, Hatano M, Torii K and Otani T 2008 *Tech. Dig. 2008 IEEE International Electron Device Meeting* 1

[4] Hayashi R, Sato A, Ofuji M, Abe K, Yabuta H, Sano M, Kumomi H, Nomura K, Kamiya T, Hirano M and Hosono H 2008 *SID Symp. Dig. Tech. Papers* **39** 621

[5] Hayashi R, Ofuji M, Kaji N, Takahashi K, Abe K, Yabuta H, Sano M, Kumomi H, Nomura K, Kamiya T, Hirano M and Hosono H 2007 *J. SID* **15/11** 915

[6] Ofuji M, Abe K, Shimizu H, Kaji N, Hayashi R, Sano M, Kumomi H, Nomura K, Kamiya T

and Hosono H 2007 *IEEE Elec. Dev. Lett.* **28** 273
[7] Yabuta H, Kaji N, Hayashi R, Kumomi H, Nomura K, Kamiya T, Hirano M and Hosono H 2010 *Appl. Phys. Lett.* **97** 072111
[8] Sunamura H, Kaneko K, Furutake N, Saito S, Narihiro M, Hane M and Hayashi Y 2013 *2013 Symp. VLSI Tech. Dig. Tech. Papers* T250
[9] Abeles B and Tiedje T, 1983 *Phys. Rev. Lett.* **51** 2003
[10] Hirose M and Miyazaki S, 1984 *J. Non-Cryst. Solids* **66** 327
[11] Sze S M and Ng K K 2007 *Physics of Semiconductor devices,* 3rd ed (Hoboken: Wiley)
[12] Hishikawa Y, Nakamura N, Tsuda S, Nakano S, Kishi Y and Kuwano Y 1991 *Jpn. J. Appl. Phys.* **30** 1008
[13] Thornton J A 1986 *J. Vac. Sci. Technol.* **A 4** 3059
[14] Ross R C and Messier R 1981 *J. Appl. Phys.* **52** 5329
[15] Kang D, Song I, Kim C, Park Y, Kang T D, Lee H S, Park J -W, Baek S H, Choi S -H and Lee H 2007 *Appl. Phys. Lett.* **91** 091910
[16] Yamazoe N, Shimanoe K and Sawada C 2007 *Thin Solid Films* **515** 8302
[17] Yamazoe N and Shimanoe K 2008 *Sensors and Actuators B* **128** 566
[18] Yamazaki T, Furuta T, Shen Y, Jin C, Kikuta T and Nakatani N 2006 *Jpn. J. Appl. Phys.* **45** 9180
[19] Nespolo M, Sato A, Osawa T and Ohashi H 2000 *Cryst. Res. Technol.* **35** 151
[20] Cho D -Y, Song J, Na K D, Hwang C S, Jeong J H, Jeong J K and Mo Y –G 2009 *Appl. Phys. Lett.* **94** 112112
[21] Nogawa H, Hitosugi T, Chikamatsu A, Nakao S, Hirose Y, Shimada T, Kumigashira H, Oshima M and Hasegawa T 2010 *Jpn. J. Appl. Phys.* **49** 041102
[22] Yamaguchi N, Taniguchi S, Miyajima T and Ikeda M 2009 *J. Vac. Sci. Technol. B* **27** 1746
[23] Martins R, Barquinha P, Pimentel A, Pereira L and Fortunato E 2005 *Phys. Stat. Sol. (a)* **202** R95
[24] Leenheer A J, Perkins J D, van Hest M F A M, Berry J J, O'Hayre R P and Ginley D S 2008 *Phys. Rev. B* **77** 115215
[25] Shinozaki B, Makise K, Funaki M, Yano K, Inoue K and Nakamura H 2009 *J. Phys: Conf. Series* **150** 022076
[26] Takagi A, Nomura K, Ohta H, Yanagi H, Kamiya T, Hirano M and Hosono H 2005 *Thin Solid Films* **486** 38
[27] Nomura K, Kamiya T, Ohta H, Shimizu K, Hirano M and Hosono H 2008 *Phys. Stat. Sol. (a)* **205** 1910
[28] Satoh K, Kakehi Y, Okamoto A, Murakami S, Moriwaki K and Yotsuya T 2008 *Thin Solid Films* **516** 5814
[29] Hosono H 2006 *J. Non-Cryst. Solids* **352** 851
[30] Kamiya T, Nomura K and Hosono H 2009 *J. Disp. Tech.* **5** 462
[31] Kamiya T, Nomura K and Hosono H 2010 *Appl. Phys. Lett.* **96**, 122103
[32] Adler D, Flora L P and Senturia S D 1973 *Solid State Commun.* **12** 9
[33] Ng H M, Doppalapudi D and Moustakas T 1998 *Appl. Phys. Lett.* **73** 821
[34] Jeong P S, Kim H R, Choi S -H and Lee Y -S 2009 *J. Korean Phys. Soc.* **54** 2378
[35] Kumomi H, Yaginuma S, Omura H, Goyal A, Sato A, Watanabe M, Shimada M, Kaji N, Takahashi K, Ofuji M, Watanabe T, Itagaki N, Shimizu H, Abe K, Tateishi Y, Yabuta H, Iwasaki T, Hayashi R, Aiba T and Sano M 2009 *J. Disp. Tech.* **5** 539
[36] Omura H, Kumomi H, Nomura K, Kamiya T, Hirano M and Hosono H 2009 *J. Appl. Phys.* **105** 093712
[37] Mott N F 1990 *Metal - Insulator Transitions*, 2nd ed. (Taylor & Francis)
[38] Hamberg I and Granqvist C G 1986 *J. Appl. Phys.* **60** R123
[39] Hamberg I and Granqvist C G 1984 *Appl. Phys. Lett.* **44** 721
[40] Bellingham J R, Phillips W A and Adkins C J 1992 *J. Mater. Sci. Lett.* **11** 263

[41] Seto J Y W 1975 *J. Appl. Phys.* **46** 5247
[42] Orton J W and Powell M J 1980 *Rep. Prog. Phys.* **43** 81
[43] Werner J H 1994 *Solid State Phenomena* **37** 213
[44] Chung H, Jeong J H, Ahn T K, Lee H J, Kim M, Jun K, Park J, Jeong J K, Mo Y and Kim H D 2008 *Electrochem. Solid-State Lett.* **11** H51
[45] Saitoh T and Matsubara S 1977 *J. Electrochem. Soc.* **124** 1065
[46] Tsurumi T, Nishizawa S, Ohashi N and Ohgaki T 1999 *Jpn. J. Appl. Phys.* **38** 3682
[47] Orton J W, Goldsmith B J, Powell M J and Chapman J A 1980 *Appl. Phys. Lett.* **37** 557
[48] Seager C H and Pike G E 1979 *Appl. Phys. Lett.* **35** 709
[49] Shimakawa K, Narushima S, Hosono H and Kawazoe H 1999 *Philos. Mag. Lett.* **79** 755
[50] Shinozaki B, Makise K, Shimane Y, Nakamura H and Inoue K 2007 *J. Phys. Soc. Jpn.* **76** 074718
[51] Anderson P W, Abrahams E and Ramakrishnan T V 1979 *Phys. Rev. Lett.* **43** 718
[52] Shinozaki B, Hidaka K, Ezaki S, Makise K, Asano T, Tomai S, Yano K and Nakamura H 2013 *J. Appl. Phys.* **113** 153707
[53] Tauc J 1974 *Amorphous and Liquid semiconductors*, ed. J Tauc (London: Plenum) p 159.
[54] Kasap S, Koughia C, Singh J, Ruda H and O'Leary S K 2006 *Springer Handbook of Electronic and Photonic Materials*, eds S Kasap and P Capper (New York: Springer) p 47
[55] Jellison Jr G E and Modine F A 1996 *Appl. Phys. Lett.* **69** 371
[56] Jellison Jr G E and Modine F A 1996 *Appl. Phys. Lett.* **69** 2137
[57] Fujiwara H and Kondo M 2005 *Phys. Rev. B* **71** 075109
[58] Rovira P I and Collins R W 1999 *J. Appl. Phys.* **85** 2015
[59] Kamiya T, Nomura K and Hosono H 2009 *Phys. Stat. Sol. (a)* **860** 1910
[60] Tiwald T E, Thompson D W, Woollam J A, Paulson W and Hance R 661 *Thin Solid Films* **313-314** 661
[61] Nomura K, Kamiya T, Yanagi H, Ikenaga E, Yang K, Kobayashi K, Hirano M and Hosono H 2008 *Appl. Phys. Lett.* **92** 202117
[62] Nakazawa H, Ito Y, Matsumoto E, Adachi K, Aoki N and Ochiai Y 2006 *J. Appl. Phys.* **100** 093706
[63] Nakamura M, Kimizuka N and Mohri T 1991 *J. Solid State Chem.* **93** 295
[64] Gupta V, Sreenivas K and Fahim M 1998 *Proceedings of the Eleventh IEEE International Symposium on Applications of Ferroelectrics, ISAF* (Montreux, Switzerland, August 24-27) p 325
[65] Medvedeva J E 2007 *EPL* **78** 57004
[66] Kamiya T, Nomura K and Hosono H 2009 *J. Display Technol.* **5** 273
[67] Orita M, Takeuchi M, Sakai H and Tanji H 1995 *Jpn. J. Appl. Phys.* **34** L155

Journal of Physics: Conference Series **518** (2014) 012002 doi:10.1088/1742-6596/518/1/012002

Deposition of crystalline Ge nanoparticle films by high-pressure RF magnetron sputtering method

D Ichida[1], G Uchida[2], H Seo[1], K Kamataki[3], N Itagaki[1,4], K Koga[1] and M Shiratani[1]

[1] Graduate School of Information Science and Electrical Engineering, Kyushu University, 744 Motooka, Nishi-ku, Fukuoka, 819-0395, Japan
[2] Jointing and Welding Reserch Institute, Osaka University, 11-1 Mihogaoka, Ibaraki, Osaka, 567-0047, Japan
[3] Faculty of Arts and Science, Kyushu University, 744 Motooka, Nishi-ku, Fukuoka, 819-0395, Japan
[4] PRESTO Japan Science and Technology Agency, Gobancho, Chiyoda-ku, Tokyo, 102-0076, Japan

E-mail: d.ichida@plasma.ed.kyushu-u.ac.jp

Abstract. We report here deposition of crystalline Ge nanoparticle films using a radio frequency magnetron sputtering method in argon and hydrogen gas mixture under a high pressure condition. The size of Ge nanoparticles is deduced to be 6.3-6.4 nm from the peak frequency shift of Raman spectra. Raman and X-ray diffraction spectra show that the films are crystalline. The film crystallinity strongly depends on substrate temperature (T_s). Highly crystalline Ge nanoparticle films are successfully fabricated at $T_s = 180^oC$.

1. Introduction

Semiconductor quantum-dot (QD) films have attracted much attention because of their application to high efficiency solar cells [1-3]. The unique characteristics of QDs such as tunable band gap and high multiple exciton generation (MEG) yield are expected to enhance the energy conversion efficiency above the Shockley and Queisser limit of 33% [4]. MEG can produce more than two excitons by one incident photon and represents a promising route to increase solar conversion efficiencies in single-junction photovoltaic cells. Our interest has been motivated by QD solar cells using Si and Ge nanoparticles [5-9], because Si and Ge are important materials of current electronics and solar cell industries. Ge QDs have more outstanding quantum confinement effects than Si QDs due to the larger excitonic Bohr radius of 24.3 nm for Ge QDs than 4.9 nm for Si ones [10-12]. Ge has also an inherent advantage over Si since Ge is known to exhibit much larger absorption coefficient in infrared light region than Si, and a high power conversion efficiency is expected for high solar light absorption.

In this study, we report fabrication of Ge QD films using Ge nanoparticles, where a high pressure rf magnetron sputtering method is applied to the deposition of Ge QD films with high crystallinity. First, we describe optical emission spectroscopy (OES) in argon and hydrogen gas mixture discharge plasma. Then, we present Raman and X-ray diffraction spectra of deposited Ge nanoparticle films deposited using a high pressure magnetron sputtering method.

2. Experimental

Figure 1 shows a schematic of experimental apparatus. Ge nanoparticle films were fabricated using 13.56 MHz RF magnetron sputtering process in an Ar and H_2 gas mixture under a high pressure condition of 200 Pa. At such high pressure, the mean free pass of Ge atoms is as short as the order of micrometer, and gas-phase Ge nanoparticle formation is possible. The fabricated Ge nanoparticles were transported to the quartz glass substrate located at 50 mm from the cathode by neutral gas flow. Ar and H_2 gas mixture were supplied at total flow of 133-250 sccm in the direction from the cathode to the substrate. H_2 dilution ratio $R_H = H_2/(H_2+Ar)$ was set in 20 %. The RF magnetron sputtering discharge plasma was generated by applying 13.56 MHz voltage to the powered electrode. The discharge RF power (P_{RF}) was 10-50 W. The sputtering target was a poly-crystalline Ge disk (1 inch) with a purity of 99.99 %.

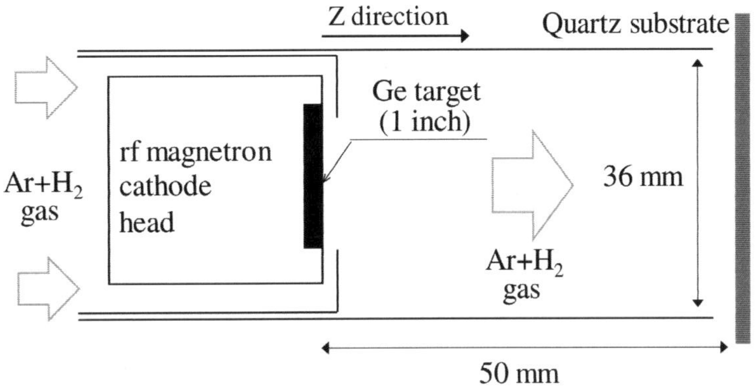

Figure 1. Experimental apparatus of high pressure RF sputtering.

3. Results and discussion

3.1. Optical emission spectroscopy of high pressure Ar/H_2 gas mixture sputtering discharge plasmas

First, we observed optical emission spectroscopy (OES) to study Ge nanoparticle production in Ar/H_2 sputtering discharge plasma. In our deposition system, Ge nanoparticles are nucleated and grow in Ar/H_2 plasma, and they are transported to the downstream region by neutral gas flow [13-15]. Therefore, the number density of Ge atoms in the plasma is an important parameter to control the particle size. Figure 2 shows a typical emission spectrum of Ar/H_2 discharge plasma. The collimator lens of 1-inch diameter was set to face to the Ge target for collecting emission light, and the intensity of light integrated in the z-direction was detected with spectrometer without setting a quartz substrate. The gas flow velocity was 0.27 m/s for the total gas flow rate of 133 sccm, and P_{RF} was 50 W. Ge 303 nm line ($I_{Ge:\ 303nm}$) is clearly detected in addition to Ar and H lines. As can be seen in figures 3 (a) and (b), Ge emission intensity increases linearly with P_{RF}, which is well consistent with the dependence of Ar emission intensity on P_{RF} because Ge target is mainly sputtered by Ar-ion bombardment. This results indicates that the number density Ge atoms in Ar/H_2 plasma is controlled by P_{RF}.

The emission intensity ratio of H_β 486 nm to H_α 656 nm ($I_{H:\ 486nm}/I_{H:\ 656nm}$), which corresponds to the effective electron temperature [16], is almost constant, although $I_{H:\ 486nm}/I_{H:\ 656nm}$ sharply increases at P_{RF} = 10 W (see figure 3(c)). The observed increase of $I_{H:\ 486nm}/I_{H:\ 656nm}$ at P_{RF} = 10 W is related to the shrink of plasma volume induced by lowering input power, because the electron temperature is determined by gas pressure and plasma volume [17].

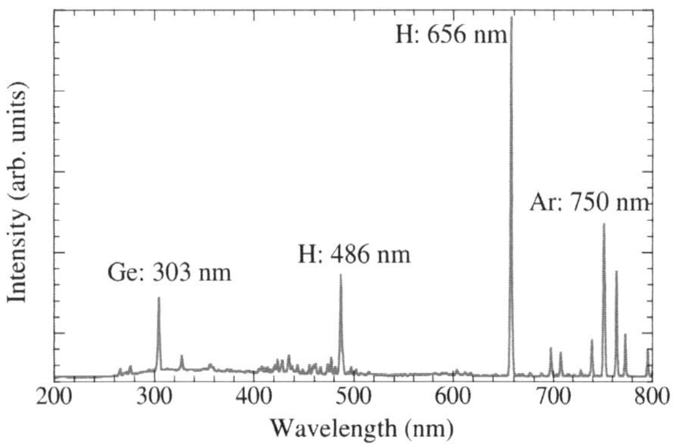

Figure 2. Typical optical emission spectrum from Ar/H_2 RF discharge plasma with the Ge sputtering target.

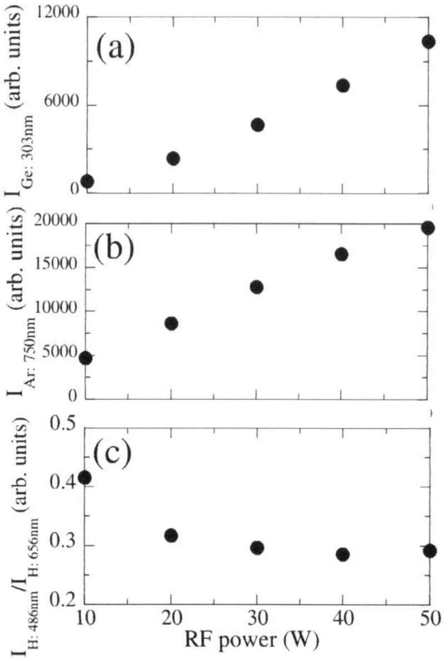

Figure 3. Discharge RF power (P_{RF}) dependence of emission intensity of (a) Ge 303 nm and (b) Ar 750 nm, and (c) emission intensity ratio of H_β 486 nm to H_α 656 nm.

3.2. Deposition of Ge nanoparticle films

We deposited Ge nanoparticle films using the high pressure rf magnetron sputtering method. The gas flow velocity was 0.50 m/s for the total gas flow rate of 250 sccm, and H_2 dilution ratio was 20 %. Figure 4 shows Raman spectra obtained for Ge nanoparticle films at P_{RF} = 10, 30 and 50 W, together with that of single crystalline Ge (c-Ge). Substrate temperature (T_s) was room one. Deposition rate increases linearly from 0.0039 to 0.042 nm/s with increasing P_{RF} from 10 to 50 W, which is well consistent with the P_{RF} dependence of $I_{Ge:303nm}$, as shown in figure 3(a). The peak frequency and the full width at half maximum (FWHM) of each peak are also given in figure 4. All films show a sharp peak at around 300 cm^{-1}, and these peaks are assigned to Ge crystals. Melting point of bulk Ge is as high as 938°C, which is much higher than gas temperature. However, the melting point of Ge nanoparticles is expected to decrease with decreasing nanoparticle size. The lower melting point for nanoparticles plays important roles to produce crystalline Ge nanoparticles. The shift of peak frequency with respect to that of reference single c-Ge depends on the nanoparticle size. There was no significant change in particle size as RF power was increased from 30 to 50 W. The particle size was deduced to be 6.3 - 6.4 nm from the relative Raman peak-shift frequency.

We also investigated the crystal orientation of Ge nanoparticle films. Figure 5 shows X-ray diffraction (XRD) pattern of Ge nanoparticle films deposited at T_s = room one and 180°C. P_{RF} was 50 W. X-ray diffraction spectra of Ge nanoparticle films show crystalline structure, where the diffraction peaks appear at $2\theta = 27^\circ$, 45°, and 53° corresponding to the (111), (220), and (311) crystal planes of Ge, respectively. A sharp peak with narrow width of 0.17 deg was observed for T_s = 180°C, and high crystallinity Ge nanoparticle films were successfully fabricated by the high pressure RF sputtering process. The H_2 dilution ratio, gas flow velocity, RF power, and substrate temperature are important parameters for realizing higher crystallinity, and further experimental investigations will be carried out to obtain deep understanding of the mechanism of Ge-nanoparticle crystallization.

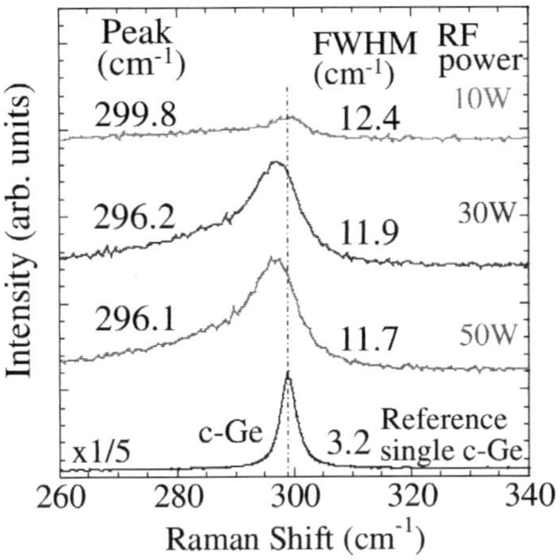

Figure 4. Raman spectra of Ge nanoparticle films showing crystalline structure. H_2 dilution ratio = 20 %, gas flow rate = 250 sccm, and substrate temperature (T_s) = room one.

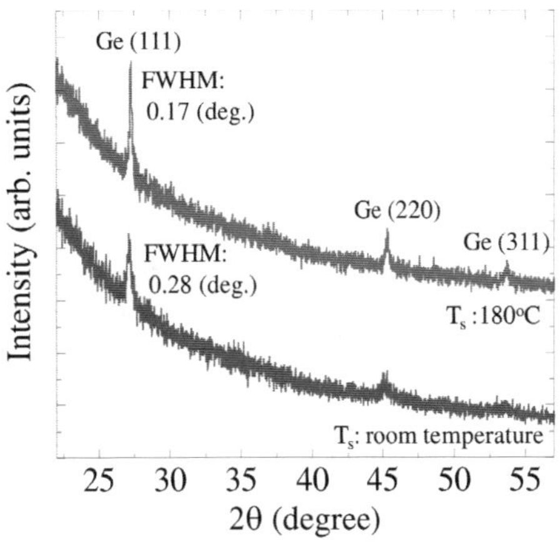

Figure 5. X-ray diffraction spectra of Ge nanoparticles deposited at substrate temperature (T_s) = room one and 180°C. Discharge RF power (P_{RF}) = 50 W, gas flow rate = 250 sccm, and H_2 dilution ratio = 20 %.

4. Conclusions

We deposited crystalline Ge nanoparticle films by the high pressure rf magnetron sputtering method. The peak frequency of Raman spectra shifts to high frequency with increasing P_{RF}, and the particle size is deduce to be 6.3 - 6.4 nm. The film crystallinity depends on T_s, and high crystalline Ge nanoparticle films are successfully fabricated at T_s = 180°C.

Acknowledgments

This work was partly supported by the New Energy and Industrial Technology Development Organization (NEDO).

References

[1] Nozik A J, Beard M C, Luther J M, Law M, Ellingson R J and Johnson J C 2010 *Chem. Rev.* **110** 6873
[2] Green M A 2006 *Third Generation Photovoltaics: Advanced Solar Energy Conversion* (Berlin: Springer)
[3] Conibeer G, Green M, Cho E, König D, Cho Y, Fangsuwannarak T, Scardera G, Pink E, Huang Y, Puzzer T, Huang S, Song D, Flynn C, Park S, Hao X and Mansfield D 2008 *Thin Solid Films* **516** 6748
[4] Shockley W and Queisser H J 1961 *J. Appl. Phys.* **32** 510
[5] Uchida G, Yamamoto K, Kawashima Y, Sato M, Nakahara K, Kamataki K, Itagaki N, Koga K, Kondo M and Shiratani M 2011 *Phys. Status Solidi* C **8** 3017
[6] Uchida G, Yamamoto K, Sato M, Kawashima Y, Nakahara K, Kamataki K, Itagaki N, Koga K and Shiratani M 2012 *Jpn. J. Appl. Phys.* **51** 01AD05
[7] Uchida G, Sato M, Seo H, Kamataki K, Itagaki N, Koga K and Shiratani M 2013 *Thin Solid Films* **544** 93
[8] Seo H, Wang Y, Uchida G, Kamataki K, Itagaki N, Koga K and Shiratani M 2013 *Electrochimi. Act.* **87** 213
[9] Seo H, Wang Y, Sato M, Uchida G, Kamataki K, Itagaki N, Koga K and Shiratani M 2013 *Jpn. J. Appl. Phys.* **52** 01AD05
[10] Fujii M, Hayashi S and Yamamoto K 1990 *Appl. Phys. Lett.* **57** 2692

[11] Zhang B, Shrestha S, Green M A and Conibeer G 2010 *Appl. Phys. Lett.* **96** 261901
[12] Maeda Y, Tsukamoto N, Yazawa Y, Kanemitsu Y and Masumoto Y 1991 *Appl. Phys. Lett.* **59** 3168
[13] Shiratani M, Koga K, Iwashita S, Uchida G, Itagaki N and Kamataki K 2011 *J. Phys. D* **44** 174038
[14] Kaneya T, Koga K, Shiratani M, Watanabe Y and Kondo M 2006 *Thin Solid Films* **506–507** 288
[15] Koga K, Inoue T, Bando K, Iwashita S, Shiratani M and Watanabe Y 2005 *Jpn. J. Appl. Phys.* **44** 1430
[16] Takai M, Nishimoto T, Kondo M and Matsuda A 2000 *Appl. Phys. Lett.* **77** 2828
[17] Lieberman M A and Lichtenberg A J 1994 *PRINCIPLES OF PLASMA DISCHARGE AND MATERIALS PROCESSING* (New York: John and Wiley and Sons)

Characteristics of plasma parameters in sputtering deposition using a powder target

H Kawasaki, D Taniyama, S Takeichi, T Ohshima, Y Yagyu, T Ihara, Y Tanaka and Y Suda

Department of Electrical and Electronic Engineering, Sasebo National College of Technology, 1-1 Okishin, Sasebo, Nagasaki 857-1193, Japan

E-mail: h-kawasa@sasebo.ac.jp

Abstract. Fabrication mechanisms of functional thin films prepared using a sputtering deposition method with metal-based powder targets were examined and profiles of electron density and emission intensity in the processing plasma were investigated. The electron density and electron temperature were 0.5×10^9 to 3×10^9 cm^{-9} and 0.5 to 1.5 eV, respectively. The radial profiles of the electron density were almost uniform, but the electron density and temperature at the edge of the electrode were higher than those at the center near the powered electrode. Axial profiles suggested that the electron density was the highest at the plasma/sheath boundary. The profile shape using a powder target was almost the same as that using a bulk target.

1. Introduction

Sputtering deposition is a method for preparation of thin films using an ion sputtered material from a bulk target; that is, deposition onto a substrate. This method has been used to prepare uniform amorphous and crystalline thin films with large areas. Based on the versatility of the sputtering method, we have developed several kinds of functional thin films, including those made from tungsten carbide, silicon carbide, chromium carbide, titanium carbide, cubic boron nitride, carbon nitride and silicon nitride [1-7]. High density bulk targets (>3 g/cm^3 and >95% density) are generally used in this method. Therefore, when producing a thin film with certain elemental ingredients using this method, it is necessary to make a new target compared to that used for other methods, such as spark plasma sintering. However, target preparation can be time-consuming and expensive. Thus, sputtering deposition would become more attractive if powder material targets can be used.

Kajima et al. prepared ferromagnetic nanocomposite oxide sputtered films with a Bi_2O_3-Fe_2O_3-$PbTiO_3$ ternary system using powder targets [8] and suggested that the prepared films worked well as thin film capacitors. We have also prepared TiO_2 films using a Ti powder target [1-2] and found that the deposition rate, Ti atom emission intensity and electron density in the processing plasma using the powder target were higher than those found using a bulk target. These parameters also slightly increased with an increase in the powder size. In contrast, electron temperatures in the plasma using a powder target were almost the same as those using a bulk target. These results suggest that the net-cathode surface area of the powder target is larger than that of the bulk target. However, spatial profiles of the plasma parameters, such as the electron density, electron temperature and emission species, were not investigated.

In this paper, functional thin films were prepared by a sputtering deposition method using metal-based powder and bulk targets. The characteristics of the sputtering plasma were studied based on measurements of electron density, electron temperature, and Ti atoms emission intensity in the plasma. Based on the results, the mechanisms of thin film deposition using powder targets were explored.

2. Experimental

A schematic illustration of the experimental apparatus has been presented elsewhere [2]. The deposition chamber was fabricated from stainless steel with a diameter of 400 mm and a length of 370 mm. The metal powder targets were set up on a stainless steel target holder. Rutile Ti powders of 45 μm diameter were used as the powder targets and a Ti metal target was used as a bulk target. The purity of the powder target in the target holder was similar to that in the bulk target (99.9%). The chamber was evacuated to 1 Pa using a rotary pump and then to a base pressure of 1×10^{-2} Pa using a turbomolecular pump. A deposition total pressure (P) of 10 Pa was produced using a mixture of oxygen (O_2: 99.99%) and argon (Ar: 99.99%) at a flow rate of 10 sccm. RF non-magnetron sputtering plasmas were generated at 13.56 MHz discharge frequency and 30-70 W discharge RF power. The Si(100) substrates were located 5.0 cm from the target. The substrates were cleaned by repeated bathing in an ultrasonic agitator and then rinsed in high-purity deionized water prior to loading into the deposition chamber. The electron density and temperature of the processing plasma were measured using a Langmuir probe of size 0.3 mmϕ and a 0.5 mm tungsten wire. The optical emission intensity of Ti atoms and the laser light scattering intensity were recorded through a spectrometer (Hamamatsu C5095) equipped with an ICCD camera (Hamamatsu C7164-03) after integrating 16 times.

3. Results and discussion

Radial distributions of electron density in the processing plasma using powder and bulk targets at different RF powers are shown in figure 1. Measurements were made using a Langmuir probe at 10 mm above the cathode surface. The electron density ranged from 0.5×10^9 to 3×10^9 cm^{-9} and increased with increasing RF power. The profiles suggest that the electron density at the edge of the electrode (-5 and +5 cm in figure 1) was higher than that at other positions. The radial electron density profile using a Ti bulk target at 70 W RF power (figure 1) was almost the same as that obtained with the powder target at the same RF power. However, the electron density using a powder target was almost 3 times higher than that using a bulk target. In our previous work, the deposition rate increased with increasing RF power using a powder target. These results are consistent with those in figure 1.

Radial distributions of electron temperature 10 mm above the substrate at different RF powers are shown in figure 2. The electron temperatures ranged from 0.5 to 1.5 eV and were almost independent of the RF power. The radial electron temperature profile using the Ti bulk target at 70 W RF power (figure 2) and the absolute electron temperature with the bulk target were almost the same as those obtained with the powder target.

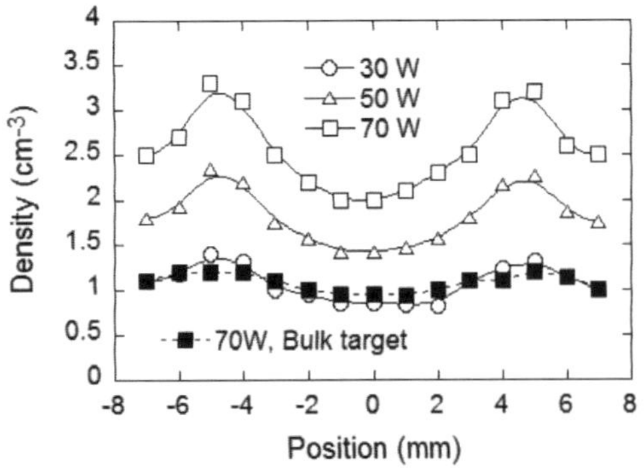

Figure 1. Radial distributions of electron density 5 mm above the substrate at different RF powers.

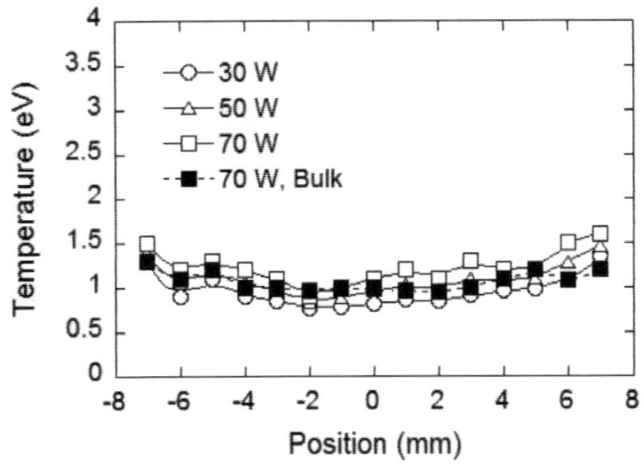

Figure 2. Radial distributions of electron temperature 5 mm above the substrate at different RF powers.

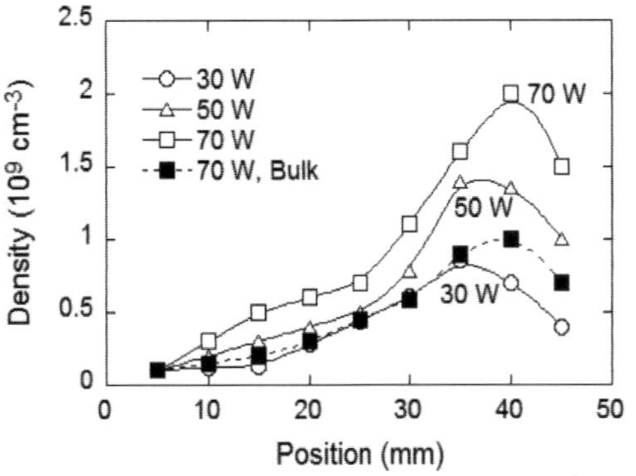

Figure 3. Axial distributions of electron density at the center of the electrode.

Axial distributions of electron density in sputtering plasma using a powder target at different RF powers are shown in figure 3. The electron density ranged from 0.5×10^9 to 3×10^9 cm^{-9}. The maximum electron density occurred at 35 to 40 mm from the anode surface, at the plasma/sheath boundary region.

Emission profiles of Ti atoms (337 nm) measured by a monochromator using a powder target at different RF powers are shown in figure 4. The profiles of Ti emission intensity were similar to the profiles of the electron density.

Axial profiles of the electron density, Ti emission intensity and laser light scattering (LLS) intensity using a powder target are shown in figure 5. LLS intensity was measured by 532 nm laser light and light scattering was measured using a photodetector. All the profiles have a sharp peak at the plasma/sheath region. The peak position of the LLS profiles was 5 mm shifted to the cathode. Small particle generation and growth were related to the high energy electrons near the sheath area. No LLS signal in the sputtering plasma was obtained using the bulk target.

In our previous work, the quality of films prepared by plasma processing was decreased by particle generation, which was detected by LLS methods [6]. LLS intensity was detected at the plasma/sheath region, but not on the anode surface.

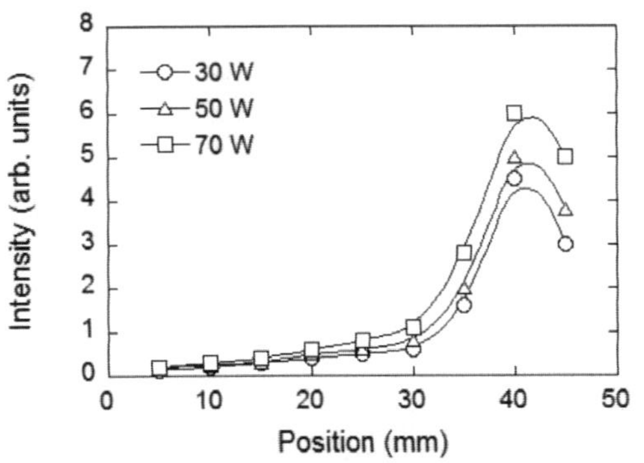

Figure 4. Axial distributions of Ti (337nm) emission at the center of the electrode.

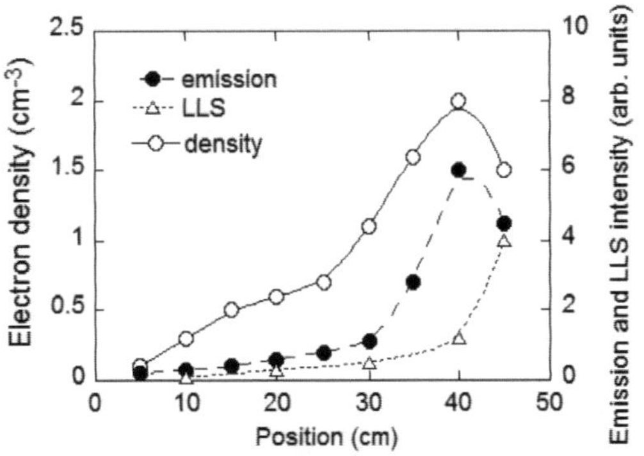

Figure 5. Axial distributions of electron density, Ti (337 nm) emission and LLS intensity at the center of the electrode.

4. Conclusion

Radial and axial profiles of electron density, electron temperature and LLS intensity were investigated in sputtering plasma using a powder target. The electron density and electron temperature were 0.5×10^9 to 3×10^9 cm^{-9} and 0.5 to 1.5 eV, respectively. The profiles of these parameters were uniform, which suggests that uniform thin films with large areas can be prepared. Small particle generation and growth were related to the high energy electrons near the sheath area at a particle size of 45 μm.

Acknowledgments

This study was supported in part by a Grant-in-Aid for Scientific Research in Priority Areas (B) (No.23340181), a Grant-in-Aid for Scientific Research on Innovative Areas (No.22110519), the Nippon Sheet Glass Foundation for Materials Science and Engineering, and the Circle for Promotion of Science and Engineering. The authors thank Prof. H. Fujiyama of Nagasaki University, Prof Shiratani of Kyusyu University and Prof. Inoue of Toyohasi University of Technology for their helpful discussions.

References

[1] Kawasaki H, Ohshima T, Arafune K, Yagyu Y and Suda Y 2012 *Trans. Mater. Res. Soc. Jpn* **37** 147
[2] Kawasaki H, Kanazawa T, Aoqui S, Ohshima T, Yagyu Y and Suda Y 2011 *Trans. Mater. Res. Soc. Jpn* **36** 455
[3] Kawasaki H, Shigematsu T, Imasaka K, Ohshima T, Yagyu Y and Suda Y 2011 *Trans. Mater. Res. Soc. Jpn* **36** 479
[4] Ohshima T, Murakami Y, Kawasaki H, Suda Y and Yagyu Y 2011 *Jpn. J. Appl. Phys.* **50** 08JD09
[5] Kawasaki H, Shibahara K, Ohshima T, Yagyu Y and Suda Y 2010 *Jpn. J. Appl. Phys.* **49** 08JF01
[6] Kawasaki H, Ohshima T, Yagyu Y and Suda Y 2008 *Trans. Mater. Res. Soc. Jpn* **33** 655
[7] Kawasaki H, Matsunaga T, Guan W, Ohshima T, Yagyu Y and Syda Y 2009 *J. Plasma Fusion Res. Series* **8** 1431
[8] Kajima A, Arita T, Tsuji Y, Inoue M and Fujii T 2006 *J. Magn. Soc. Jpn.* **30** 174

Journal of Physics: Conference Series **518** (2014) 012004 doi:10.1088/1742-6596/518/1/012004

Chemical vapor deposition of a-CN_x:H films for electron field emission using band supermagnetron plasma

H Kinoshita and S Yagi

Department of Electrical and Electronic Engineering, Graduate School of Engineering, Shizuoka University, 3-5-1 Johoku, Naka-ku, Hamamatsu 432-8561, Japan

E-mail: rdhkino@ipc.shizuoka.ac.jp

Abstract. A supermagnetron plasma apparatus with two band magnetron cathodes, which can form uniform magnetron plasma under a stationary magnetic field (about 160 G), was used to deposit a-CN_x:H films at N_2 gas concentrations of 0 and 70%. A high deposition rate of about 85±5 nm/min was obtained at a low dc self-bias voltage, between -20 and -62 V. The optical band gap could be controlled between 1.1 and 3.8 eV. The double-layer a-CN_x:H films with optical band gaps of 2.2 eV (upper layer; N_2 0%) and 1.5 eV (lower layer; N_2 70%) formed on p-Si substrate showed a low-threshold-emission electric field of 9 V/μm.

1. Introduction

The excellent electron transport characteristics of amorphous carbon films have attracted considerable attention due to their promising applicability to cold-cathode materials in the next generation of high-performance electronic devices, such as display devices and microelectronics [1-3]. Hydrogenated amorphous carbon (a-C:H) and hydrogenated amorphous carbon nitride (a-CN_x:H) films are of considerable interest due to their unique properties such as a high degree of hardness, low stress and chemical inertness [4]. Carbon nanotubes are also the recent leading candidate for use as field emitters [5, 6]. However, they must be deposited at high temperatures of 700−1000°C, which is too high to deposit on low-cost glass substrates [6, 7]. a-CN_x:H films can be deposited at temperatures below about 400°C [8, 9], and show excellent electron field emission characteristics from their flat surfaces [2, 10-12]. For cold cathode operation of devices such as flat-panel displays and panel light sources, a large-area electron emission with a low working voltage is required. Recent studies have shown that the field-emission behavior of a-CN_x:H planar structures is largely influenced by many factors relating to electron transport, such as the sp^2/sp^3 bonding ratio, surface morphology, and CC, CN, and CH bonding configurations. Many efforts have been made to reveal the relationship between field emission characteristics and preparation methods [11-14].

In the preparation of a-CN_x:H films by plasma chemical vapor deposition (CVD), high-speed and low-temperature deposition is preferable. For a high deposition rate, high-density plasma such as supermagnetron plasma (up to an order of 10^{11} cm^{-3}) is preferable [15, 16]. The supermagnetron plasma becomes very dense at an rf phase difference of 180°, which is suited to uniform deposition by rotating a magnetic field parallel to the cathode or by using a band-type magnetron cathode under a stationary magnetic field [17]. Using these plasma CVD apparatuses, the high-rate and uniform deposition of a-CN_x:II films can be achieved.

The addition of N atoms to a-C:H films, i.e. a-CN_x:H films, supports a lower electrical resistivity and a narrower optical band gap [15]. This could be caused by the increase in the sp^2/sp^3 ratio, the increase in the sp^2 cluster size or the connection of sp^2 clusters by N atoms [13, 18]. The emission threshold field is reported to be dependent on the distribution of sp^2 clusters. It gives rise to the high internal field enhancement within a-CN_x:H films [12]. Using supermagnetron plasma apparatuses, a-CN_x:H layers with low resistivity and/or high N atom concentration have been realized [14, 19]. These layers render excellent field emission characteristics [8].

In this study, a-CN_x:H films were prepared using i-C_4H_{10} as a hydrocarbon source and N_2 as the mixing gas. At the upper and lower electrode rf powers (UPRF/LORF) of 50–500/50–500 W (UPRF = LORF), i-C_4H_{10}/(none or N_2) mixed gas was used for the deposition of the a-C:H or a-CN_x:H planar structures. The factors relating to electron transport, such as the size or number of CC sp^2 clusters and a three-dimensional configuration of sp^3 CC bonds of these a-CN_x:H films, were controlled by the choice of rf powers and N_2 gas ratios, respectively. The field emission characteristics of these layers with single- or double-layer structure were evaluated and the mixing effect of N_2 gas was evaluated.

2. Experimental procedure

a-CN_x:H films were formed using the band supermagnetron plasma CVD apparatus shown in Figure 1. Two cathodes were arranged in parallel. The shape of each cathode was an electrode with rounded side surface, like a pillow [20]. By generating electron E×B drift with a closed loop over the front and back surface of the magnetron cathode, a uniform magnetron plasma was easily formed over the front surface of the magnetron cathode, i.e. the band magnetron cathode. By placing these two cathodes face-to-face in parallel at an electrode spacing of 30 mm, uniform supermagnetron plasma was generated over the cathode, as shown in the figure. The size of each flat cathode front surface was about 110×110 mm in square. Two different rf power sources with the same rf frequency (13.56 MHz) were supplied to two electrodes inserted into a grounded metal chamber. The front surface of the upper electrode was covered with a graphite plate to avoid the contamination of wafers by sputtering from upper metal electrode. The phase difference between the two rf voltages was controlled to be approximately 180° using a phase shifter, and a stationary magnetic field of approximately 160 G was applied parallel to the two electrode surfaces. The lower electrode (substrate holder) was heated to 30°C. Isobutane (i-C_4H_{10}; 15 sccm)/N_2 (0 or 35 sccm) was used as a deposition gas, and the gas pressure was controlled to be 4 Pa. The film thicknesses were measured by profilometry (ULVAC DECTAK-3). Using optical absorption data measured with a UV/vis/NIR spectrometer (Shimadzu

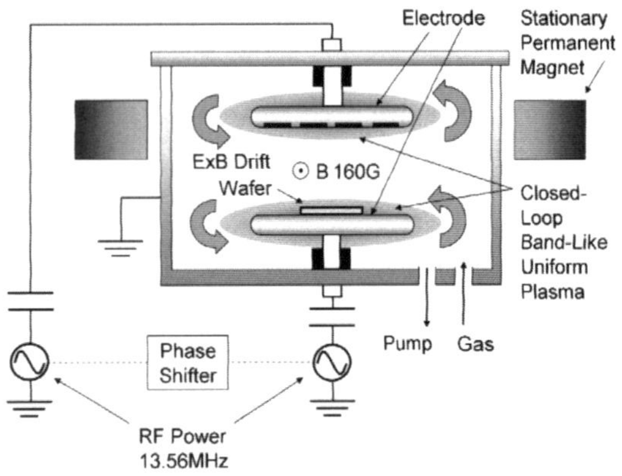

Figure 1. Schematic of the band super-magnetron plasma CVD apparatus with a statinary permanent magnet.

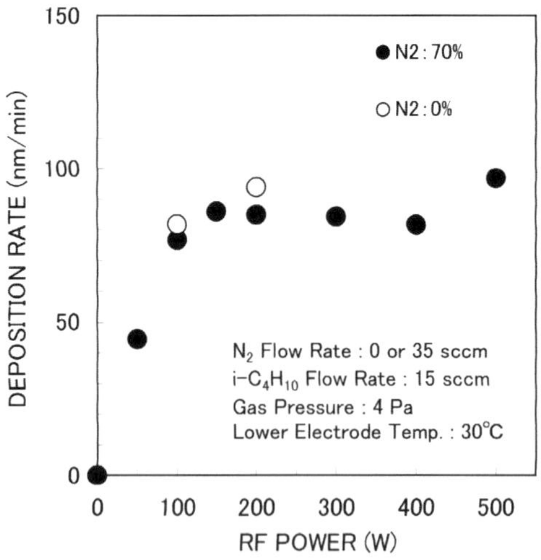

Figure 2. Deposition rate of a-CN_x:H films (N_2 of 0 or 70%) measured as a function of rf power (UPRF/LORF, UPRF = LORF).

UV-3100PC), the optical band gap was estimated from the Tauc plot based on the relation $(\alpha E)^{1/2} = A(E - E_G)$, where A is a constant, α is the absorption coefficient, E is the photon energy, and E_G is the optical band gap [21]. The bonding configurations of a-CN_x:H were measured using a Fourier transform infrared (FTIR) spectrometer (JIR WINSPEC-50). The thickness of the a-CN_x:H film used in FTIR analysis was controlled to be about 450 nm for all the samples.

3. Results and Discussion

The deposition rate of the a-CN_x:H films was measured as a function of rf power (UPRF = LORF), as shown in Figure 2. The deposition rate increased with rf power from 0 to 150 W. With the further increase of rf power from 150 to 500 W, the deposition rate changed little and was about 85±5 nm/min. At all rf powers, the nitrogen gas mixture had little influence on the deposition rate. From this experiment, it seems that the deposition rate increased with the increase of rf power from 0 to 150 W. With the further increase of rf power from 150 to 500 W, the formed a-CN_x:H films would be sputter-etched by positive ions generated in the plasma, and the deposition rate would increase very little. To investigate the sputter effect of this high-density plasma, the dc self-bias voltage (V_{DC}) of the lower electrode was measured, as shown in Figure 3. The absolute value of V_{DC} increased monotonously from 0 to -131 V with the increase of rf power from 0 to 500 W. The increase of the absolute value of V_{DC} from -41 to -131 V would cause high sputter-etching of deposited a-CN_x:H films.

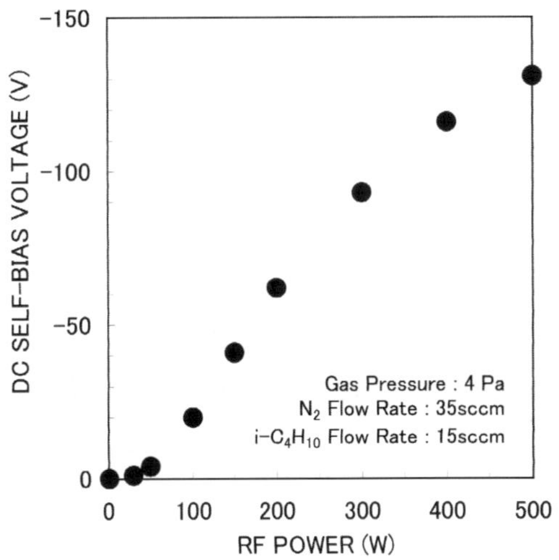

Figure 3. V_{DC} of lower electrode measured as a function of rf power.

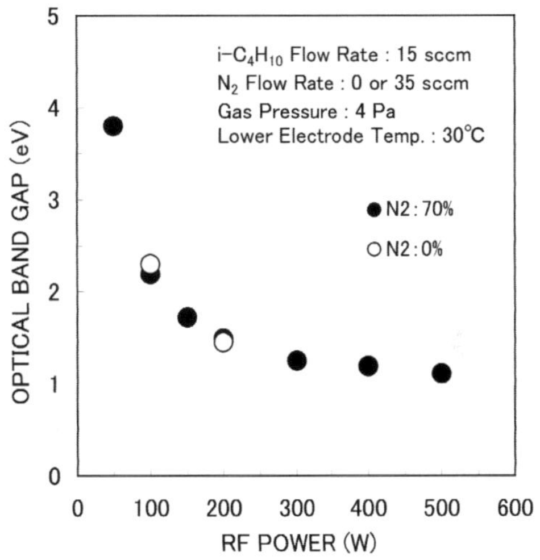

Figure 4. Optical band gap of a-CN_x:H films (N_2 of 0 or 70%) measured as a function of rf power.

The optical band gap of a-CN_x:H films was measured as a function of rf power, as shown in Figure 4. It decreased from 3.8 to 1.1 eV with the increase of rf power from 50 to 500 W, independent of the N_2 gas concentration (0 or 70%). This large decrease of the optical band gap was caused by the large increase of V_{DC}, i.e., ion bombardment energy. With the increase of ion bombardment energy, the size or number of CC sp^2 clusters would be increased [13].

The hardness of a-CN_x:H films was measured as a function of rf power, as shown in Figure 5. Hardness increased largely from 10 to about 24 GPa with the increase of rf power from 50 to 200 W. The hardness of a-C:H films was measured at 100 and 200 W, and was higher than that of a-CN_x:H films (N_2 gas concentration of 70%). Between the rf power of 200 to 500 W, that of a-CN_x:H films was almost constant (23±1 GPa). These values were slightly higher than that of vitreous silica (SiO_2),

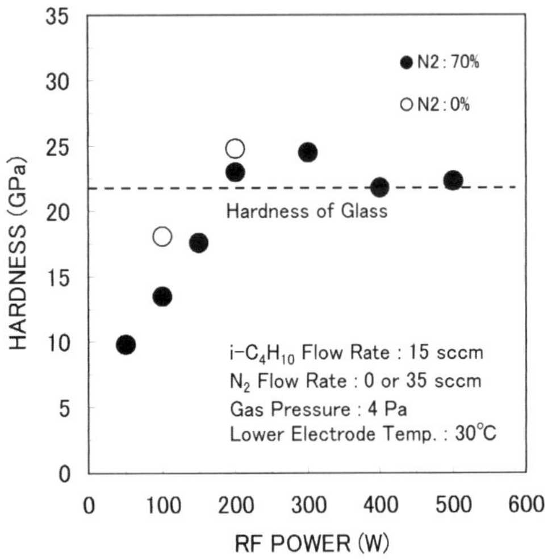

Figure 5. Hardness of a-CN_x:H films (N_2 of 0 or 70%) measured as a function of rf power.

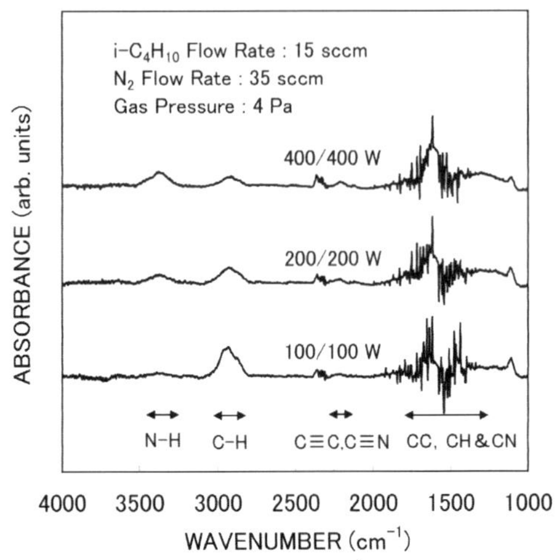

Figure 6. FTIR absorption spectra of a-CN_x:H films (N_2 of 70%) measured as a function of rf power.

which we measured as 22 GPa. Such a high hardness was ascribed by the inclusion of a three-dimensional configuration of sp^3 CC bonds and also sp^2 CC bonds. In spite of such a low V_{DC}, between -60 (200 W) and -130 V (500 W), high hardness was achieved using this uniform band supermagnetron plasma CVD apparatus. This plasma CVD apparatus is suited to synthesizing hard a-CN_x:H films under very low V_{DC} (< -130 V).

FTIR absorption spectra were measured for a-CN_x:H films deposited at rf powers of 100, 200 and 400 W, as shown in Figure 6. The intensities of the absorption bands at 3300 cm^{-1} (NH bonds) increased with rf power. This increase in intensity was caused by the inclusion of nitrogen atoms in the films [22]. The intensities of absorption bands at 2930 cm^{-1} (CH_3, CH_2, and CH bonds) were decreased with the increase in rf power. Those at 1100 to 1700 cm^{-1} (C=C, C=N, and C=N-H bonds) increased slightly with rf power. The FTIR absorption spectrum of a-C:H (100 or 200 W) was also measured, and was similar to that of a-CN_x:H film deposited at 100 W. In the spectrum, no peaks at 1100 to 1700 cm^{-1} (C=C, C=N, and C=N-H bonds) were observed.

a-CN_x:H films of about 200-nm thickness with single- or double-layer structures were deposited at an rf power of 100–500 W on p-Si substrates to investigate the influence of the rf power and layer structure on their field emission characteristics, as shown in Figures 7 and 8. The surface morphology of the 200-nm thick a-CN_x:H film with single-layer structure (N_2 of 70%, 500 W) deposited for field emission measurement was analyzed by AFM. The surface of the a-CN_x:H film was flat with a surface roughness less than 1 nm over an area of 500 × 500 nm^2 or 5 × 5 μm^2, as shown in Figure 9 [23]. In the case of the double-layer structure, the layer thickness of each a-CN_x:H film (N_2 of 0 or 70%) was about 100 nm. Fowler-Nordheim (FN) plots of their field emission characteristics and the constructions of the field emitters are shown in the figure insets. The threshold emission electric field (E_{TH}) was defined as the applied electric field at the field emission current of 0.01 μA. In the case of single-layer structure (N_2 of 70%), the minimum E_{TH} of 13 V/μm was obtained at rf powers of 200, 300 and 400 W, and the maximum E_{TH} 17 V/μm was obtained at an rf power of 500 W. This means that the fitting between the two energy levels of a-CN_x:H and p-Si is important to ensure the flow of electrons from p-Si to a-CN_x:H layer, i.e., electrons in the valence band of p-Si (0.01 Ω cm; about 1×10^{19} cm^{-3}) are injected into the conduction band (including CC sp^2 clusters) of a-CN_x:H by tunneling with a high electric field applied between p-Si and n-type a-CN_x:H [24]. The electron channel formed in a-CN_x:H film may be CC sp^2 clusters. The optimal rf powers in this experiment for the energy level fitting were 200, 300 and 400 W. In the case of the double-layer structure [upper layer (N_2 of 0 and 70%), 100 W; lower layer (N_2 of 70%), 200 W], E_{TH} took small values of 9 and 12 V/μm, respectively. The FN plots of these a-CN_x:H thin films with single- and double-layer structures were almost linear, as shown in the figure insets. This means that the electron field emissions examined obeyed the FN relationship for field emission. From this experiment, it was found that a double-layer structure with wide optical band gap (upper layer) is effective to lessen E_{TH}.

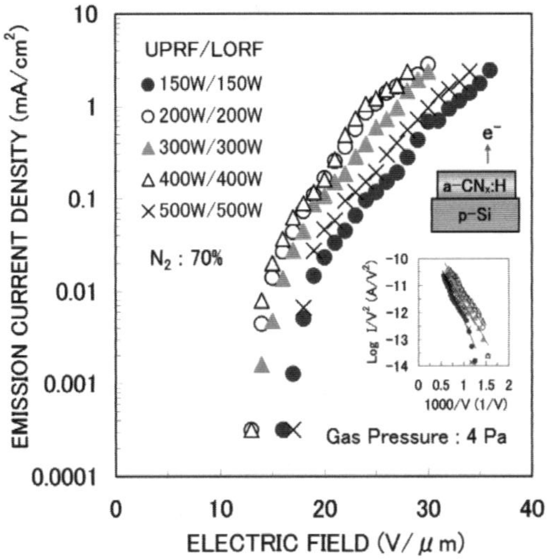

Figure 7. Emission current-voltage characteristics of a-CN_x:H films (single-layer) as a function of rf power. The inset shows their FN plots and their constructions.

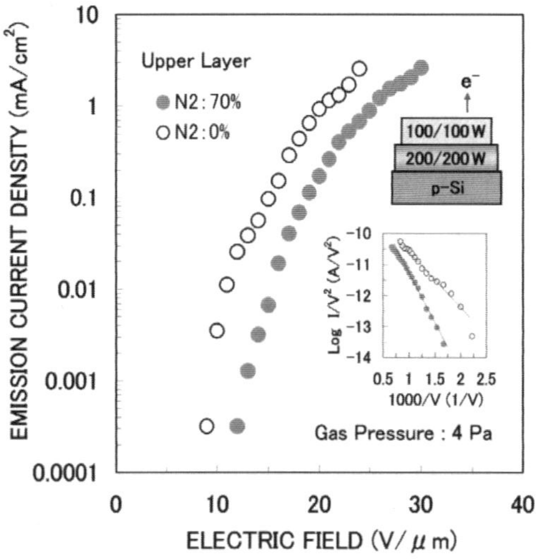

Figure 8. Emission current-voltage characteristics of a-CN_x:H films (double-layer) as a function of rf power. The inset shows their FN plots and their constructions.

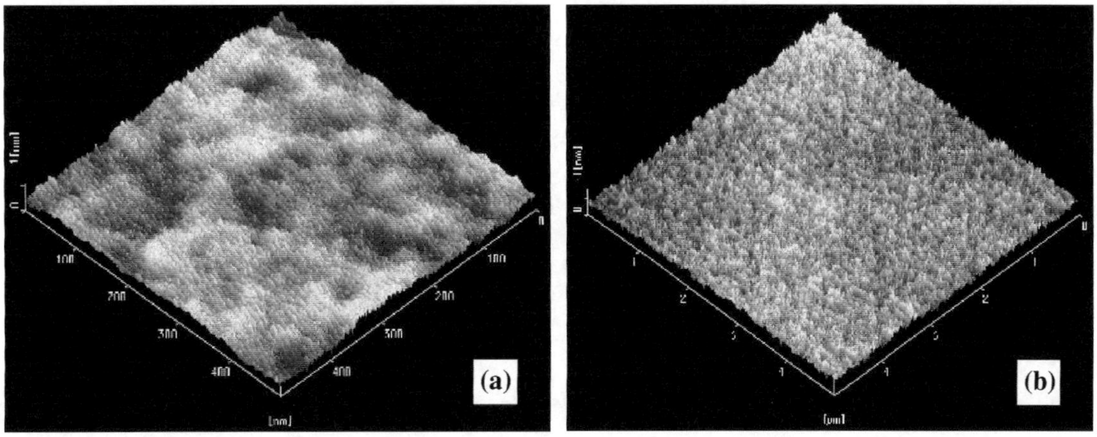

Figure 9. AFM images of the a-CN_x:H film with single-layer structure (N_2 of 70%, 500 W, 200-nm thickness). x-y axes (maximum): (a) 500 nm; (b) 5 μm, and z axes (maximum): 1 nm.

4. Conclusions

A uniform supermagnetron plasma CVD apparatus operating under a stationary magnetic field was used for a-CN_x:H film (N_2 of 0 or 70%) formation. A high deposition rate of about 85±5 nm/min was obtained at a low V_{DC}, between -20 and -62 V (rf power of between 100 and 200 W). Hard (19 ± 6 GPa) and flat (surface roughness less than 1 nm) a-CN_x:H films were successfully formed. The a-CN_x:H films with single- or double-layer structures were applied to electron field emission. The a-CN_x:H films with a double-layer structure [optical band gap: 2.2 eV (upper layer) and 1.5 eV (lower layer)] on p-Si substrate showed a low E_{TH} of 9 V/μm at 100 W for the upper layer (N_2 of 0%) and at 200 W for the lower layer (N_2 of 70%). Wide optical band gap of the upper layer is effective for reducing E_{TH} values.

References

[1] Jung J H, Ju B K, Kim H, Oh M H, Chung S J and Jang J 1998 *J. Vac. Sci. Technol. B* **16** 705
[2] Lin C M, Chang S J, Yokoyama M, Chuang F Y, Tsai C H, Wang W C and Lin I N 1999 *Jpn. J. Appl. Phys.* **38** 890
[3] Xu N S, Wu Z S, Deng S Z and Chen J 2001 *J. Vac. Sci. Technol. B* **19** 1370
[4] Lettington A H 1993 *Philos. Trans. R. Soc. London Ser. A* **342** 287
[5] Knapp W and Schleussner D 2003 *Vacuum* **69** 333
[6] Küttel O M, Groening O, Emmenegger C and Schlapbach L 1998 *Appl. Phys. Lett.* **73** 2113
[7] Semet V, V. Binh V T, Vincent P, Guillot D, Teo K B K, Chhowalla M, Amaratunga G A J, Milne W I, Legagneux P and Pribat D 2002 *Appl. Phys. Lett.* **81** 343
[8] Kinoshita H, Ikuta R and Murakami S 2004 *J. Vac. Sci. Technol. A* **22** 1857
[9] Kinoshita H and Suzuki H 2011 *J. Mod. Phys.* **2** 398
[10] Amaratunga G A J and Silva S R P 1996 *Appl. Phys. Lett.* **68** 2529
[11] Cui J B, Robertson J and Milne W I 2001 *J. Appl. Phys.* **89** 3490
[12] Uemura Y, Murai S, Koide Y and Murakami M 2002 *Diamond Relat. Mater.* **11** 1429
[13] Ilie A, Ferrari A C, Yagi T, Rodil S E, Robertson J, Barborini E and Milani P 2001 *J. Appl. Phys.* **90** 2024
[14] Panwar O S, Kumar S, Rajput S S, Sharma R and Bhattacharyya R 2004 *Vacuum* **72** 183
[15] Kinoshita H and Fukushima F 2000 *J. Appl. Phys.* **88** 2263
[16] Kinoshita H and Otaka N 2002 *J. Vac. Sci. Technol. A* **20** 1481
[17] Kinoshita H, Yagi S and Sakurai M 2013 *J. Mod. Phys.* **4** 587
[18] Wood P, Wydeven T and Tsuji O 1995 *Thin Solid Films* **258** 151
[19] Kinoshita H and Yamaguchi A 2010 *Jpn. J. Appl. Phys.* **49** 08JF07
[20] Narasimhan M, Sasserath J and Ghanbari E 1992 *J. Vac. Sci. Technol. A* **10** 1100
[21] Tauc J 1974 *Amorphous, Liquid Semiconductors* (Plenum Press, New York)
[22] Kinoshita H and Yamashita M 2007 *Thin Solid Films* **515** 5142
[23] Kinoshita H, Yamashita M and Yamaguchi T 2006 *Jpn. J. Appl. Phys.* **45** 8401
[24] Kinoshita H, Ninomiya Y and Kato T 2013 *Jpn. J. Appl. Phys.* **52** 116201

Journal of Physics: Conference Series **518** (2014) 012005 doi:10.1088/1742-6596/518/1/012005

Dependence of ion current on magnetic unbalancing in high power pulsed magnetron glow discharge

K Azuma and Y Inoue

Department of Electrical Engineering and Computer Sciences, University of Hyogo, 2167 Shosha, Himeji, Hyogo, Japan

E-mail: easter11_azuma@mem.iee.or.jp

Abstract. A high-power pulsed sputtering (HPPS) discharge is able to generate high ionization metallic plasma by applying a negative high voltage pulse-train with low duty ratio and high current capacity to a sputtering target. If HPPS discharge technique is applied to unbalanced magnetron sputtering (UBMS) system which is one of the ionized sputtering, we will get the metallic plasma ionized more. The ion current of direct-current UBMS increases with increase in magnetic un-balancing. However, high-power pulsed UBMS does not necessarily have the same characteristics as dc-UBMS. We investigated a connection between the ion current to a collector electrode and the number of peripheral magnets in the sputter target. The collector current increased with increase in the number of the magnets, but the ratio of the collector current to the target current depended more strongly on average input power than that for dc-UBMS. It was found that the current ratio indicated the maximum of 18.8% at the input power of 400 W in the hpp-UBMS with the peripheral 10-magnets.

1. Introduction

The sputtering deposition is one of the technologies that cover the surface of a substrate with functional thin film. It is known as the deposition process to able to form dense thin film. Especially a magnetron sputtering process has high deposition rate. The planar-type magnetron sputtering method was originally developed in 1974. A closed magnetic field is formed near the planar magnetron target by arranging permanent magnets on the backside of the target. High-density plasma is then only formed near the target by capturing secondary electrons in the magnetic field. The ions generated in the plasma are then accelerated toward the target to sputter the target material. Planar magnetron sputtering has a faster deposition rate than sputtering in the absence of a magnetic field. However, there is a drawback of low ionization-rate of the sputtered metallic species. A UBMS (unbalanced magnetron sputtering) [1, 2] can be used to improve the ionization rate compared to the traditional planar magnetron sputtering system. Another technique of raising the ionization rate of metallic plasma is a method of applying a high-power pulsed sputtering (HPPS) discharge. The HPPS discharge is able to generate high ionization metallic plasma by applying a negative high voltage pulse-train with low duty ratio and high current capacity to a sputtering target. This can be applied to the various sputter targets. A magnetron sputter target applied with HPPS technique is called HPPMS (high-power pulsed magnetron sputtering) or HiPIMS (high-power impulse magnetron sputtering) that was reported in 1999 [3]. Of course, it has been studied that HPPS discharge is applicable to the UBMS (for instance [4, 5]). If HPPS technique is applied to UBMS, we will get the metallic plasma more ionized using hpp-UBMS (high-power pulsed UBMS) than dc-UBMS (direct-current UBMS).

Commonly, the ion current of dc-UBMS increases with increase in magnetic un-balancing. However, hpp-UBMS does not necessarily have the same characteristics as dc-UBMS. In this study, we made the sputtering target electrode into which the number of the peripheral magnets was changed, and investigated a connection between the collector ion current and the number of the peripheral magnets in the sputter target.

2. The experimental setup

Figure 2 shows the experimental setup of the hpp-UBMS (high power pulsed unbalanced magnetron sputtering) system. The magnetron target consisted of the sputter target, the heat sink and the permanent magnets. The heat sink made from a circular plate of copper had the center hole in diameter of 14 mm and the torus channel in inside and outside diameters of 50 mm and 78 mm. One permanent magnet was inserted into the center hole. The others were inserted into the torus channel equiangularly. The magnets were NdFeB magnets (Niroku seisakusho Co., NE062, Ø14×17). A titanium sputter target with diameter of 70 mm and thickness of 5 mm was bolted to on the heat sink. The distance from the magnet surface to the target surface was 7 mm. The vacuum chamber had an inner diameter of 310 mm and a height of 300 mm. The sputter target was set on the upper flange of the chamber and the collector electrode with diameter of 80 mm was inserted from the bottom flange. The distance between the sputter target and the collector electrode was 100 mm. The evacuation system consisted of a turbo molecular pump (TMP, 350 L/min) and a rotary pump. The inlet gas pressure was controlled by a mass flow controller. The argon gas was introduced from behind the sputter target. The pressure inside of the vacuum chamber was measured using a ceramic capacitance manometer (ULVAC, GM-2001/CCMT-10A) and an ionization gauge (ANELVA, MIG-430/MG-2).

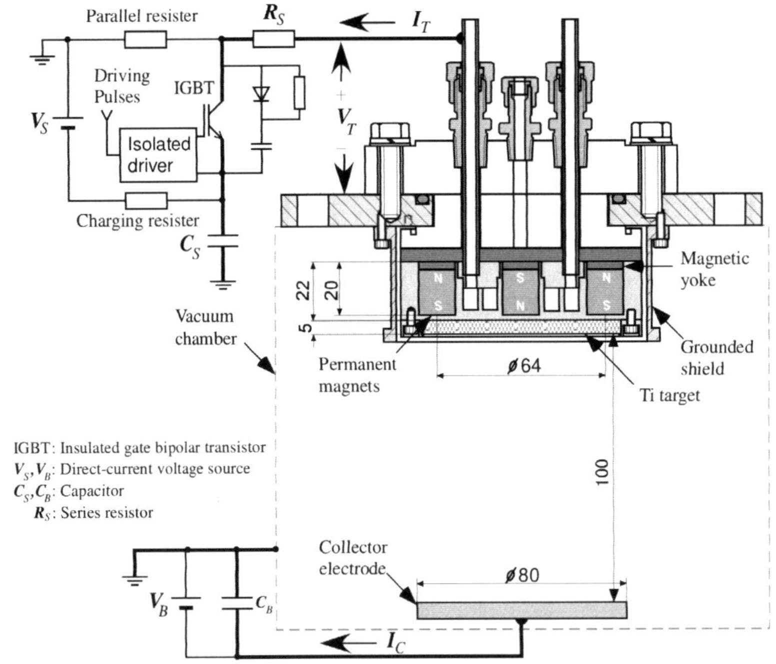

Figure 1. The experimental setup.

The negative high voltage pulse-train was generated by an IGBT (Insulated Gate Bipolar Transistor). The sputter target was connected to a high voltage capacitor C_S (200 μF, 2kV) through a series resistor R_S (4 Ω, 300 W) and the IGBT. The capacitor was connected to a negative direct-current voltage source V_S (−800V, 5 kW). The peak value of the voltage pulse-train was controlled by the voltage source, and the pulse duration and repetition rate were controlled by the driving signals to the

IGBT. The collector electrode was connected to a negative bias voltage source (−50V, 3A) with parallel-connected capacitor (40 μF, 275 V). The target current, I_T, through the circuit was observed using a current transformer (Pearson type 110A). The target voltage, V_T, was observed using a voltage probe (Tektronix, type 5100). The collector current I_C was observed using a current transformer (Tektronix, P6021). These waveforms were monitored with an oscilloscope (Tektronix, TDS3034B). The skew of the current signal caused by the frequency property of the current transformer was corrected by software processing. In the case of driving the sputter target with only the dc power source, the target voltage, the target current and the ion current were measured by three digital multi-meters.

The plasma was generated at the gas pressure of 0.5 Pa and the gas flow rate of 15 sccm. The negative voltage pulse-train (a set-up voltage of −800 V, a pulse duration of 100 μs and a repetition rate of 500 Hz) was applied the titanium sputter target. The bias voltage of −50 V was applied to the collector electrode. A floating voltage of the collector electrode was −10 V to −20 V. Therefore, the collector current observed was the ion current.

3. Results

3.1. Electrical characteristics of dc-UBMS

Figure 2 shows the current-voltage characteristics of the sputter target by driving the direct-current power source at the various number n of the peripheral magnets. *I-V* curves of n = 6 and n = 7 were completely different from those of n = 8 and n = 10. It is thought that the non-uniformity of the magnetic field in the circumferential direction has influenced the discharge at $n < 8$. The *I-V* curves of n = 6 and n = 7 were saturated at the range greater than V_T = 460 V and V_T = 500 V, respectively. When eight or more magnets were arranged on the outside zone, the target current I_T grew rapidly with increasing the target current V_T. In the case of n = 8 and n = 10, I_T was expressed by empirical composition with base V_T and exponent 7.34 ±0.07 and 8.55 ±0.23 respectively.

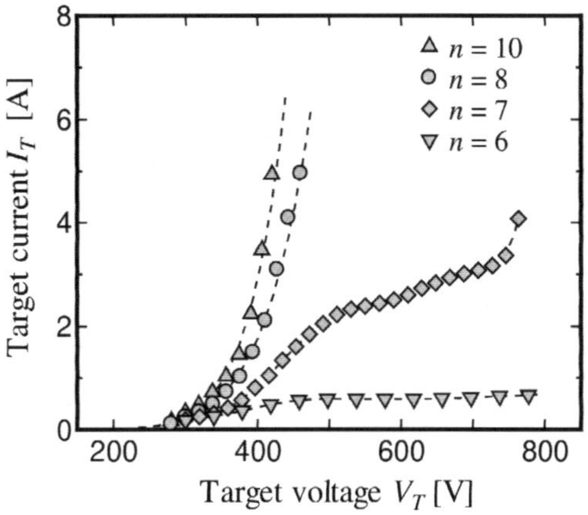

Figure 2. The current-voltage characteristics of the sputter target with n-magnets by driving the direct-current power source.

3.2. Electrical characteristics of hpp-UBMS

Figures 3 show waveforms of (a) target voltage, V_T, (b) target current, I_T, and (c) collector ion current, I_C. The setup voltage $-V_S$, the pulse duration τ and repetition rate f of the applied negative voltage pulse-train were −800 V, 100 μs and 500 Hz respectively. In the case of 8-magnets, the target voltage $-V_T$ reached −800 V just after a pulse voltage was applied to the target and began to decrease after less than a several microseconds. The target current I_T grew rapidly, but its growth rate slowed down after 14 μs. The peak value of was 72 A at t = 28 μs after the turn-on time. The target voltage gradually decreased, followed by a stationary state. This is for a voltage drop caused by series resistance R_S. When the target current reached a peak, the target voltage indicated about −470 V. Then, the target current decreased slowly. At the turn-off time of 100 μs, I_T and V_T became 51 A and −540 V, respectively. The collector ion current I_C continued to grow until about the time of 57 μs and was saturated with 5.6 A. After the turn-off of the voltage pulse, the collector ion current was decayed more slowly than the target current. In the case of 8-magnets, the target current grew up more rapidly than in case of 10-magnets and reached the peak of 80 A at t = 26 μs. At the turn-off time, I_T and V_T became 69 A and −460 V, respectively. The collector ion current I_C reached the peak of 9.7 A at t = 68 μs. After this, the ion current decreased slowly. At the turn-off time, I_C was 8.6 A.

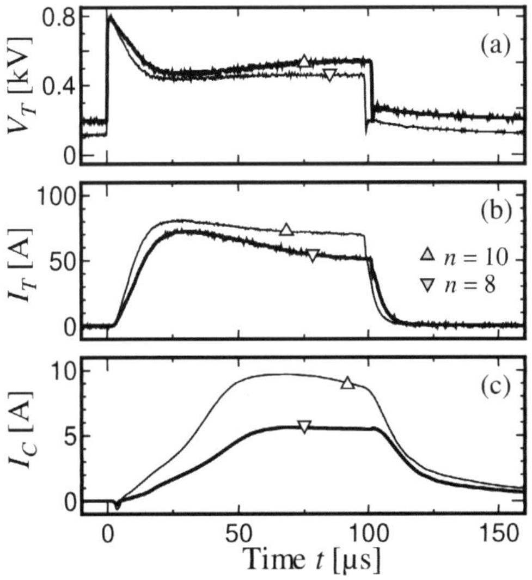

Figure 3. The typical time response of (a) target voltage $-V_T$, (b) target current I_T and (c) collector ion current I_C on the collector electrode by driving the pulse voltage source of −800 V, 100 μs and 500 Hz under the ambient argon gas pressure of 0.5 Pa. The heavy lines and narrow lines indicate waveforms observed at the 8-magnets and 10-magnets respectively.

3.3. Dependence on the magnetic unbalancing

Figure 4 shows the unbalance coefficient dependence of (a) the peak target current $I_{T,peak}$ and the peak collector ion current $I_{C,peak}$ and (b) the peak input power $p_{T,peak}$ and the average input power P_T at the setup voltage of −800 V. The coefficient of unbalance K is the ratio of magnetic fluxes from the central and peripheral magnets on the target surface [5]. In our system, K is nearly equal to the number n of the peripheral magnets because all of the central 1-magnet and peripheral n-magnets are the same material and the same proportion. The distance from the target surface to a magnetic null point at n = 6, 8 and 10 is 18.9 mm, 16.6 mm and 13.5 mm, respectively.

The average input power P_T is the value which carried out the time average of the instant electric power $p_T(t)= V_T(t)\times I_T(t)$ supplied to a plasma source, and is denoted by the following formula:

$$P_T = f\int_0^{1/f} p_T(t)dt \tag{1}$$

where f is the repetition frequency of high-voltage pulses. At $n \geq 8$, The $I_{T,peak}$, $p_{T,peak}$ and P_T all barely depended on n, and were about 80 A, 35 kW and 1.5 kW respectively. In contrast, the collector ion current $I_{C,peak}$, increased from 5.6 A to 9.7 A with increase in n from 8 to 10-magnets. It was confirmed that larger ion current can be obtained by raising the magnetic unbalancing.

Since hpp-UBMS is pulsed discharge, the electrical properties of it cannot be directly compared with dc-UBMS. Then, we introduce the amount of charges which passes an electrode per unit time, Q. Q is given by the following formula:

$$Q = f\int_0^{1/f} i(t)dt \tag{2}$$

where $i(t)$ is the instantaneous current detected the electrode and f is the repetition rate of the pulse-train. This is exactly the average current of $i(t)$. Q is not the electric current strictly in electric engineering. Here, since we would like to discuss the quantity of the ions which reaches a collector electrode, we think it is appropriate to compare Q of hpp-UBMS with I of dc-UBMS.

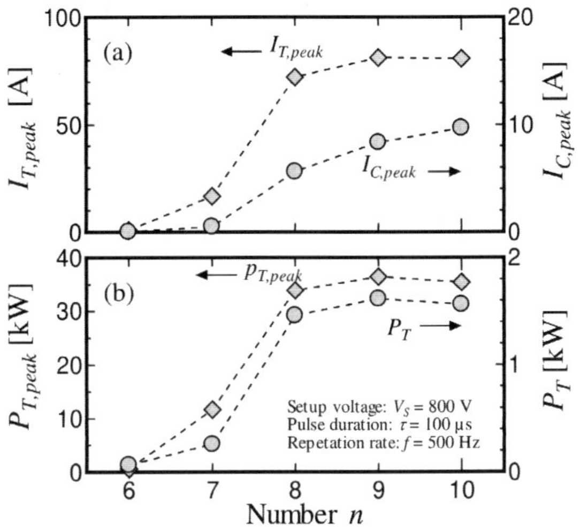

Figure 4. The dependence of (a) the peak target current $I_{T,peak}$ and the peak collector ion current $I_{C,peak}$ and (b) the peak power $p_{T,peak}$ and the input power P_T on the n -magnets.

The input power dependence of the ratio of I_C to I_T is shown in Figure 5. Figure 5 (a) and (b) show I_C/I_T in dc-UBMS and and Q_C/Q_T in hpp-UBMS. The subscripts T and C express the target and collector electrode, respectively. I_C/I_T reached the peak at $P_T \approx 100$ W and decreased with increase in P_T to about 1.3 kW. I_C/I_T decreased with increase in P_T and increased with increase in n. But, I_C/I_T became independent from P_T with increase in n. The plasma generated by sputtering is high density near the target, and lower density near the collector electrode. When electric power becomes low, an ion sheath on the collector electrode will become thick and a potential difference of the sheath becomes large. For this reason, when electric power is small, I_C becomes large relative to I_T in the case of low P_T, otherwise it becomes smaller. When n becomes large, the density of the plasma near the collector electrode will become higher and the ratio of I_C to I_T stops being dependent on P_T. On the other hand, the curve of Q_C/Q_T-P_T reached the peak at $P_T \approx 400$ W and approached more precipitously with increase in n. When n becomes large, a ratio of Q_C to Q_T not only becomes large but it will be strongly dependent on P_T. Such a characteristic of Q_C/Q_T may be related to how depending on which plasma spreads. The pressure of plasma will become high when P_T becomes large enough. The plasma pressure increases the cross section of a plasma flow. When the cross section of plasma flow becomes larger than a collector electrode, the ratio of Q_C to Q_T will fall. This guess is backed up by the fact that P_T at maximum point of Q_C/Q_T - P_T curve decreases with increase in n. At n = 10, the peaks of I_C/I_T and Q_C/Q_T were 17.2% and 18.8%, respectively. Generally, dc-MS is operated with the power flux density of 10 W/cm^2 on the sputter target [7]. When it is converted by our system, the input power is about 200 W. In terms of the improvement of the ionization due to the ion current by applying HPPS technique, hpp-UBMS should be operated with several-fold power of dc operation. These experimental results have suggested that the behavior of the ion current differs considerably in dc-UBMS and hpp-UBMS.

4. Conclusions

We investigated a connection between the ion current detected at the collector electrode and magnetic unbalancing of hpp-UBMS. The magnetic unbalancing was realized by changing the number of peripheral magnets in the sputter target. The ion current increased with increase in the number of the magnets, but, the ratio of the collector ion current to the target current depended more strongly on the average input power than dc-UBMS. When the number of peripheral magnets is eight or more, the peak input power, the average input power and the peak of the target current all barely depended on the number of magnets and indicated about 35 kW, 1.5 kW and 80 A. In contrast, the peak value of the collector current increased from 5.6 A to 9.7 A with increase in 8 the number of magnets from to 10. The ratios of the collector current to the target current with 10-magnets in dc-UBMS and hpp-UBMS had the peak of 17.2% at the input power of 100 W and 18.8% at 400 W respectively. It is suggested that the behavior of the ion current differs considerably by dc-UBMS and hpp-UBMS.

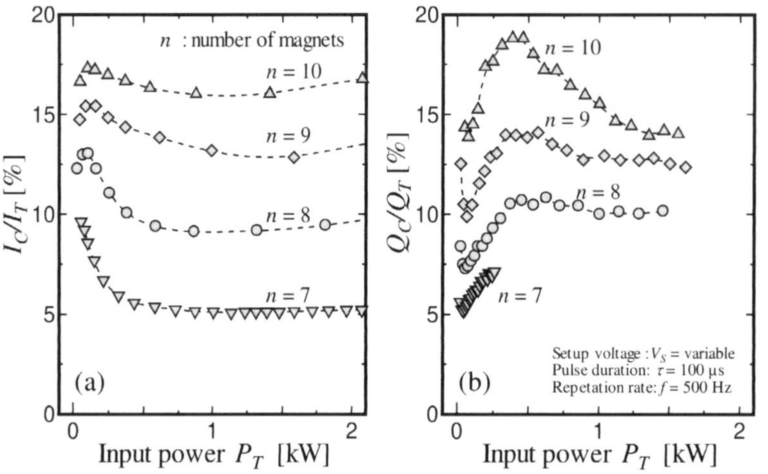

Figure 5. The input power dependence of the ratio of the collector ion current to the target current I_C/I_T and Q_C/Q_T in the cases of (a) dc-MS and (b) hpp-MS.

References

[1] Window B and Savvides N 1986 Unbalanced dc magnetrons as sources of high ion fluxes *Journal of Vacuum Science & Technology A* **4** (3) 453–456

[2] Savvides N and Window B 1986 Unbalanced magnetron ion-assisted deposition and property modification of thin films *Journal of Vacuum Science & Technology A* **4** (3) 504–508

[3] Kouznetsov V, Macák K, Schneider J M, Helmersson U and Petrov I 1999 A novel pulsed magnetron sputter technique utilizing very high target power densities *Surface and Coatings Technology* **122** (2–3) 290–293

[4] Vlček J, Kudláček P, Burcalová K and Musil J 2007 High-power pulsed sputtering using a magnetron with enhanced plasma confinement *Journal of Vacuum Science and Technology A* **25** (1) 42–47

[5] Mishra B, Moore J J, Lin J and Sproul W D 2010 Advances in Thin Film Technology through the Application of Modulated *Materials Science Forum* **638–642** 208–213

[6] Svadkovski I V, Golosov D A and Zavatskiy S M 2002 Characterisation parameters for unbalanced magnetron sputtering systems *Vacuum* **68** (4) 283–290

[7] Anders A 2010 High power impulse magnetron sputtering and related discharges: Scalable plasma sources for plasma-based ion implantation and deposition *Surface and Coatings Technology* **204** (18–19) 2864–2868

Deposition of SiO_2 in a SiH_4/O_2 inductively coupled plasma

L Z Tong

Keisoku Engineering System Co., Ltd., 1-9-5 Uchikanda, Chiyoda-ku, Tokyo, Japan 101-0047

E-mail: tong@kesco.co.jp

Abstract. Plasma enhanced chemical vapour deposition (PECVD) process in a SiH_4/O_2 inductively coupled plasma (ICP) is studied in this paper. The detailed chemical reactions in plasma and surface reactions on the wafer are included in the simulation using finite element method (FEM), so that the formation of bulk species of SiO_2 is examined. The study is performed for 10% SiH_4/90% O_2 gas mixtures operating at the rf frequency of 13.56 MHz and the total gas pressure of 25 mTorr with the variation of gas flow rate from 20 to 500 sccm. The gas temperature is 300 K and the input power is 700 W. It is found that the deposition rate near the rim of wafer is much lower than that in the centre. The deposited film thickness near the rim of wafer is 70% of that in the centre after the discharge of 0.2 sec. The study shows that the oxygen atom density, which directly contributes to the deposition of SiO_2, decreases when the gas flow rate increases. The deposition rate on the wafer is reduced by around one order and the deposited film thickness is reduced to 12% of the original when the gas flow rate is increased from 20 to 500 sccm.

1. Introduction

High density plasma deposition of silicon dioxide (SiO_2) is an important process in integrated circuit manufacturing [1-3]. To deposit SiO_2 on the wafer, typically a SiH_4/O_2 or SiH_4/N_2O mixture is applied [4]. SiH_4 is the precursor gas, while an oxidizing agent (e.g., O_2) is added to grow a SiO_2 film. SiO_2 can be deposited on the wafer at low temperatures of 100-300°C, using PECVD [1]. Low-pressure SiH_4/O_2 reaction in chemical vapour deposition (CVD) has been found to be a fast gas phase reaction. However, the hydroxyl reaction products could be imbedded into the growing oxide and lead to low-quality electrical properties of the film. PECVD holds the potential to overcome the limitation by the selective excitation of gas phase species to control reaction pathways [2]. The numerical study of PECVD in SiH_4/O_2 mixtures would provide us direct insights into the correlation between the deposition of films and the reaction of gas phase species.

Plasma conditions determine the film properties such as the dielectric constant and stress which in turn affect the device performance and reliability. Although many experimental researches have been carried out, model investigations are still limited to the estimation of fluxes of precursor species to the wafer and deposition rates [4,5] due to the lack of detailed data for all the surface reactions on the wafer. Recently, the investigation of $SiH_4/O_2/Ar$ inductively coupled plasmas used for the filling of microtrenches in shallow trench isolation (STI) was reported [6], in which the plasma chemistry of $SiH_4/O_2/Ar$ discharges as well as surface processes such as deposition were presented. An extra analytical module was used to predict changes in the composition of the surface layers.

The gas flow rate, used as an operating parameter for experiments in plasmas, has been found to affect the plasma properties in $N_2/H_2/Ar$ and Ar/N_2 inductively coupled plasma (ICP) reactors [7-9]. It has been indicated that electron density increases and electron temperature decreases with a rise in gas flow rate for the different Ar fractions in Ar/N_2 discharges. However, the effect of gas flow rate on surface reactions such as deposition is still not clear.

In order to estimate the deposited film thickness, the discharge time considered for PECVD will be longer than that commonly used for plasma simulation, *e.g.*, 1~10 msec. It is known that although for the regime of operation a kinetic description based on the particle scheme can provide accurate electron energy distribution, the fluid approximation can be applicable in low pressure plasma discharges [6,9]. It is noted that the calculation time can be dramatically reduced using fluid models. In this work, COMSOL Multiphysics, a finite element analysis simulation software, is used to combine the surface reactions on the wafer with the chemical reactions occurred in plasma so as to estimate the formation of bulk species of SiO_2 in a SiH_4/O_2 PECVD. The deposition rate and deposited film thickness of SiO_2 are obtained. The effect of gas flow rate on plasma properties as well as deposition of SiO_2 is investigated.

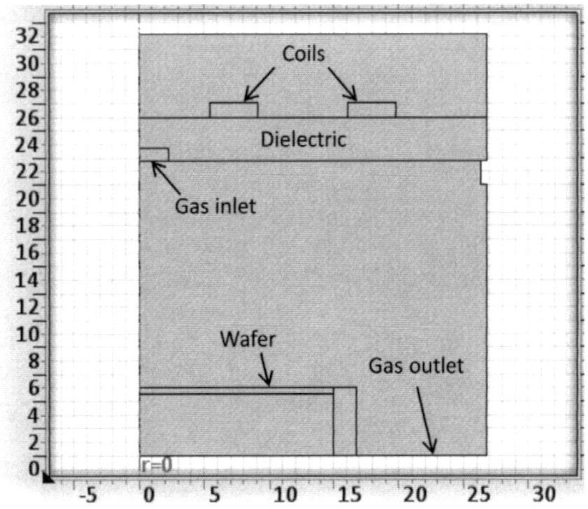

Figure 1. Model geometry of PECVD.

2. Description of the model

Figure 1 shows a model geometry used in this work, which has been described in the earlier publication [10]. The calculations are performed for the case of 10% SiH_4/90% O_2 PECVD operating at the rf frequency of 13.56 MHz and the total gas pressure of 25 mTorr. The gas temperature is assumed to be 300 K and the input power is 700 W. The gas flow rate is varied from 20 to 500 sccm.

The species taken into account include electrons, neutrals (O_2, O_2^*, O, O^*, H_2, H, H_2O, OH, SiH_2, SiH_3, SiH_4, SiHO, SiH_2O, SiH_3O, SiO, SiO_2), ions (O_2^+, H_2^+, SiH_2^+, SiH_3^+, O^-, SiH_2^-, SiH_3^-), as well as surface species (Si(s), SiH(s), SiH_2(s), SiH_3(s), SiO(s), SiHO(s), SiH_2O(s), SiH_3O(s), SiO_2(b)). 88 types of chemical reactions occurred in plasma and 29 types of surface reactions appeared on the wafer are taken into account. The electron impact cross-section data and chemical reaction rates for the above-described reactions are taken from the published literature [1,5,6]. The basic equations for the plasma simulation used in this research include a pair of drift diffusion equations for the electrons, a modified Maxwell-Stefan equation for the ion and neutral species, a Possion's equation for the space charge electric field, and an electromagnetic field equation for the induced electric field. In this work, the plasma simulations are coupled with the fluid equations, which include the continuity equation and momentum equation. The detailed information for the plasma simulation can be found in our previous work [8,9].

In this work, the surface reactions on the wafer are calculated to obtain the deposition rate and deposited film thickness, which are combined with the above-described plasma simulation. The surface rate expression for species k is defined as

$$R_k = \sum_{i=1}^{I} \nu_{ki} q_i \tag{1}$$

where $\nu_{ki} = \nu_{ki}^r - \nu_{ki}^f$ is the stoichiometric coefficient, *f* and *r* mean the forward and reverse reaction, respectively, *I* is the number of surface reactions, q_i is the rate of the *i*th reaction governed by the law of mass action

$$q_i = k_{f,i} \prod_{k=1}^{K} c_k^{\nu_{ki}^f} - k_{r,i} \prod_{k=1}^{K} c_k^{\nu_{ki}^r} \tag{2}$$

where *K* is the number of chemical species, c_k is the concentration of the *k*th species, $k_{f,i}$ and $k_{r,i}$ are the rate coefficients for the *i*th forward and reverse reactions, respectively. The rate coefficient $k_{f,i}$ is related to the sticking coefficient γ_i via the Motz-Wise correction [11]

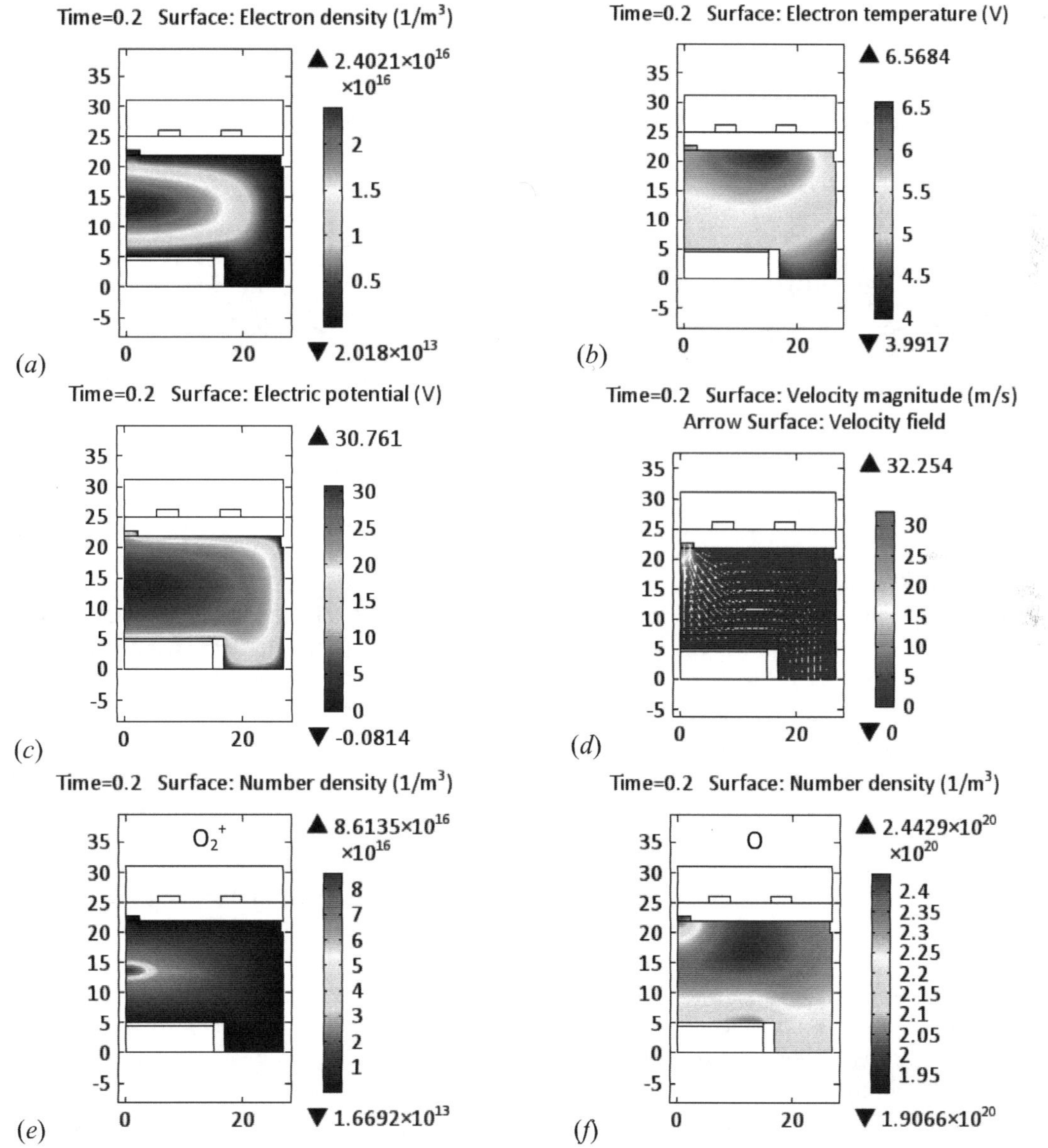

Figure 2. Discharge structure in a 10% SiH_4/90% O_2 ICP plasma (p = 25 mTorr, P = 700 W, T = 300 K, Q = 50 sccm, t = 0.2 sec). (*a*) Electron density, (*b*) Electron temperature, (*c*) Electric potential, (*d*) Fluid field, (*e*) Oxygen ion: O_2^+ density and (*f*) Oxygen atom density.

$$k_{f,i} = \left(\frac{\gamma_i}{1-\gamma_i/2}\right)\frac{1}{(\Gamma_{tot})^m}\left(\frac{1}{4}\right)\sqrt{\frac{8RT}{\pi M_n}} \tag{3}$$

where Γ_{tot} is the total surface site concentration, m is the sum of all of the surface species' stoichiometric coefficients, M_n is the mean molecular weight, R is the universal gas constant, T is the temperature.

For the bulk surface species, the following equation is solved to obtain the deposited film thickness h_k

$$\frac{dh_k}{dt} = -\frac{R_k M_k}{\rho_k} \tag{4}$$

where M_k is the molecular weight, and $\boldsymbol{\rho_k}$ is the density of the bulk species.

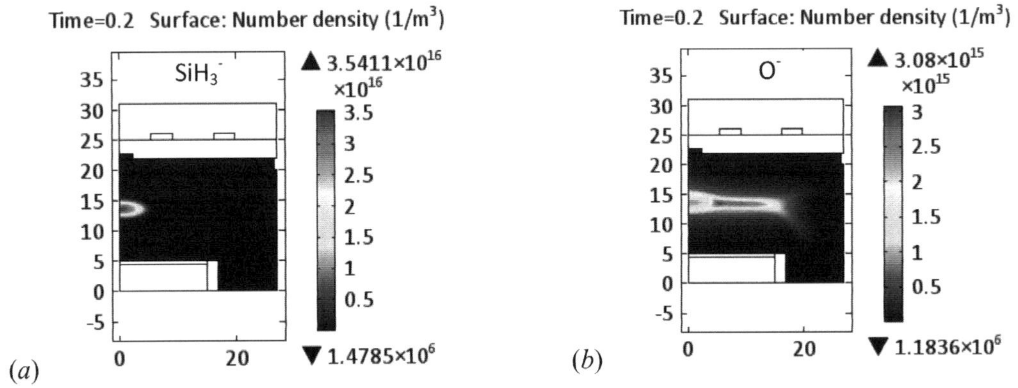

Figure 3. Negative ion density in a 10% SiH_4/90% O_2 ICP plasma (p = 25 mTorr, P = 700 W, T = 300 K, Q = 50 sccm, t = 0.2 sec). (*a*) SiH_3^- ion and (*b*) O^- ion.

3. Results and discussion

Figure 2 shows the calculated results for a 10% SiH_4/90 %O_2 PECVD at a gas flow rate of 50 sccm. The maximum density of electron arrives at 2.40 × 10^{16} m^{-3} and the electron temperature is in a range from 3.99 to 6.57 eV with the low electron temperature near the outlet of gas flow and the high electron temperature beneath the coil domain. The fast gas flow appears around the inlet of the gas flow and the maximum velocity reaches 32.25 $m \cdot s^{-1}$. The governing ion is O_2^+, being concentrated in the centre of the reactor. It is known that O may react with SiO(s) on the wafer to form the bulk species of SiO_2. Figure 2 (*f*) shows that the density of O spreads from the high electron temperature area, which could be deduced due to electron impact collisions with O_2 in the area.

Since both SiH_4 and O_2 are electronegative gases, the plasma considered in this research is electronegative. In electronegative discharges, the negative ions occurred in discharges are held in the core of the plasma, where they can acquire energy from the high frequency electromagnetic field. The calculations show that SiH_3^- and SiH_2^- possess a similar distribution. Thus in this paper, only SiH_3^- and O^- ion densities are shown in figure 3. The densities of SiH_3^- and SiH_2^- are so high in the core of plasma that O^- ions are excluded from the area with high densities of SiH_3^- and SiH_2^-. It should be noted that the maximum densities of SiH_3^- and SiH_2^- have reached 3.54 × 10^{16} m^{-3} and 2.62 × 10^{16} m^{-3}, respectively, both being higher than electron density.

The distributions of surface species on the wafer are shown in figures 4 (a) and (b). The governing surface species is SiO. The concentration of SiO is found to be over one order of magnitude higher than the others, which is consistent with the previous research of Tinck and Bogaerts [6]. The obtained deposition rate and deposited film thickness are given in figures 4 (c) and (d). The deposition rate near the rim of wafer is 55% of that in the centre. As a result, the deposited film thickness near the rim of wafer is 70% of that in the centre.

Results of the effect of gas flow rate on SiO_2 deposition are shown in figures 5 and 6. The density of O shows a decreasing tendency when the gas flow rate is increased. The decrement of O could be considered as a major factor to cause the reduction of deposition rate and deposited film thickness. The surface reaction of O to form SiO_2 can be simplified as [11]

$$SiO(s)+O \rightarrow SiO_2(b) \quad (5)$$

where SiO(s) is the surface species of SiO on the wafer and SiO_2(b) is the bulk species of SiO_2. As seen in figure 6, the deposition rate is reduced by nearly one order when the gas flow rate is increased from 20 to 500 sccm and the deposited film thickness is reduced to 12% of the original.

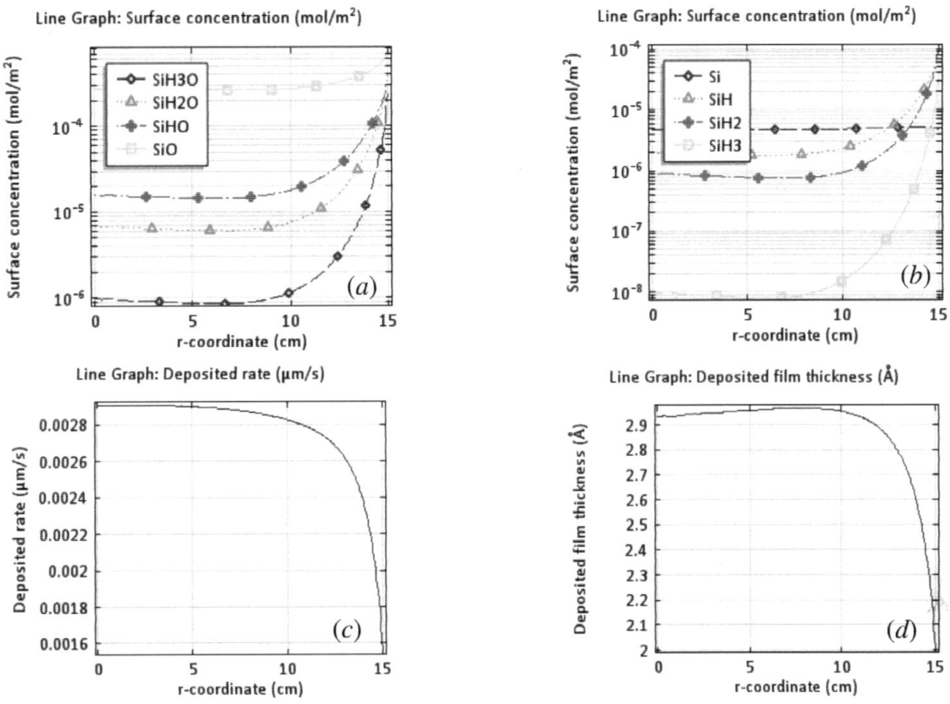

Figure 4. Concentration of surface species, deposition rate and deposited film thickness on the wafer in a 10% SiH_4/90% O_2 ICP plasma (p = 25 mTorr, P = 700 W, T = 300 K, Q = 50 sccm, t = 0.2 sec). (*a*) and (b) Concentration of surface species, (c) Deposition rate and (*d*) Deposited film thickness.

4. Conclusion

In this work, the deposition of SiO_2 in a SiH_4/O_2 inductively coupled plasma was investigated at the gas flow rates of 20-500 sccm. The spatiotemporal distributions of plasma species in plasma as well as on wafer surface were obtained. It was found that the deposition rate and deposited film thickness near the rim of wafer were much lower than those in the centre.

The effect of gas flow rate on SiO_2 deposition was presented. The increase of gas flow rate caused the reduction of oxygen atom density. The deposition rate on the wafer is reduced by around one order and the deposited film thickness is reduced to 12% of the original when the gas flow rate is increased from 20 to 500 sccm. The obtained results provide insight on the gas flow control in the PECVD process, which could be used to improve industrial applications.

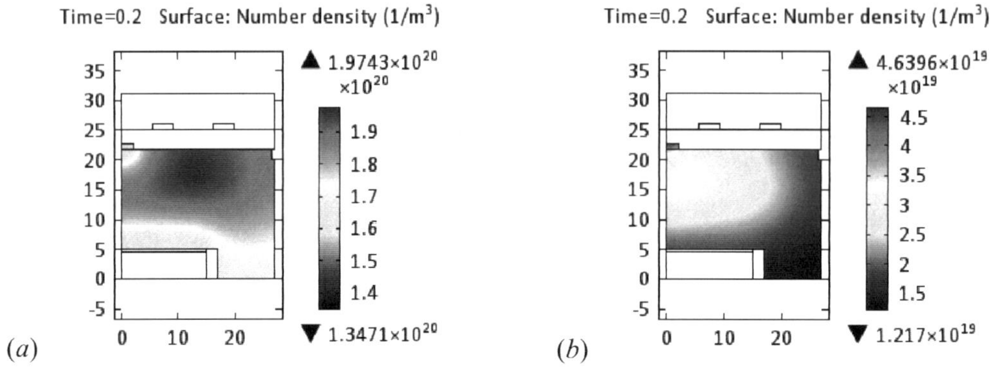

Figure 5. Oxygen atom density at the gas flow rates of (*a*) 100 sccm and (*b*) 500 sccm in a 10% SiH_4/90% O_2 ICP plasma (p = 25 mTorr, P = 700 W, T = 300 K, t = 0.2 sec).

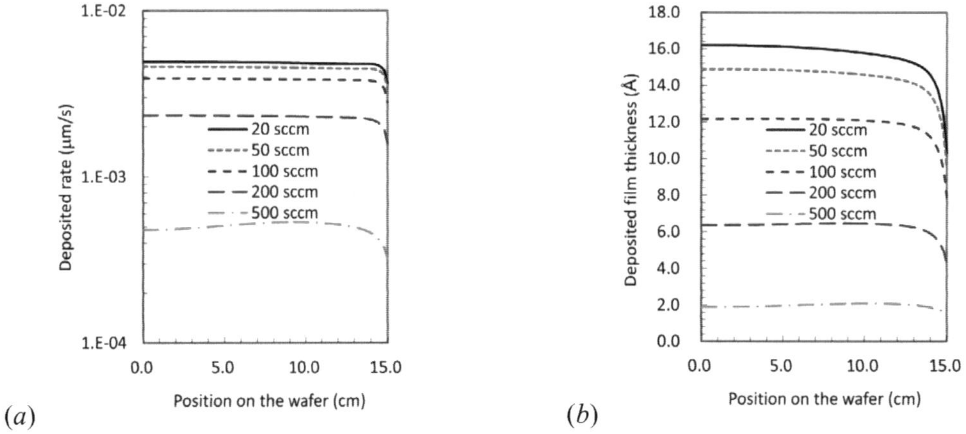

Figure 6. Deposition rate and deposited film thickness on the wafer in a 10% SiH_4/90% O_2 ICP plasma after t = 0.5 sec (p = 25 mTorr, P = 700 W, T = 300 K).

5. References

[1] Lieberman M A and Lichtenberg, A J 1994 *Principles of Plasma Discharges and Materials Processing* (New York: John Wiley & Sons)
[2] Liehr M and Cohen S A 1992 *Appl. Phys. Lett.* **60** 198
[3] Moravej M, Babayan S E, Nowling G R, Yang X and Hicks R F **2004** *Plasma Sources Sci. Technol.* **13** 8
[4] Kushner M J 1993 *J. Appl. Phys.* **74** 6538
[5] Meeks E, Larson R S, Ho P, Apblett C, Han S M, E Edelberg and Aydil E S 1998 *J. Vac. Sci. Technol. A* **16** 544
[6] Tinck S and Bogaerts A 2012 *Plasma Process. Polym.* **9** 522
[7] Lee H J, Kwon B S, Park Y R, Ahn J, Kim J S and Lee N 2010 *J. Korean Phys. Soc.* 56 1441
[8] Tong L Z 2012 *Cent. Eur. J. Phys.* **10** 888
[9] Tong L Z 2013 *Jpn. J. Appl. Phys.* **52** 05EA03
[10] Tinck S, Boulart W and Bogaerts A 2008 *J. Phys. D: Appl. Phys.* **41** 065207
[11] Kee R J, Coltrin M E and Glarborg P 2003 *Chemically Reacting Flow: Theory and Practice* (New Jersey: John Wiley & Sons)

Effects of filter gap of cluster-eliminating filter on cluster eliminating efficiency

Y Hashimoto, S Toko, D Yamashita, H Seo, K Kamataki, N Itagaki, K Koga and M Shiratani

Graduate School of Information Science and Electrical Engineering, Kyushu University, 744 Motooka, Nishi-ku, Fukuoka 819-0395, Japan

E-mail: y.hashimoto@plasma.ed.kyushu-u.ac.jp

Abstract We have developed a cluster-eliminating filter which reduces amount of amorphous silicon nanoparticles (clusters) incorporated into a-Si:H films. We have measured film thickness dependence of a ratio of a hydrogen content associated with Si-H_2 bonds (C_{SiH2}) to that with Si-H bonds (C_{SiH}) as a parameter of the filter gap (d_{gap}). The cluster removal efficiency increases with decreasing d_{gap}. C_{SiH2} and the ratio of C_{SiH2} to C_{SiH} decrease with increasing the film thickness, suggesting the amount of incorporated clusters is high near the interface between the film and the substrate.

1. Introduction

Hydrogenated amorphous silicon (a-Si:H) is most widely employed as a top cell material of thin film Si tandem solar cells [1-9]. The light-induced degradation of a-Si:H limits the efficiency of the tandem solar cells, because light exposure initially reduces the efficiency owing to the degradation [10-20]. For a-Si:H film deposition, SiH_3, high order silane (HOS) radicals in a size range below 0.5 nm, and amorphous silicon nanoparticles (clusters) in a size range between 0.5nm and 10 nm are generated in silane discharge plasmas and they form films [21, 22]. SiH_3 radicals are the main deposition precursor for high quality films. HOS related radicals and clusters are minor deposition precursors and are incorporated into films during the deposition and may degrade film qualities. The degradation tends to be slightly reduced using H_2 dilution and VHF discharge frequency [23-25]. Although the mechanism of the degradation is still unclear, large polymerized molecules in plasmas are an important factor of the degradation [26-28]. In our recent studies, a-Si:H films with a less volume fraction of clusters in the films have been found to show less light induced degradation. We have succeeded in depositing highly stable a-Si:H films using a multi-hollow discharge plasma chemical vapor deposition (CVD) method by which the volume fraction of clusters in films deposited in the upstream region is significantly reduced by driving clusters towards the downstream region [27-38]. Using this method, we have successfully deposited cluster-free a-Si:H films with a stabilized defect density of 4.7×10^{15} cm^{-3} at a rate of 3.0 nm/s [30]. The films show high stability against light exposure.

In this paper, the cluster-eliminating filter was installed in the upstream region of multi-hollow discharge plasma CVD reactor. This cluster-eliminating filter is developed to reduce a volume fraction (V_F) of amorphous silicon nanoparticles above approximately 1 nm in size incorporated into a-Si:H films. By employing the filter together with the multi-hollow discharge plasma CVD reactor, the V_F value is reduced below 1/180 compared to that for the conventional device quality films [39].

Here, we report effects of the filter gap on the cluster eliminating efficiency.

2. Experiment

The cluster-eliminating filter reduces the contribution of the clusters to deposition of films using the difference between the sticking probability of the clusters and the surface reaction probability of the SiH_3 radicals [39]. Cluster removal in the filter takes place by collision and attachment to the surface of the filter plates, since the sticking probability of clusters above 2 nm in size to the surface is nearly 100%, whereas the surface reaction probability of SiH_3 is 30%. The filter was composed of two stainless-steel plates of 0.1 mm in thickness with 8 or 9 slits of 1 mm in width. The slits of one plate were covered by the other plate as shown in figure 1 (top view) and figure 2 (side view) and these plates were placed at a distance of below 0.5 mm, being shorter than the mean free path of SiH_3 and that of clusters.

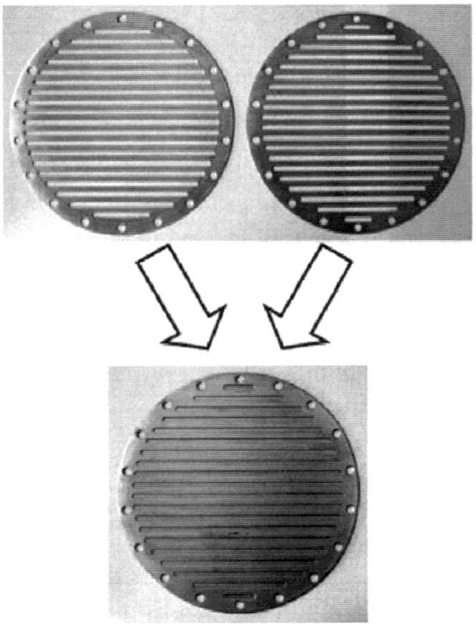

Figure 1. Cluster-eliminating filter image.

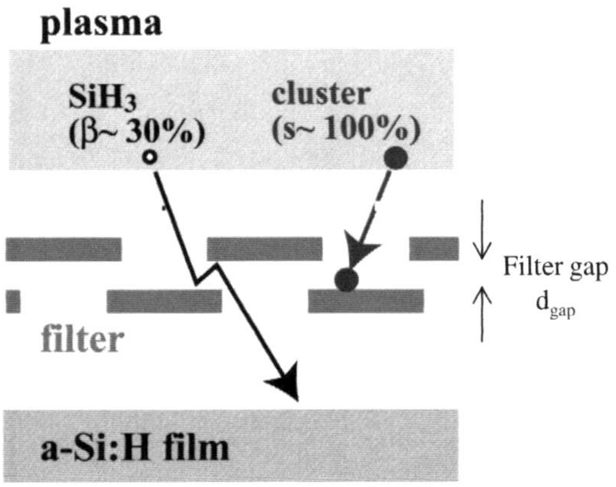

Figure 2. Cluster-eliminating filter schematic sectional view. β is surface reaction probability of SiH_3 radicals. s is sticking probability of clusters.

These plates were placed at a filter gap distance (d_{gap}) of 0.2-0.5 mm. The distance from cluster-eliminating filter to one of the grounded electrode was 8 mm. The distance from substrate to cluster-eliminating filter was 2 mm. We applied this cluster-eliminating filter to deposition a-Si:H films.

For deposition experiments, a-Si:H films were deposited by a multi-hollow discharge plasma CVD method as shown in Figure 3. For the method, clusters, which were generated in the discharge region, were driven toward the downstream region by gas flow, because the gas flow velocity in the discharge region was much faster than the diffusion velocity of clusters. Therefore, cluster-free a-Si:H films were deposited on a n-type c-Si substrate of 20x20 mm^2 in the upstream region. Gas of SiH_4 was supplied at a flow rate of 5 sccm. The total pressure was 0.1 Torr. Discharge frequency and power were 60 MHz and 40 W. Under these condition, a-Si:H films with 100-500nm thickness were deposited at T_s= 180°C. The hydrogen content in films was determined by Fourier-transform infrared spectroscopy (FT-IR; JASCO FT-IR6200).

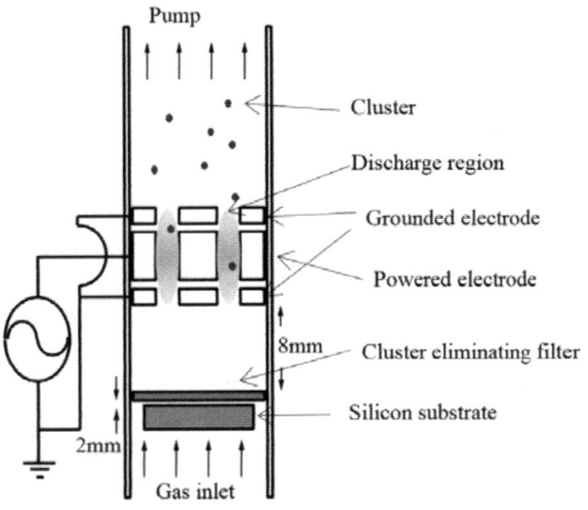

Figure 3. Schematic of the multi-hollow discharge plasma CVD reactor.

3. Results and discussion

The gap between two plates (d_{gap}) is an important parameter of the filter. d_{gap} relates to the transmittance of clusters. First, we studied dependence of deposition rate on d_{gap} as shown in Figure 4.

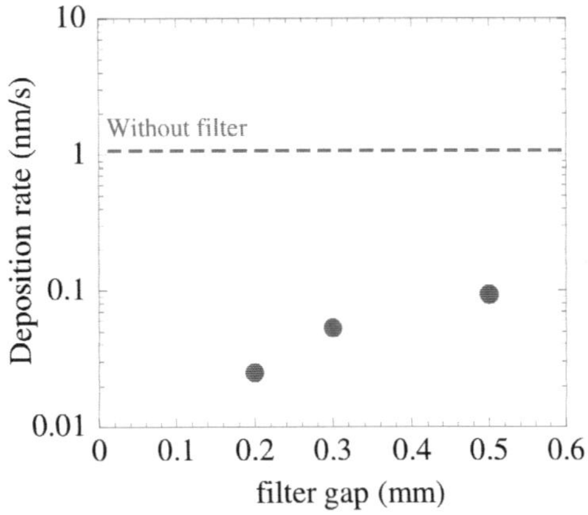

Figure 4. Dependence of deposition rate on filter gap (d_{gap}).

The deposition rate decreases with decreasing d_{gap}. The decrease of the deposition rate indicates that the filter transmittance of SiH_3 decreases. To optimize the filter structure, we have measured film thickness dependence of a hydrogen content associated with Si-H_2 bonds (C_{SiH2}) and a ratio of C_{SiH2} to that with Si-H bonds (C_{SiH}) as a parameter of d_{gap} as shown in figures 5 and 6, respectively. For d_{gap}= 0.5 and 0.3 mm, the ratio considerably decreases with increasing the film thickness, whereas the ratio gradually decrease with increasing the thickness for d_{gap}= 0.2 mm. The decrease of the ratio indicates that large amount of Si-H_2 bonds exists on the surface of the films and/or Si-H_2 bonds generates during the initial deposition phase. Previously, we have found that incorporation of clusters in the initial deposition phase is much higher than that in the steady state [40]. C_{SiH2} in clusters is much higher than that in bulk a-Si:H films. These results indicate that the cluster removal efficiency increases with decreasing d_{gap}.

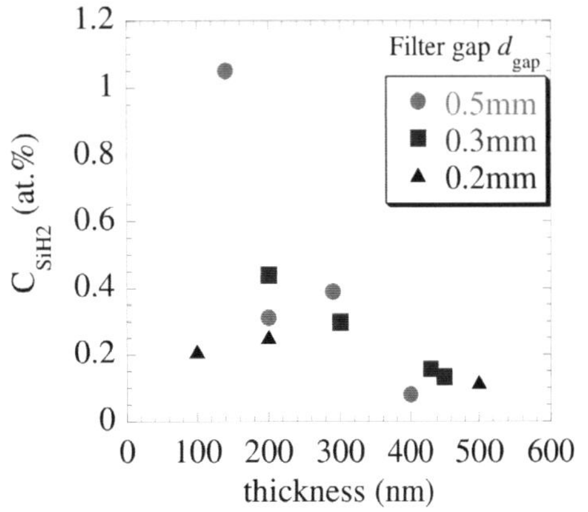

Figure 5. Film thickness dependence of C_{SiH2} as a parameter of the d_{gap}.

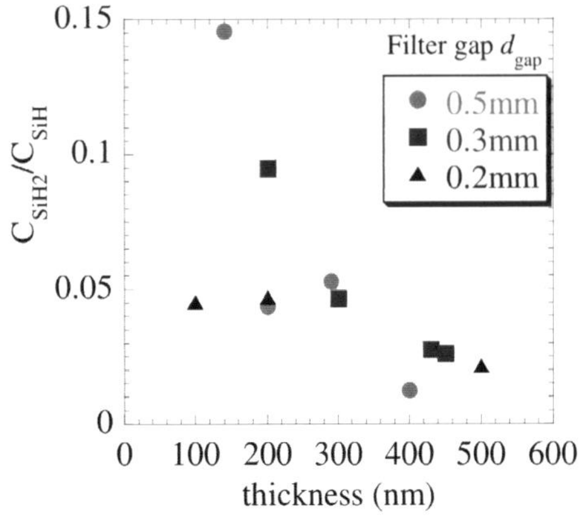

Figure 6. Film thickness dependence of ratio C_{SiH2} and C_{SiH} as a parameter of the d_{gap}.

4. Conclusions

We have applied the cluster-eliminating filter to deposit a-Si:H films. The cluster removal efficiency increases with decreasing d_{gap}. The incorporation of clusters in the initial deposition phase is much higher than that in the steady state.

Acknowledgements

This work was partly supported by the New Energy and Industrial Technology Development Organization (NEDO) and Photovoltaic Power Generation Technology Research Association (PVTEC).

References

[1] Meier J, Dubail S, Fluckiger R, Keppner H, and Shah A 1994 *Proc. IEEE 1st World Conference on Photovoltaic Energy Conversion* 409
[2] Meier J, Torres P, Platz R, Dubail S, Kroll U, Selvan J A A, Vaucher N P, Hof C, Fischer D, Keppner H, and Shah A, 1996 *Mater. Res. Soc. Symp. Proc.* **420** 3
[3] Meier J, Dubail S, Cuperus J, Kroll U, Plats R, Torres P, Selvan J A, Pernet P, Beck N, Pellaton Vaucher N, Hof Ch, Fischer D, Keppner H, and Shah A 1998 *J. Non-Cryst. Solids* **227** 1250
[4] Keppner H, Meier J, Torres P, Fischer D, and Shah A 1999 *Appl. Phys.* A **69** 169
[5] Yamamoto K, Yoshimi M, Tawada Y, Okamoto Y, and Nakajima A 2000 *J. Non-Cryst. Solids* **266** 1082
[6] Zeman M and 2010 *Electr. J Eng.* **61** 71
[7] Konagai M 2011 *Jpn. J. Appl. Phys.* **50** 030001
[8] Matsui T, Sai H, Saito K and Kondo M 2011 *21st International Photovoltaic Science and Engineering Conference* **4A-2O** 03
[9] Matsuda A 2004 *Jpn. J. Appl. Phys.* **43** 7909
[10] Schropp R E I and Zeman M Amorphous and Microcrystalline Silicon Solar Cells 1998 *Kluwer Academic Publishers* 47 99
[11] Sonobe H, Sato A, Shimizu S, Matsui T, Kondo M and Matsuda A 2006 *Thin Solid Films* **502** 306
[12] Kondo M and Matsuda A 2004 *Thin Solid Films* **457** 97
[13] Nishimoto T, Takai M, Miyahara H, Kondo M and Matsuda A 2002 *J. Non-Cryst. Solids* **299** 1116
[14] Staebler D L and Wronski C R 1977 *Appl. Phys. Lett.* **31** 292
[15] Zeman M, van Elzakker G, Sutta P, Isabella O, and Krc J 2009 *J. Microelectronic Components*

and Mater. **39** 223
[16] Chowdhury A, Stradins S, and Ray S 2009 *Sol. Energ. Mater. Sol. C.* **93** 597
[17] Stradins P 2010 *Proc. 35th IEEE Photovoltaic Specialists Conference* 142
[18] Inoue H, Tanaka K, Sano Y, Nishimura T, Teramoto A, Hirayama M, and Ohmi T 2011 *Jpn. J. Appl. Phys.* **50** 0365502
[19] Koga K, Matsuoka Y, Tanaka K, Shiratani M and Watanabe Y 2000 *Appl. Phys. Lett.* **77** 196
[20] Watanabe Y, Harikai A, Koga K, and Shiratani M 2002 *Pure Appl. Chem.* **74** 483
[21] Matsuda A 2000 *J. Plasma Fusion Res.* **76** 760
[22] Shiratani M and Koga K 2010 *J. Plasma Fusion Res.* **86** 33
[23] Yang J, Xu X, and Guha S 1994 *Mater. Res. Soc. Symp. Proc.* **336** 687
[24] Curtins H, Wyrsch N, Favre M, and Shah A 1987 *Plasma Chem. Plasma Process.* **7** 267
[25] Oda S, Noda J, and Matsumura M 1990 *Jpn. J. Appl. Phys.* **29** 1889
[26] Strahm B and Hollenstein Ch 2010 *J. Appl. Phys.* **107** 023302
[27] Takai M, Nishimoto T, Kondo M, and Matsuda A 2000 *Appl. Phys. Lett.* **77** 2828
[28] Koga K, Kaguchi N, Shiratani M, and Watanabe Y 2004 *J. Vac. Sci. Technol. A.* **22** 1536
[29] Koga K, Inoue T, Bando K, Iwashita S, Shiratani M, and Watanabe Y 2005 *Jpn. J. Appl. Phys.* **44** L1430
[30] Nakamura W M, Matsuzaki H, Sato H, Kawashima Y, Koga K, and Shiratani M 2010 *Surface and Coating Technologies* **205** S241
[31] Nakamura W M, Miyahara H, Koga K, and Shiratani, M 2008 *Journal of Physics: Conference Series* **100 8** *082018*
[32] Shiratani M, Koga K, Ando S, Inoue T, Watanabe Y, Nunomura S, and Kondo M 2007 *Surface and Coatings Technology* **201 9** 5468
[33] Nakamura W M, Miyahara H, Sato H, Matsuzaki H, Koga K, and Shiratani M 2008 *Plasma Science, IEEE Transactions on*, **36 4** 888
[34] Shiratani M, Koga K, Iwashita S, Uchida G, Itagaki N, and Kamataki K 2011 *Journal of Physics D: Applied Physics*, **44 17** 174038
[35] Uchida G, Yamamoto K, Kawashima Y, Sato M, Nakahara K, Kamataki K, and Shiratani M 2011. *physica status solidi (c)* **8 10** 3017
[36] Koga K, Matsunaga T, Nakamura W M, Nakahara K, Kawashima Y, Uchida G, and Shiratani M 2011 *Thin Solid Films*, **519 20** 6896
[37] Shiratani M, Iwashita S, Bando K, Inoue T, and Koga K 2006 *Photovoltaic Energy Conversion, Conference Record of the 2006 IEEE 4th World Conference on* **2** 1596
[38] Shiratani M, Nakamura W M, Miyahara H, and Koga K 2007 *Journal of Physics: Conference Series* **86 1** 012021
[39] Koga K, Kaguchi N, Bando K, Shiratani M, and Watanabe Y 2005 *Rev. Sci., Instrum.* **76** 113501
[40] Kim Y, Hatozaki K, Hashimoto Y, Uchida G, Kamataki K, Itagaki N, Seo H, Koga K, and Shiratani M 2013 *Jpn. J. Appl. Phys.* **52** 01AD01

Contribution of ionic precursors to deposition rate of a-Si:H films fabricated by plasma CVD

S Toko, Y Hashimoto, Y Kanemitu, Y Torigoe, H Seo, G Uchida, K Kamataki, N Itagaki, K Koga and M Shiratani

Graduate School of Information Science and Electrical Engineering, Kyushu University, 744 Motooka, Nishi-Ku, Fukuoka, 819-0395, Japan

E-mail: s.toko@plasma.ed.kyushu-u.ac.jp

Abstract. We have studied contribution of ionic precursors to deposition rate of a-Si:H films in the downstream region of a multi-hollow discharge plasma CVD reactor using a DC bias grid and QCMs. The deposition rate decreases from 1.1 to 0.93 by applying negative bias voltage to the bias grid. The ionic precursors contribute to 7% of the total deposition rate and the dominant ionic precursors are considered to be negatively charged clusters.

1. Introduction

Hydrogenated amorphous silicon (a-Si:H) films deposited by plasma CVD have been widely used for a top cell of thin film silicon tandem solar cells[1-7]. A-Si:H films show metastability against light exposure, the Stabler-Wronski effect, which reduces significantly its initial photovoltaic conversion efficiency[8]. In SiH_4 discharge plasmas employed for a-Si:H deposition, there are three size groups of species: SiH_x ($x \leq 3$) radicals, higher order silane (HOS) molecules Si_mH_n ($m \leq 4$), and a-Si:H particles in a size range below 10 nm (clusters)[9-13]. SiH_3 radicals are main deposition precursors, which dominate the film deposition rate. HOS molecules are generated due to polymerization reactions between SiH_2 and SiH_4. HOS radicals are generated from HOS molecules due to electron impact dissociation and they can react on surfaces as well as in gas phase. Clusters are nucleated from HOS molecules of Si_nH_x ($n{\sim}4$) and grow due to coagulation between clusters and/or deposition of SiH_x radicals to the surface of clusters. HOS radicals and clusters are minor deposition precursors and are pointed out to lead to the light-induced degradation[14, 15].

We have previously reported that suppression of cluster incorporation into films is the key to realizing highly stable a-Si:H films[16, 17]. Based on the results, we have successfully deposited highly stable a-Si:H films of 4.7×10^{15} cm^{-3} in stabilized defect density by using a multi-hollow discharge plasma CVD method, by which the volume fraction of clusters in films can be reduced significantly[18]. Although ESR measurements of our 0.6 μm thick a-Si:H films show high stability, Schottky cells using such "stable a:Si:H films" still show modest light-induced degradation. We also have developed a real-time in-situ monitor of the cluster volume fraction in films by employing three quartz crystal microbalances (QCMs)[19, 20]. The real time measurements reveal that large amount of clusters deposit in the initial 60 s after turning on the discharges[20]. To reduce cluster incorporation further, we need information on minor deposition precursors in detail. Here, we report contribution of ionic precursors to the total deposition rate, obtained by using QCMs equipped with a bias grid mesh.

2. Experimental

Experiments were carried out using a multi-hollow discharge plasma CVD reactor with a DC bias grid and QCMs as shown in Fig. 1. The detail of multi-hollow discharge plasma CVD reactor is described elsewhere[17, 18]. DC bias grid of 10 cm in diameter was set at 10 mm below the lower electrode and 10 mm above the QCMs. The aperture of the DC bias grid was 1.8 × 1.8 mm^2, the wire was 0.8 mm, and the aperture ratio was 47 %. SiH_4 was supplied to the reactor at 30 sccm. The working pressure was 0.5 Torr. Discharge plasmas were sustained by applying 60 MHz peak to peak voltage V_{pp} of 34 V between the powered and grounded electrodes.

To obtain information on electron impact generation rate of radicals from SiH_4, optical emission intensities of Si* 288 nm and SiH* 414 nm were measured.

Deposition rate *DR* was obtained with QCMs which were set at 20 mm under the lower grounded electrode. Resonance frequency of quartz crystal decreases with increasing mass of films deposited on the crystal. Relation between mass change Δm and frequency change Δf is given by Sauerbrey equation[21],

$$\Delta f = -\frac{2f_0^2}{A_1(\mu_q \rho_q)^{1/2}} \Delta m \cdot \tag{1}$$

where f_0 is the initial resonance frequency of crystal, μ_q and ρ_q is the density of crystal and the shear modulus of the quartz, A_1 is the area of crystal. From the Δm, *DR* was deduced by

$$\Delta m / \Delta t = \rho A_2 \cdot DR \tag{2}$$

where ρ is the weight density, Δt is the discharge duration, and A_2 is the area of deposition. In this study, we employed $\rho = 2.3$ g/cm^3 which is the mass density of a-Si:H.

DC bias voltage V_{grid} was applied to the grid set above QCMs to obtain information on contribution of ionic precursors to deposited films. V_{grid} was set in a range from -100 V to +100 V.

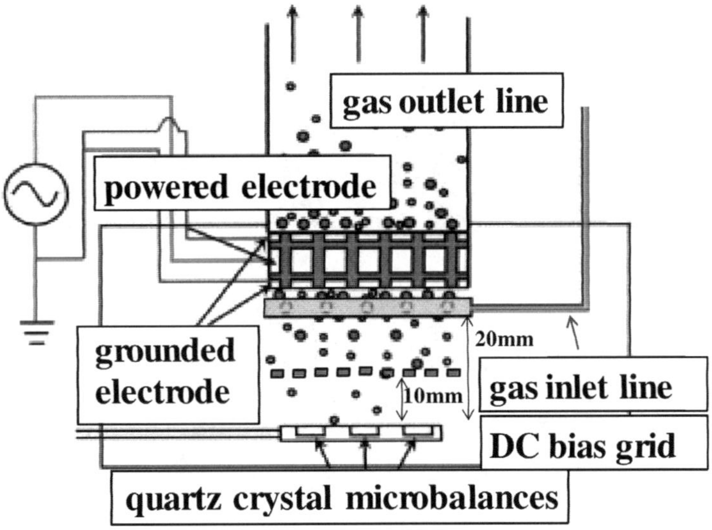

Figure 1. Multi-hollow discharge plasma CVD reactor with QCMs and a DC bias grid.

3. Results and Discussion

First, we have examined influence of V_{grid} on the discharges. Figure 2 shows V_{grid} dependence on space potential V_{space} which is deduced by

$$V_{space} = (\omega / 2\pi)(kT_e / e) \int_0^{2\pi/\omega} \ln\{[1 + A_c / A_a \times \exp(eV(t)/kT_e)]/[1 + (A_c / A_a)]\} dt + V_0, \quad (3)$$

where ω is excitation frequency, k is Boltzmann coefficient, T_e is electron temperature, A_c is cathode area, A_a is anode area, V_0 is plasma potential relation to floating objects, and $V(t)$ is the VHF voltage on the powered electrode[22]. V_{space} tends to increase with V_{grid}, in particular, it increases sharply above V_{grid} = 30V, and V_{space} = 34V for V_{grid} = 100V. These results suggest that the bias grid is effective to control the flux of ionic species towards QCMs.

Second, we have measured Si* 288 nm emission intensity I_{Si*} and SiH* 414 nm emission intensity I_{SiH*} to obtain information on influence of V_{grid} on radical generation rates. Figure 3 shows the V_{grid} dependence of I_{Si*} and I_{SiH*}. These intensities are normalized by those for V_{grid} = 0 V. I_{Si*} and I_{SiH*} are almost constant between V_{grid} = -100 V and V_{grid} = +30 V then sharply increase for V_{grid} > +30 V. Figure 4 shows V_{grid} dependence of the ratio I_{Si*}/I_{SiH*} which indicates the slope of high energy tail of electron energy distribution in the plasma[23-25]. The results suggest the generation rate ratio of SiH_2 and SiH_3 is irrelevant to V_{grid}, because I_{Si*}/I_{SiH*} is nearly constant for the whole range of V_{grid} = -100 – +100 V. The sharp increases in I_{Si*} and I_{SiH*} for V_{grid} > +30 V indicate that the radical generation rates increase sharply for V_{grid} > +30 V, and hence it is difficult to discuss contribution of ionic precursors to deposition. Therefore, hereafter, we discuss contribution of ionic precursors for V_{grid}< 30 V.

Figure 5 shows dependence of *DR* on V_{grid}. *DR* is normalized by that for V_{grid}= 0 V. DR increases slightly from 0.93 for V_{grid}= -100 V to 1.09 for V_{grid} = 30 V. From the results in figures 3 and 4, generation rates of radicals and clusters are almost constant for V_{grid}< 30 V, and hence composition of precursors in the plasmas is almost the same for V_{grid} < 30 V. Therefore, the increase of *DR* in figure 5 shows that the flux of ionic precursors increases with increasing V_{grid}. Effect of sputtering by positive ion was evaluated using Ar discharge, because mass of Ar^+ is closed to that of SiH_3^+ and the sputtering rate depends on mass of positive ion[26]. The results show that sputtering contribute little to V_{grid} dependence of DR in figure 5. These results suggest that major ionic deposition precursors are negatively charged such as SiH_3^-, $Si_mH_x^-$, and negatively charged clusters and the contribution of them to the total deposition rate for V_{grid}< -40 – +20 V is 7%. Since the electron attachment rate of clusters is much higher than that of radicals[27-32], the negatively charged precursors are considered to be mainly clusters. These results suggest that contribution of negatively charged clusters to films can be reduced by using electrostatic force exerted on the clusters[33-37].

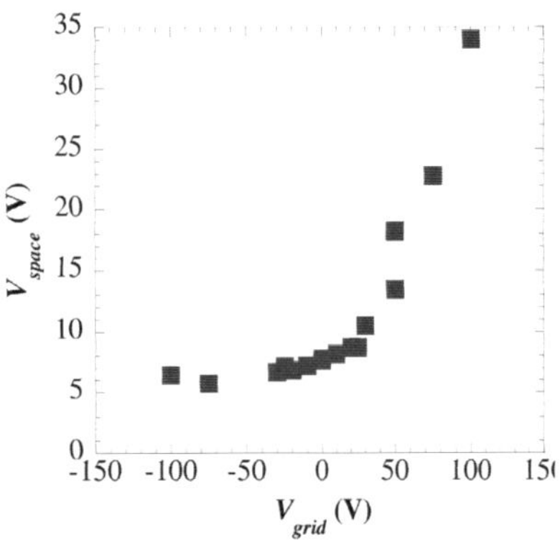

Figure 2. V_{grid} dependence of V_{space}.

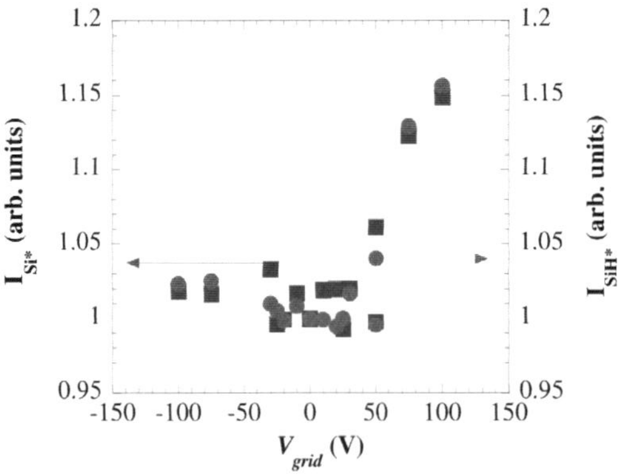

Figure 3. V_{grid} dependence of I_{Si*} and I_{SiH}.

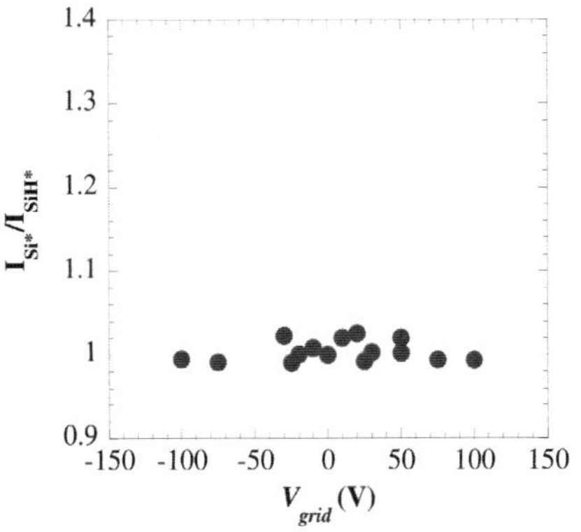

Figure 4. V_{grid} dependence of the ratio of I_{Si*}/I_{SiH*}.

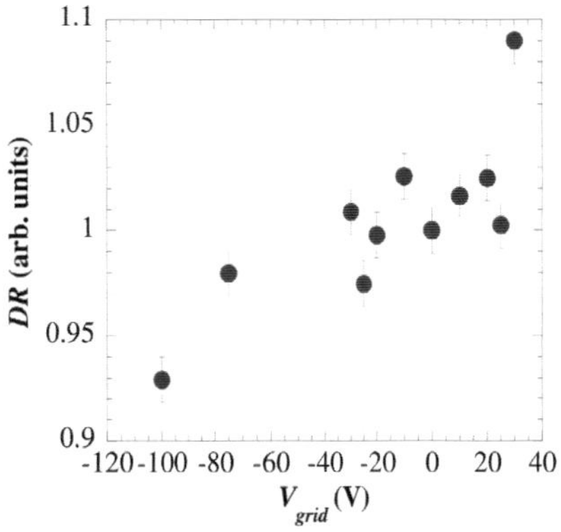

Figure 5. V_{grid} dependence of DR.

4.Conclusions

We have studied contribution of ionic deposition precursors in deposition region of multi-hollow discharge plasma CVD using the bias grid and QCMs. The contribution of ionic precursors is 7% of the total deposition rate. The dominant ionic precursors are considered to be negatively charged clusters. These results suggest that contribution of negatively charged clusters to films can be reduced by using electrostatic force exerted on the clusters.

Acknowledgments

This work was partly supported by NEDO, PVTEC, and MEXT KAKENHI Grant Number 21110005.

References

[1] Matsui T and Kondo K 2013 *Solar Energy Materials & Solar Cells* **119** 156
[2] Terakawa A 2013 *Solar Energy Materials & Solar Cells* **119** 204
[3] Kim S, Kasashima S, Sichanugrist S, Kobayashi T, Nakada T, Konagai M 2013 *Solar Energy Materials & Solar Cells* **119** 214
[4] Tanaka K 1991 *J. Non. Cyst. Solids* **137-138** 1
[5] Osborne I S, Hata N, Ganguly G, and Matsuda A 1994 *Jpn. J. Appl. Phys.* **33** 5663
[6] Takai T, Nishimoto T, Kondo M, and Matsuda A 2000 *Appl. Phys. Lett.* **77** 2828
[7] Ganguly G, Newton J, Carlson D E, and Arya R R 2002 *J. Non. Cyst. Solids* **299-302** 53
[8] Stabler D L and Wronski C R 1980 *J. Appl. Phys.* **51** 3262
[9] Koga K, Matsuoka Y, Tanaka K, Shiratani M, and Watanabe Y 2000 *Appl. Phys. Lett.* 77 196
[10] Kawasaki H, Kida J, Sakamoto K, Fukuzawa T, Shiratani M, and Watanabe Y 1998 *J. Appl. Phys.* **83** 5665
[11] Matsuoka Y, Shiratani M, Fukuzawa T, Watanabe Y, and K. S. Kim 1999 *Jpn. J. Appl. Phys.* **38** 4556
[12] Nunomura S, Kita M, Koga K, Shiratani M, Watanabe Y 2006 *J. Appl. Phys.* **99** 83302
[13] Shiratani M, Koga K, Kaguchi N, Bando K, Y. Watanabe 2006 *Thin Solid Films* **506-507** 17
[14] Takagi T, Hayashi H, Ganguly G, Konda M, and Matsuda A 1999 *Thin Solid Films* **345** 75
[15] Koga K, Kaguchi N, Shiratani M, and Watanabe Y 2004 *J. Vac. Sci. Technol.* **A 22** 1536
[16] Koga K, Kai M, Shiratani M, Watanabe Y, and Shikatani N 2002 *Jpn. J. Appl. Phys.* **41** L168
[17] Koga K, Inoue T, Bando K, Iwashita S, Shiratani M, and Watanabe Y 2005 *Jpn. J. Appl. Phys.* **44** L1430
[18] Nakamura W M, Matsuzaki H, Sato H, Kawashima Y, Koga K, and Shiratani M 2010 *Surf. Coat. Technol.* **205** S241
[19] Shiratani M 2012 *Proc. Mater. Res. Sympo.* **1426** 307

[20] Kim Y, Hatozaki K, Hashimoto H, Uchida G, Kamataki K, Itagaki N, Seo H, Koga K, and Shiratani M 2013 *Jpn. J. Appl. Phys.* **152** 01AD01
[21] Sauerbrey G 1959 *Z. Phys.* **155** 206 [in German]
[22] Song Y P, Field D, and Klemperer D F 1990 *J. Phys. D Appl. Phys.* **23** 673
[23] Perrin J and Aarts J F M 1983 *Chem. Phys.* **80** 351
[24] Takai M, Nishimoto T, Kondo M, and Matsuda A 2001 *Sci. Technol. Adv. Mater.* **2** 495
[25] Matsuda A, Takai M, Nishimoto T, and Kondo M 2003 *Sol. Energy Mater. Sol. Cells* **78** 3
[26] Min Wu S, Kruijs R, Zoethout E, and Bijkerk F 2009 *J. Appl. Phys.* **106** 054902
[27] Shiratani M, Fukuzawa T, Eto K, and Watanabe Y 1992 *Jpn. J. Appl. Phys.* **31** L1791
[28] Perrin J, Leroy O, and Bordage M C 1996 *Contrib. Plasma. Phys.* **36** 3
[29] Fukuzawa T, Kawasaki H, Shiratani M, and Watanabe Y 1994 *Jpn. J. Appl. Phys.* **33** 4212
[30] Kono A, Haverlag M, Kroesen G M W, and Hoog F J 1991 *J. Appl. Phys.* **70** 2939
[31] Haverlag M, Kono A, Passchier D, Kroesen G M W, Goedheer W J, and Hoog F J 1991 *J. Appl. Phys.* **70** 3472
[32] Shindo M, Uchino S, Ichiki R, Yoshimura S, and Kawai Y 2001 *Rev. Sci. Instrum.* **72** 2288
[33] Shiratani M, Koga K, Iwashita S, Uchida G, Itagaki N, and Kamataki K 2011 *J. Phys. D: Appl. Phys.* **44** 174038
[34] Koga K, Iwashita S, and Shiratani M 2007 *J. Appl. Phys. D* **40** 2267
[35] Shiratani M, Koga K, Iwashita S, and Nunomura S 2008 *Faraday Discuss.* **137** 127
[36] Iwashita S, Schungel E, Schulze J, Hartmann P, Donko Z, Uchida G, Koga K, Shiratani M, Czarnetzki U 2012 *Plasma Sources Sci. Technol.* **21** 032001
[37] Iwashita S, Schungel E, Schulze J, Hartmann P, Donko Z, Uchida G, Koga K, Shiratani M, Czarnetzki U 2013 *J. Phys. D: Appl. Phys.* **46** 245202

Contribution of H_2 plasma etching to radial profile of amount of dust particles in a divertor simulator

M Tateishi[1], K Koga[1], D Yamashita[1], K Kamataki[2], H Seo[1], N Itagaki[1,3], M Shiratani[1], N Ashikawa[4], S Masuzaki[4], K Nishimura[4], A Sagara[4] and the LHD Experimental Group[4]

[1] Faculty of Information Science and Electrical Engineering, Kyushu University, 744 Motooka, Nishi-ku, Fukuoka 819-0395, Japan.
[2] Fuculty of Arts and Science, Kyushu University, 744 Motooka, Nishi-ku, Fukuoka 819-0395, Japan.
[3] PRESTO Japan Science and Technology Agency, 5 Sanban-cho Chiyoda-ku Tokyo, 102-0075, Japan.
[4] National Institute for Fusion Science, 322-6 Oroshi-cho Toki-city Gifu, 509-5292, Japan.

E-mail: m.tateishi@plasma.ed.kyushu-u.ac.jp

Abstract. We have studied contribution of H_2 plasma etching to radial profile of amount of dust particles generated due to interactions between H_2 plasmas and graphite target in a divertor simulator. Dust fluxes of spherical particles and flakes are the maximum at the distance r = 100 mm and 120 mm from the centre axis of the plasma column, respectively. From ion density and dust flux, we have deduced etched volume of deposited dust particles due to H_2 plasma irradiation. Sum of the etched volume and measured volume of spherical dust particles is almost constant for r < 120 mm and decreases with increasing r for r > 120 mm, whereas that of flakes is the maximum at r = 120 mm. H_2 plasma etching significantly reduces size of dust particles for r smaller than 100 mm.

1. Introduction

In fusion devices, dust particles generated due to plasma-wall interactions are pointed out to become safety hazard due to tritium retention and radioactive content in dust particles. Dust particles also may cause deterioration of plasma confinement [1-8]. Dust particles can easily react with oxygen and water when the reactor is exposed to air during maintenance, leading to an explosion hazard [9]. Hydrogen isotopes such as tritium trapped in them and on their surface can also react with air. Their vaporization in the edge of fusion plasmas can be a source of impurity, leading to deterioration of plasma confinement [10]. Therefore, their maximum quantity in International Thermonuclear Experimental Reactor is restricted below 6 kg of carbon or 11 kg of beryllium and 230 kg tungsten [11, 12]. To overcome these issues, it is important to understand their generation and transport mechanisms in fusion devices. So far, we have developed a divertor simulator using helicon discharges to study generation and transport mechanisms of dust particles. Using the discharges, we have simulated generation of dust particles in nanometer size due to interaction between plasmas and graphite target, which is employed as a divertor material in Large Helical Device (LHD) [13-17]. For the dust collection experiments, collected dust particles are exposed by irradiation of H_2 plasma, leading to etching of dust particles. The contribution of the H_2 plasma etching may be important to evaluate the amount of collected dust particles. Here we estimated such contribution to radial profile of amount of dust particles. The results give insight on the etching effect on dust particles re-deposited in nuclear fusion devices.

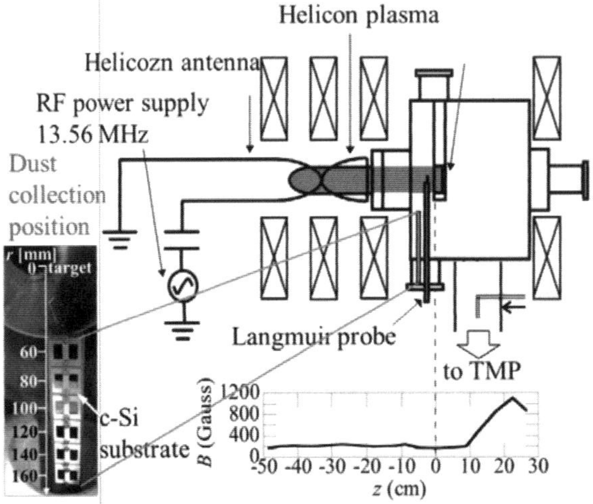

Figure 1. Experimental set up of the divertor simulator, c-Si substrates and spatial profile of magnetic flux density on center axis of the reactor. c-Si substrates are set on a Ti holder placed at 25 mm from the target and 60, 80, 100, 120, 140 and 160 mm from the center of plasma column. The surface of substrates faces the graphite target. The inset shows the magnetic flux density in center axis of the discharge tube. The graphite target was set at z = 0 mm.

2. Experimental

Experiments were carried out with the divertor simulator as shown in figure 1. The reactor is composed of a stainless vessel with 267 mm maximum inner diameter and 294 mm length, a quartz tube of 50 mm diameter and 200 mm length, as well as an antenna for the m = 1 helicon mode excitation of 170 mm length placed around the tube [18]. The uniform magnetic field of 0.015 T (150 Gauss) was applied along the center axis of the discharge tube with the four magnetic coils. Gas of pure H_2 was supplied at 13 sccm in flow rate at pumping port to reduce effects of gas flow on dust transport. The gas pressure was 5 mTorr. H_2 plasmas were generated by applying pulsed RF voltage of 13.56 MHz to a helicon antenna. The ion density and electron temperature at the center of the discharge and 20 mm from the target was 2.8×10^{12} cm^{-3} and 6 eV, respectively. The ion density and electron temperature are close to those of divertor plasmas in LHD [19]. The discharging period was 0.25 s and the interval was 1.0 s to avoid over heating the quartz discharge tube. In order to produce dust particles due to interactions between the carbon wall and the H_2 plasmas, a polycrystalline graphite target (Toyo Tanso IG-430U) of 35 mm in diameter and 8 mm in thickness was placed as shown in figure 1. The graphite target is employed for divertor material in LHD.

Dust particles were collected on c-Si substrates set on a Ti holder placed at 25 mm from the target and 60, 80, 100, 120, 140 and 160 mm from the center of plasma column. The surface of substrates faces the graphite target. The size and shape of the dust particles collected on the substrates were measured with a scanning electron microscope (SEM). The minimum detectable particle size in the SEM measurements was 50 nm. Their composition was obtained by energy dispersive X-ray (EDX) analysis.

3. Results and Discussion

Collected dust particles can be classified into two kinds: spherical particles and flakes. EDX analysis shows that the major compositions of the dust particles are C and Fe; C is the predominant composition of the graphite target and Fe is one of compositions of the chamber materials. Carbon is the primary composition of spherical particles and flakes. The shape of spherical particles indicates they are formed in gas phase due to deposition of carbon radicals on their surface. Flakes have an irregular shape which suggest they are formed due to peeling from the carbon films deposited on the reactor wall and/or the target [4].

Figure 2 shows size distribution of dust flux as a parameter of distance *r* from the center of the plasma column. The flux of collected dust particles towards the substrates was obtained by

$$\Gamma_d = \frac{n(d)}{t_{\text{total}}}, \tag{1}$$

where $n(d)$ and t_{total} are the area density of dust particles with a size *d* deposited on a substrate and the total discharging period (581 s), respectively. All the spherical particles are smaller than 300 nm and the flakes are in a size range from 40 nm to a few μm. The peak in the size distribution is independent of *r* and around 50 nm for both of the spherical particles and flakes. The size distributions shows similar shape for any *r*. It suggests that the radial transport of the dust particles depend little on their size. Therefore, the volume of spherical particles and flakes obtained by integrating their volume between 50 nm and 1 μm in size, respectively, are employed to discuss qualitatively radial transport of dust particles.

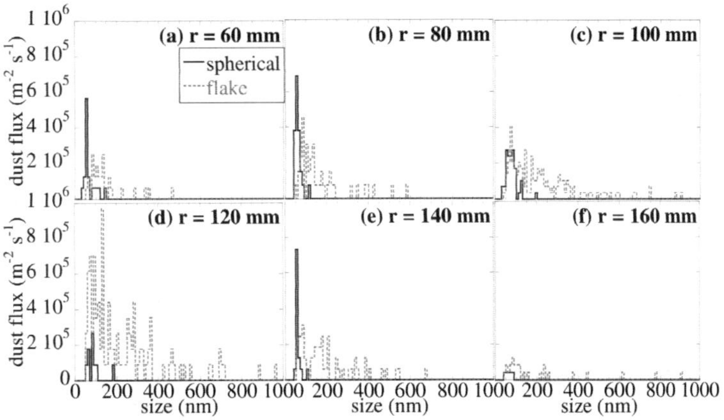

Figure 2. Size distribution of flux of collected dust particles for *r* = (a) 60 mm, (b) 80 mm, (c) 100 mm, (d) 120 mm, (e) 140 mm, and (f) 160 mm.

Volume of collected dust particles V_{col} is given by

$$V_{col} = \int \frac{4}{3}\pi \left(\frac{d}{2}\right)^3 n(d)\,\mathrm{d}d \qquad \text{for spherical particles,} \tag{2}$$

$$V_{col} = \int 100 \times 10^{-9} \frac{(d)^2}{2} n(d)dd \qquad \text{for flakes,} \tag{3}$$

where the volume is integrated in a size range between 50 nm and 1 μm. Thickness of flakes is assumed to be 100 nm.

During the dust collection, H_2 plasmas are irradiated to dust particles deposited on the substrates. To estimate the etching effects on the dust particles, we have evaluated from the erosion yield Y_{tot} of carbon films [20]. In this study, Y_{tot} has been employed as etching yield of the dust particles. Y_{tot} is a function of ion energy E_0, ion flux Γ_i and electron temperature T_e. E_0 is assumed to be the same as the difference between substrate potential and space potential of hydrogen plasmas, because of low pressure. Γ_i is obtained from the ion density n_i and electron temperature T_e measured near the substrate. Figure 3 shows *r* dependence of n_i and T_e. n_i exponentially decreases with increasing *r*. T_e is almost constant of about 4 eV between *r* = 60 mm and 110 mm. Etched thickness of dust particles t_{etch} is given by

$$t_{etch} = \frac{\Gamma_i Y_{tot} m_c}{\rho_d} t_{total}, \tag{4}$$

As shown in figure 4, etched thickness of dust particles t_{etch} decreases from 39.2 nm for r= 60 mm to 0.01 nm for r= 160 mm. H_2 plasma etching reduces size of dust particles and hence some dust particles became smaller than the lower detection limit of size.

The etched volume V_{etch} of observed dust particles is given by

$$V_{etch} = \int \frac{\Gamma_i Y_{tot} m_c}{\rho_d} S t_{total} n(d) dd, \tag{5}$$

where the volume is integrated in a size range between 50 nm and 1 μm. m_c is mass of a carbon atom. ρ_d is the mass density of the dust which is assumed to be 1.8 g/cm^3 [21], the mass density of the graphite target. d is the size of a dust particle. S is given by

$$S = \pi \left(\frac{d}{2}\right)^2 \quad \text{for spherical particles,} \tag{6}$$

$$S = \frac{(d)^2}{2} \quad \text{for flakes,} \tag{7}$$

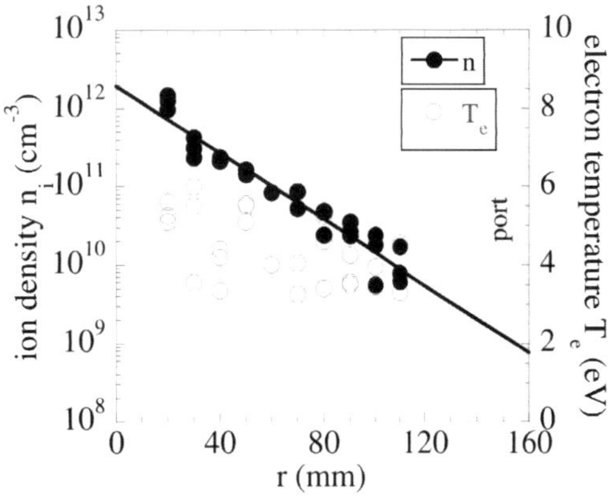

Figure 3. Distance dependence of ion density and electron temperature.

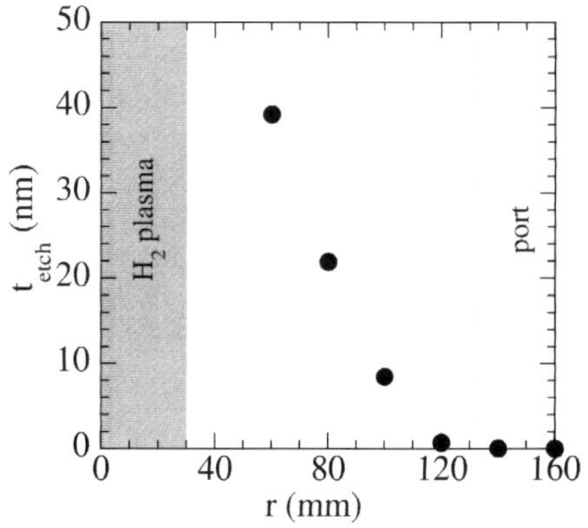

Figure 4. Distance dependence of etched thickness of dust particles.

In the eq. (4), all dust particles are assumed to be etched by H_2 plasmas during discharge period t_{total}, namely, the maximum etched volume of dust particles. Here, we neglect the growth of dust particles collected on substrates due to deposition of radicals or self-organization. We have to investigate the growth of dust particles on vessel wall in future research. Figure 5 shows r dependence of V_{col} and V_{col} + V_{etch}. V_{col} + V_{etch} of spherical particles is almost constant for r below 120 mm and decreases with increasing r for r above 120 mm. That of flakes has a maximum value for r = 120 mm, suggesting that flakes are formed by peering of re-deposited carbon films on vessel wall (r= 133.5 mm).

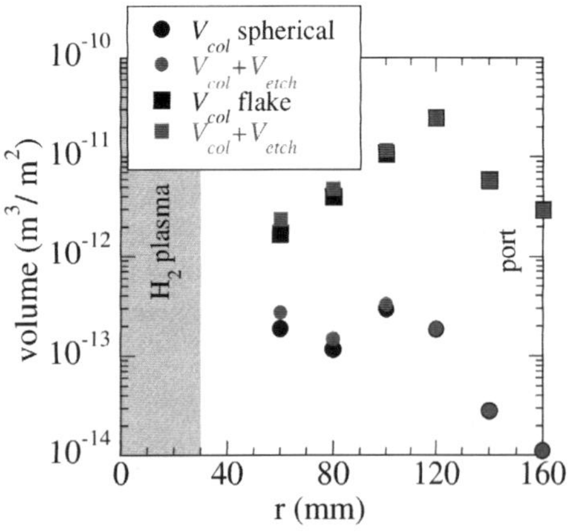

Figure 5. Distance dependence of V_{col} and V_{col} + V_{etch}.

Figures 6 (a) and (b) show a ration R of V_{etch} to V_{col} + V_{etch} versus r and ion density at 5 mm from each collection substrate, respectively. R of spherical dust particles and flakes monotonically increasing with decreasing r and increasing n_i from 0.01 and 0.00 for r= 160 mm to 31.0 and 28.2 % for r= 60 mm, respectively. R is smaller than 1 % for r larger than 120 mm that means n_i smaller than 5.39×10^9 cm^{-3}. Therefore, etching effects on dust particles are negligible for r larger than 120 mm. R is larger than 7 % for r smaller than 100 mm that corresponds to n_i larger than 1.47×10^{10} cm^{-3}. These results suggest that volume of dust particles deposited on vessel wall can be reduced by H_2 plasma irradiation.

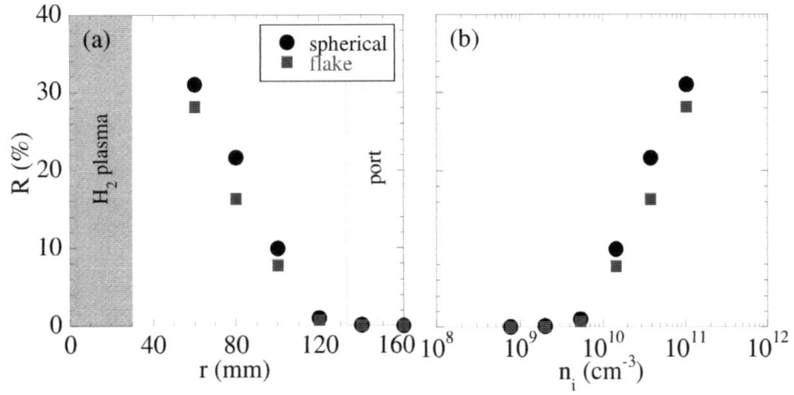

Figure 6. (a) a ration R of V_{etch} to V_{col} + V_{etch} versus distance r. (b) a ration R of V_{etch} to V_{col} + V_{etch} versus ion density n_i.

4. Conclusions

We have studied spatial profile of flux of dust particles generated due to interaction between H_2 plasmas and graphite target and evaluated etching of the collected dust particles due to H_2 plasma irradiation. We have obtained the following two conclusions:

1. Etching effects on dust particles are negligible for r larger than 120 mm. H_2 plasma etching significantly reduces size of dust particles for r smaller than 100 mm.
2. V_{col} + V_{etch} of flakes has a maximum value for r = 120 mm, suggesting that the flakes are formed by peering of re-deposited carbon films on vessel wall (r= 133.5 mm).
3. Volume of dust particles deposited on vessel wall are reduced by H_2 plasma irradiation.

Acknowledgments

This research was supported by MEXT KAKENHI Grant Number 21110005 and General Coordinated Research Grant from the National Institute for Fusion Science Grant Number NIFS12KLPF020 and NIFS12KLPF022.

References

[1] Winter J 2000 *Phys.Plasmas* **7** 3862-66
[2] Sharpe J P, Petti D A and Bartels H –W 2002 *Fusion Eng. Des.* **63-4** 153-63
[3] Muto S, Matsui T and Tanabe T 2002 *J. Nucl. Mater.* **307-11** 1289
[4] Koga K, Uehara R, Kitaura Y, Shiratani M, Watanabe Y and Komori A 2004 *IEEE Trans. Plasma Sci.* **32** 405-09
[5] Federici F, Coad J P, Haasz A A, Janeschitz G, Noda N, Philipps V, Roth J, Skinner C H, Tivey R and Wu C H 2000 *J. Nucl. Mater.* **283-7** 110-9
[6] Girard J -Ph, Garin P, Taylor N, Uzan-Elbez J, Rodriguez-Rodrigo L and Gulden W 2007 *Fusion Eng. Des.* **82** 506-10
[7] Krasheninnikov S I, Pigarov A Yu, Smirnov R D, Rosenberg M, Tanaka Y, Benson D J,

Soboleva T K, Rognlien T D, Mendis D A, Bray B D, Rudakov D L, Yu J H, West W P, Roquemore A L, Skinner C H, Terry J L, Lipschultz B, Bader A, Granetz R S, Pitcher C S, Ohno N, Takamura S, Masuzaki S, Ashikawa N, Shiratani M, Tokitani M, Kumazawa R, Asakura N, Nakano T, Litnovsky A M, Maqueda R and the LHD Experimental Group 2008 *Plasma Phys. Control. Fusion* **50** 124054

[8] Krasheninnikov S I, Smirnov R D and Rudakov D L 2011 *Plasma Phys. Control. Fusion* **53** 083001

[9] Denkevtis K, 2010 *Fusion Eng. and Des.* **85** 1059

[10] Shinomura Y, 2007 *J. Nucl. Mater.* **363-365** 467

[11] Roth J, et al., 2009 *J. Nucl. Mater.* **390** 1

[12] Rosanvallon S, et al., 2009 *J. Nucl. Mater.* **390** 57

[13] Koga K, Iwashita S, Kiridoshi S, Shiratani M, Ashikawa N, Nishimura K, Sagara A, Komori A and the LHD Experimental Group 2009 *Plasma Fusion Res.* **4** 34

[14] Iwashita S, Miyata H, Koga K, Shiratani M, Ashikawa N, Nishimura K, Sagara A and the LHD Experimental Group 2009 *J. Plasma Fusion Res. SERIES* **8** 308-1

[15] Iwashita S, Nishiyama K, Uchida G, Seo H, Itagaki N, Koga K, Shiratani M 2013 *Fusion Eng. Des.* **88** 28-32

[16] Koga K, Nishiyama K, Morita Y, Uchida G, Yamashita D, Kamataki K, Seo H, Itagaki N, Shiratani M, Ashikawa N, Masuzaki S, Nishimura K, Sagara A and the LHD Experimental Group 2013 *J. Nucl. Mater.* **438** 727-30

[17] Nishiyama K, Morita Y, Uchida G, Yamashita D, Kamataki K, Seo H, Itagaki N, Koga K, Shiratani M, Ashikawa N, Masuzaki S, Nishimura K, Sagara A, the LHD Experimental Group, Bornholdt S and Kersten H 2013 *J. Nucl. Mater.* **438** 788-91

[18] Aramaki M, Kato K, Goto M, Muto S, Morita S and Sasaki K 2004 *J. Appl. Phys.* **43** 1164-65

[19] Masuzaki S, Kobayashi M, Akiyama T, Ohno N, Morisaki T, Shoji M, Tokitani M, Tanaka H, Peterson B J, Yoshimura S, Narihara K, Yamada I, Yasuhara R, Murakami A, Miyazawa J, Murase T, Kobuchi T, Yonezu H, Kawamura G, Murakami I, Takeiri Y, Yamada H, Komori A and LHD experiment group 2013 J. Nucl. Mater. **438** S133-S138

[20] Roth J 1999 *J. Nucl. Mater.* **266-9** 51-7

[21] Koga K, Nishiyama K, Morita Y, Yamashita D, Kamataki K, Uchida G, Seo H, Itagaki N, Shiratani M, Ashikawa N, Masuzaki S, Nishimura K, Sagara A and the LHD Experimental Group 2012 Proc. 24th IAEA Fusion Energy Conf. **266** 51

Emission spectroscopy of Ar + H_2+ C_7H_8 plasmas: C_7H_8 flow rate dependence and pressure dependence

X Dong[1], K Koga[1], D Yamashita[1], H Seo[1], N Itagaki[1], M Shiratani[1], Y Setsuhara[2], M Sekine[3] and M Hori[3]

[1] Graduate School of Information Science and Electrical Engineering, Kyushu University, 744 Motooka, Nishi-ku, Fukuoka 819-0395, Japan
[2] Joining and Welding Research Institute, Osaka University, 11-1 Mihogaoka, Ibaraki, Osaka 567-0047, Japan
[3] Department of Electrical Engineering and Computer Science, Nagoya University, Furo-cho, Chikusa-ku, Nagoya 464-8603, Japan

E-mail: x.dong@plasma.ed.kyushu-u.ac.jp

Abstract. We have studied toluene flow rate dependence and pressure dependence of Ar emission intensities at 425.9 nm, 750.4 nm and 811.5 nm for Ar + H_2 + C_7H_8 plasmas. Electron density and the electron impact excitation rate coefficients are nearly the same for Ar(100%) and Ar+H_2(33%) plasmas. Toluene addition leads to a decrease in the high energy electron density, whereas the effective electron temperature and low energy electron density remain almost the same. The effective electron temperature, and Ar metastable density and/or low energy electron density decrease significantly with increasing the total pressure from 0.1 to 5 Torr.

1. Introduction

Carbon films, due to their hardness, wear resistance, chemical inertness, and biological compatibility, have been widely used as protective coatings in areas of magnetic storage disk, car parts, micro-electromechanical systems (MEMS) [1-4]. Deposition profile control of carbon films on nano-patterned substrates is one of the concerns to realize such protective coatings.

So far, we have succeeded in controlling deposition profiles of copper films on trench substrates, and realized sub-conformal, conformal and anisotropic deposition profiles [5-10]. Then the methods of controlling deposition profiles have been applied to carbon films, and realized sub-conformal, conformal and anisotropic deposition profiles, for which carbon is deposited only to the top or without sidewall deposition [11-16]. One remaining issue of our study is the low deposition rate. To increase the deposition rate, it is important to obtain information on plasma parameters. For this purpose, we have carried out optical emission spectroscopy. In this paper, we report experimental results of toluene flow rate dependence and pressure dependence of Ar emission intensities for Ar + H_2 + C_7H_8 plasmas.

2. Experimental

Experiments were performed using the H-assisted plasma CVD reactor, in which a capacitively-coupled main discharge and an inductively-coupled discharge of H atom source were sustained as shown in figure 1 [5-16]. The main discharge, which was sustained between a mesh powered electrode of 85 mm in diameter and a plane substrate electrode of 85 mm in diameter at a distance of 33 mm, was employed to produce carbon containing radicals as precursors. The excitation frequency of the main discharge was 28 MHz and the supplied voltage was 150 V. The discharge of H atom source, which was sustained using a radio frequency (RF) induction coil of 100 mm in diameter placed at 65 mm above the substrate electrode of main discharge, generated a high flux of H atoms toward the substrate. This reactor provided independent control of generation rates of carbon containing radicals and H atoms. The H atom source was separated from the main discharge using a grounded mesh (30 meshes/inch) of 160 mm in diameter placed at 2 mm above the mesh powered electrode of main discharge. In addition, an RF bias voltage of 400 kHz was applied to the substrate to control kinetic energy of incident ions. In this study, the H atom source and RF bias voltage were not applied.

Toluene (C_7H_8), Ar and H_2 were supplied at flow rates of 0.63 - 20 sccm, 60 sccm and 30 sccm, respectively. Toluene has a low ionization potential of 8.82 eV and a methyl group. The low ionization potential of toluene is expected to bring about a high deposition rate, because the deposition rate of carbon films tends to increase with decreasing the ionization potential of precursor gas [2]. The methyl group of toluene is expected to stabilize carbon films, because thermal stability of the deposited carbon films can be enhanced due to the methyl group [17]. C_7H_8 was vaporized at 150 °C and introduced into the reactor with H_2. The total pressure was set in a range of 0.1 - 5 Torr. The substrate temperature was 100 °C.

Optical emission intensities were measured at 11 mm above the center of substrate electrode with a miniature fiber optic spectrometer (Ocean Optics USB2000+).

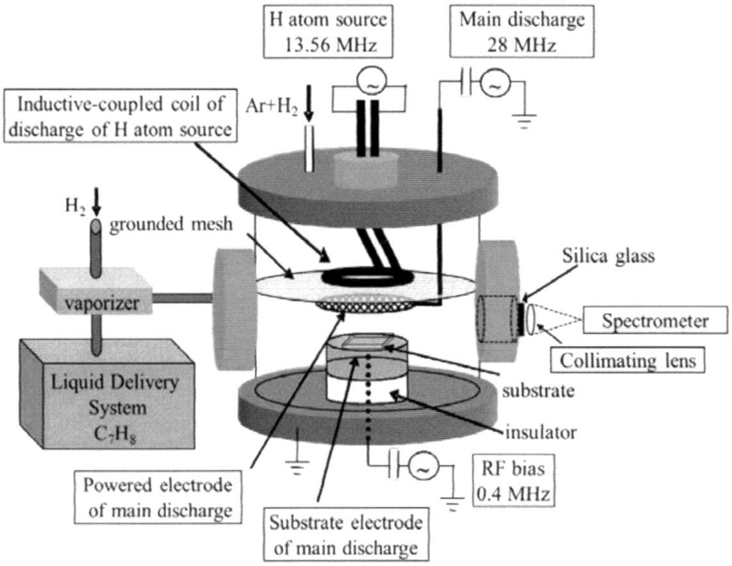

Figure 1. Schematic of H-assisted plasma CVD reactor.

3. Results and discussion

Figure 2 shows typical emission spectra for plasmas of (a) Ar(100%), (b) Ar+H_2(33%) and (c) Ar+ H_2(32.4%) + C_7H_8(0.027%) plasmas at 0.1 Torr. The emission intensities for Ar + H_2(33%) are a half of those for Ar(100%). The emission intensities for Ar + H_2(32.4%) + C_7H_8(0.027%) are a third of those for Ar(100%). Most Ar emission intensities are proportional to Ar partial pressure, electron density, and the rate coefficients for electron impact excitation from the Ar ground state. The emission intensities for Ar+H_2(33%) are weak compared to those for Ar(100%) mainly due to the decrease in the Ar partial pressure. In other words, the electron density and the excitation rate coefficients (effective electron temperature) are nearly the same for Ar(100%) and Ar+H_2(33%) plasmas. These results are consistent with the fact that Ar^+ is the predominant ion in Ar(100%) plasmas, whereas Ar^+

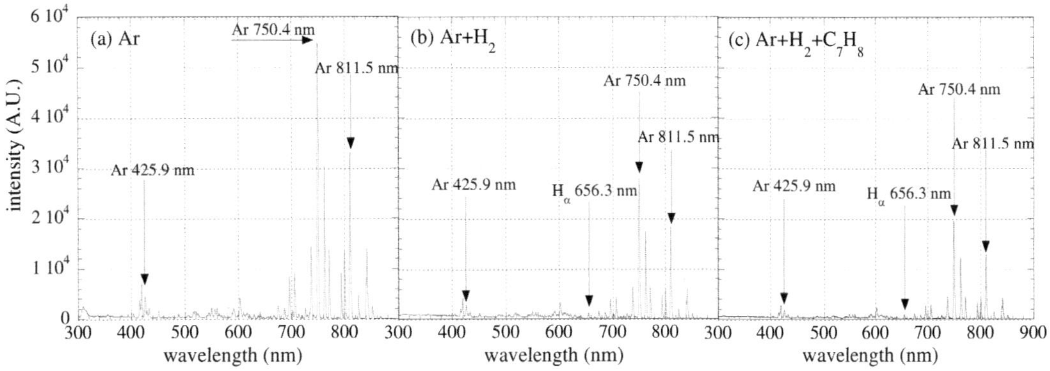

Figure 2. Emission spectra for plasmas of (a) Ar, (b) Ar + H_2(33%) and (c) Ar + H_2(32.4%) + C_7H_8(0.027%). Total pressure is 0.1 Torr.

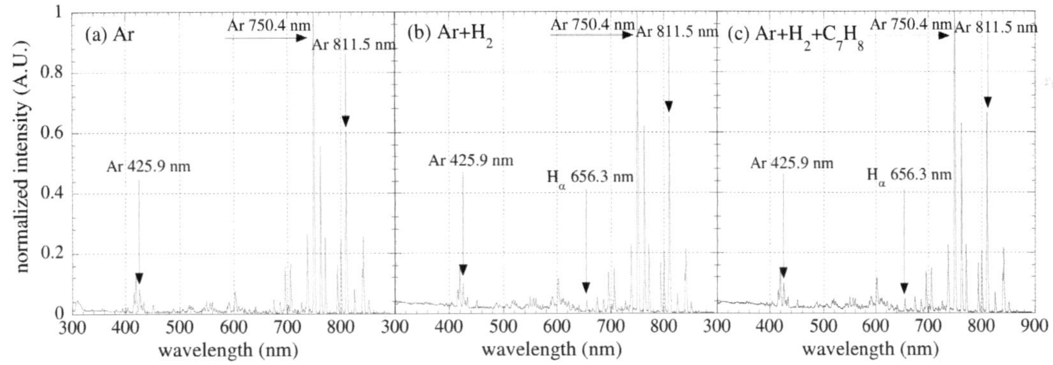

Figure 3. Normalized emission spectra for plasmas of (a) Ar, (b) Ar + H_2(33%) and (c) Ar + H_2(32.4%) + C_7H_8(0.027%). Total pressure is 0.1 Torr.

and ArH^+ are the main ions in Ar+H_2(33%) plasmas in the pressure range of 0.1 - 5 Torr [18, 19]. Because the Ar partial pressure of Ar + H_2(33%) and Ar + H_2(32.4%) + C_7H_8(0.027%) are nearly the same, the decrease in the emission intensities by adding C_7H_8 is caused by decreases in the high energy electron density ($\varepsilon \geq 13$ eV) and/or the excitation rate coefficients, which depend on electron temperature.

To obtain insight into the mechanisms of the decrease in the emission intensities by adding C_7H_8, emission spectra were normalized by the intensity at 750.4 nm $I_{750.4}$ in each subgraph of figure 2 as shown in figure 3. The normalized three spectra in figure 3 are similar with each other, indicating that effective electron temperature of these three plasmas is almost the same. Therefore, the decrease in the emission intensities by adding C_7H_8 is caused mainly by the decrease in the high energy electron density. Moreover, there exists H_α emission for Ar + H_2(33%) and Ar + H_2(32.4%) + C_7H_8(0.027%) in figure 3.

Figure 4 shows the toluene flow rate dependence of Ar emission intensities at (a) 750.4 nm, (b) 811.5 nm and (c) 425.9 nm. It can be seen a sudden drop in the $I_{425.9}$, $I_{811.5}$ and $I_{750.4}$ after introducing C_7H_8, then all of them have a linear decrease with increasing the toluene flow rate from 0.63 to 20 sccm. Ar emissions at 425.9 nm and 750.4 nm have little influence of quenching and radiation trapping, and their upper levels of $2p_1$ (excitation energy of 13.5 eV) and $3p_1$ (excitation energy of 14.7 eV) have small cross section for electron impact excitation from metastable states [20, 21]. Therefore, the ratio of emission intensity of Ar I at 425.9 nm to that at 750.4 nm, $I_{425.9}/I_{750.4}$ is used to evaluate the effective electron temperature. The Ar emission at 811.5 nm is strongly influenced by the

$1s_5$ metastable state (excitation energy of 11.5 eV) that can provide an additional channel for Ar atom excitation by low energy electrons ($1.5 \leq \varepsilon \leq 3$ eV). So the emission intensity ratio, $I_{811.5}/I_{750.4}$ is used to obtain information on the Ar metastable density and the low energy electron density [22, 23].

The Ar emission intensity ratios are shown in figure 5. Both the ratios are nearly constant irrespective of the toluene flow rate, showing that the effective electron temperature, Ar metastable density and low energy electron density remain almost the same. These results indicate that the high energy electron density decreases with increasing the toluene flow rate from 0 to 20 sccm. The total flow rate increases from 90 sccm to 110 sccm with increasing toluene flow rate from 0 to 20 sccm. The change of the total flow rate is small. In addition, the discharge voltage was kept constant in this study. Therefore the total flow rate and the discharge voltage have little influence on plasma parameters in this study.

Figure 6 shows the total pressure dependence of Ar emission intensities at 425.9 nm, 750.4 nm and 811.5 nm, normalized by each intensity for 0.1 Torr. To obtain information of plasma parameters in the bulk region, we have measured the emission intensities at 11 mm above the center of substrate. Though $I_{750.4}$ has a slight increase in the pressure range of 0.5 - 1.5 Torr, $I_{425.9}$ and $I_{811.5}$ decrease significantly with increasing the total pressure from 0.1 to 5 Torr.

The corresponding normalized Ar emission intensity ratios of figure 6 are shown in figure 7. $I_{425.9}/I_{750.4}$ and $I_{811.5}/I_{750.4}$ decrease by 84.3% and 55.6% with increasing the pressure from 0.1 to 3 Torr, and then they decrease by 24.6% and 4.8% with increasing the pressure from 3 to 5 Torr. These results indicate that the effective electron temperature, and Ar metastable density and/or low energy electron density decrease significantly with increasing the total pressure from 0.1 to 3 Torr, and then they decrease slightly with increasing the total pressure from 3 to 5 Torr.

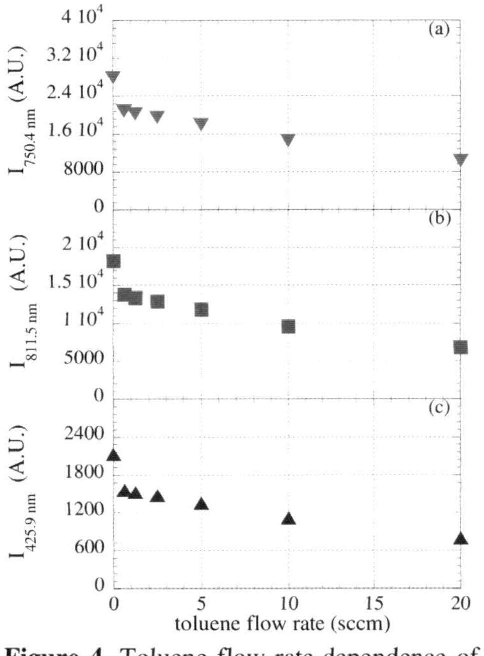

Figure 4. Toluene flow rate dependence of Ar emission intensities at (a) 750.4nm, (b) 811.5 nm and (c) 425.9 nm. Total pressure is 0.1 Torr.

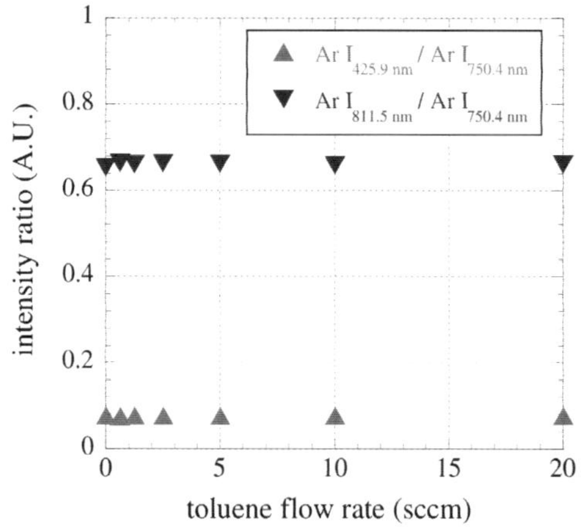

Figure 5. Toluene flow rate dependence of Ar emission intensity ratios. Total pressure is 0.1 Torr.

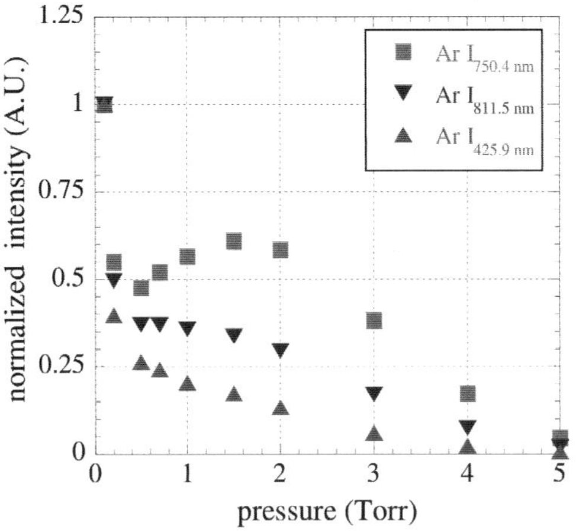

Figure 6. Pressure dependence of normalized Ar emission intensities.

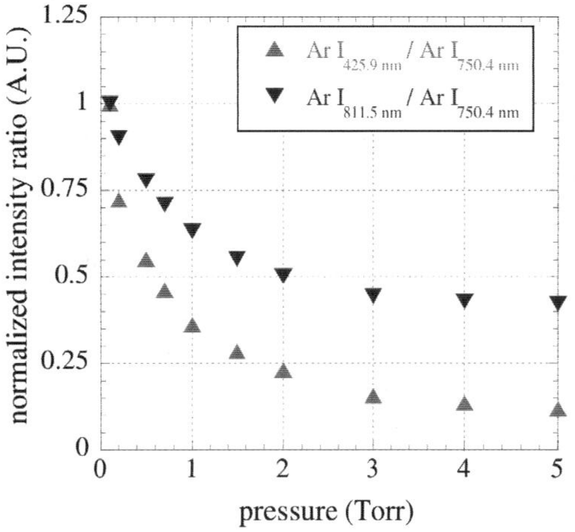

Figure 7. Pressure dependence of normalized Ar emission intensity ratios.

4. Conclusions

We studied the toluene flow rate and pressure dependence of emission intensities in Ar + H_2 + C_7H_8 plasmas. The following conclusions are obtained in this study.

(1) Electron density and the electron impact excitation rate coefficients (effective electron temperature) are nearly the same for Ar(100%) and Ar+H_2(33%) plasmas.
(2) Toluene addition leads to a decrease in the high energy electron density, whereas the effective electron temperature and low energy electron density remain almost the same.
(3) The effective electron temperature, and Ar metastable density and/or low energy electron density decrease significantly with increasing the total pressure from 0.1 to 5 Torr.

Acknowledgements

This work was partly supported by the Grant-in-Aid for Scientific Research on Innovative Areas, MEXT KAKENHI Grant Number 21110001 and 21110005.

References

[1] Voevodin A A and Donley M S 1996 *Surf. Coat. Technol.* **82** 199
[2] Robertson J 2002 *Mater. Sci. Eng. R* **37** 129
[3] Robertson J 2008 *Phys. Status Solidi A* **205** 2233
[4] Robertson J 2001 *Jpn. J. Appl. Phys.* **50** 01AF01
[5] Shiratani M, Jin H J, Takenaka K, Koga K, Kinoshita T and Watanabe Y 2001 *Sci. Technol. Adv. Mater.* **2** 505
[6] Takenanka K, Kita M, Kinoshita T, Koga K, Shiratanai M and Watanabe Y 2004 *J. Vac. Sci. Technol. A* **22** 1993
[7] Takenaka K, Shiratani M, Takashita M, Kita M, Koga K and Watanabe Y 2005 *Pure Appl. Chem.* **77** 391
[8] Takenaka K, Koga K, Shiratani M, Watanabe Y and Shingen T 2006 *Thin Solid Films* **506** 197
[9] Umetsu J, Koga K, Inoue K, Matsuzaki H, Takenaka K and Shiratani M 2008 *Surf. Coat. Technol.* **202** 5659
[10] Umetsu J, Inoue K, Nomura T, Matsuzaki H, Koga K, Shiratani M, Setsuhara Y, Sekine M and Hori M 2008 *DPS Proc.* **30** 35
[11] Umetsu J, Inoue K, Nomura T, Matsuzaki H, Koga K, Shiratani M, Setsuhara Y, Sekine M and Hori M 2009 *J. Plasma Fusion Res.* **8** 1443
[12] Nomura T, Koga K, Shiratani M, Setsuhara Y, Sekine M and Hori M 2009 *MRS Proc.* **1222**

DD05-16
[13] Nomuta T, Urakawa T, Korenaga Y, Yamashita D, Matsuzaki H, Koga K, Shiratani M, Setsuhara Y, Sekine M and Hori M 2010 *TENCON Proc.* 2213
[14] Urakawa T, Nomora T, Matsuzaki H, Yamashita D, Uchida G, Koga K, Shiratani M, Setsuhara Y, Sekine M and Hori M 2001 *ISPC 20 Proc. POL* **02**
[15] Koga K, Urakawa T, Uchida G, Kamataki K, Seo Y, Itagaki N, Shiratani M, Setsuhara Y, Sekine M and Hori M 2012 *PES Proc.* **136**
[16] Urakawa T, Matsuzaki H, Yamashita D, Uchida G, Koga K, Shiratani M, Setsuhara Y, Sekine M and Hori M 2012 *Surf. Coat. Technol.* **228** S15
[17] Joo J, Quan Y C, and Jung D 2000 *J. Mater. Res.* **15** 228.
[18] Sode M, Schwarz-Selinger T and Jacob W 2013 *J. Appl. Phys.* **113** 093304
[19] Neys E, Yan M, Bogarerts A and Gijbels R 2003 *Nuc. Instr. And Meth. In Phys. Res. B* **202** 300
[20] Moshkalyov S A, Diniz J A, Swart J W, Tatsch P J and Machida M 1997 *J. Vac. Sci. Technol. B* **15** 2682
[21] Daltrini A M, Moshkalev S A, Monteiro M J R, Besseler E, Kostryukov A and Machida M 2007 *J. Appl. Phys.* **101** 073309
[22] Czerwiec T and Graves D B 2004 *J.Phys. D: Appl. Phys.* **37** 2827
[23] Daltrini A M, Moshkalev S A, Swart L and Verdonck P B 2007 *JICS* 67

The First principle calculation of bulk modulus and Young's modulus for amorphous carbon material

A M Ito[1], A Takayama[1], Y Oda[1] and H Nakamura[1,2]

[1]Department of Helical Plasma Research, National Institute for Fusion Science, Toki, Gifu 509-5292, Japan
[2]Department of Energy Engineering and Science, Nagoya University, Nagoya 464-8602, Japan

E-mail: ito.atsushi@nifs.ac.jp

Abstract. The dependence of bulk modulus and Young's modulus on mass density for hydrogen-free amorphous carbon structures were investigated by using the first principle calculation on density functional theory. The amorphous carbon structures with various mass densities were generated by the numerical operations of structure relaxation from the random structures and volume relaxation. The bulk modulus and Young's modulus increased in proportion to the mass density. We uses the number densities of sp^2 and sp^3 atoms in place of these fractions or ratios as a classification parameter of amorphous carbon material because the number density of sp^3 atoms can fit the both of amorphous carbon and diamond by one straight line in bulk modulus and Young's modulus. By this analysis, we suggest that the increases of the bulk modulus and Young's modulus are derived from only the sp^3 carbon atom.

1. Introduction

Amorphous carbon materials include the various kinds of properties depending on the methods of fabrication. A part of them is elastic modulus and hardness close to those of a diamond. The hard amorphous carbon materials are called diamond like carbon (DLC), and then the DLC is used for industrial purposes such as coating. It is considered that the higher elastic modulus and hardness of amorphous carbon material is derived from the tetrahedral structure of sp^3 bonding of carbon atoms. Therefore, hard amorphous carbon is usually classified into tetrahedral amorphous carbon (ta-C). The relation between the elastic modulus and the sp^3 fraction, which is the ratio of number density of sp^3 carbon atoms to the number density of all kinds of carbon atoms, is often regarded as one of the most important features for the classification of amorphous carbon materials [1-6]. For example, the dependence of Young's modulus and hardness on sp^3/sp^2 ratio [7,8] was researched. In addition, because the ta-C has higher mass density close to that of a diamond, mass density is also used for the classification of amorphous carbon material. The dependence of sp^3 fraction [9,10], the Young's modulus [5,11-13] and shear modulus [12] on mass density were researched.

However, the factor to complicate the classification of amorphous carbon material is the containing of hydrogen atoms. Hydrogen atoms eliminate dangling bond, and but carbon atoms terminated by hydrogen atoms cannot construct tetrahedral carbon network structure. Namely, elastic modulus and hardness decrease as hydrogen content increases. Actually, for almost all of hydrogenated amorphous carbon, Young's modules is 250 GPa or less [7,8]. On the other hand, that Young's modulus of hydrogen free amorphous carbon which is created by using mass-selected ion beam (MSIB) deposition is 680 GPa [4].

Theoretical study on amorphous carbon material is fewer than experimental researches. Theoretical approach is necessary to solve the classification of amorphous carbon material. Especially, numerical simulation in virtual environments can simplify the complex feature of amorphous carbon material. Deposition process is often simulated by using molecular dynamics (MD) simulation [14,15]. We had also investigated the dependence of deposition rate on dependence on H/C ratio by using MD [16]. Moreover, amorphization from graphite due to hydrogen plasma irradiation was investigated with the hybrid simulation of MD and binary collision approximation (BCA) [17]. However, the problem in the MD and MD-BCA hybrid simulations is that the simulation result strongly depends on the accuracy of interatomic potential model. Especially, physical properties evaluated from MD are unreliable in amorphous carbon materials [15].

Although we had improved the quality of potential model for carbon system [18], the first principle calculation based on the density functional theory (DFT) is more suited to the evaluation of physical properties.

In this paper, we challenge to clarify mass density and sp^3 fraction dependences of bulk modulus and Young's modulus in amorphous carbon material, where we only treat hydrogen free amorphous carbon material to simplify the problem. In order to evaluate correct value of bulk modulus and Young's modulus, fully DFT calculation is employed in the present research. Simulation method and code for DFT calculation are introduced in section 2. Amorphous carbon material was created by numerical approach in section 2.2. In section 3, Bulk modulus and Young's modulus of the amorphous carbon material are calculated and then we discuss about dependence on mass density and sp^3 fraction. This paper ends with conclusions in section 4.

2. Simulation method

2.1. Density functional theory

In the present paper, the first principle calculation of the total energy and forces are executed based on the density functional theory (DFT). In the DFT, the wave function for the electrons bound by atomic nuclei is solved on the Kohn-Sham equation [19,20]. The total energy as a solution of the Kohn-Sham equation for amorphous carbon structures was calculated with Open source package for Material eXplorer (OpenMX) [21] code, which is numerical simulation software based on DFT.

The generalized gradient approximation (GGA) with a Perdew-Burke-Ernzerhof (PBE) functional [22] was employed as an exchange-correlation potential. Only the valence electrons were explicitly simulated by using norm-conserving pseudo-potentials [23]. The wave functions were numerically represented by linear combinations of pseudo-atomic orbitals (LCPAOs) [24,25]. The cutoff energy was set to 250 Rydberg.

2.2. Creation of amorphous carbon geometry

Amorphous carbon structures were prepared in numerical simulation. To put it briefly, we find the structure in which force acting on each atoms calculated by DFT balances.

Cubic simulation box of size 1.0 nm on a side with the periodic boundary condition is set. Carbon atoms are put into the simulation box at random positions. The forces acting on atoms are calculated by DFT from the current structure, and then the atoms are moved by minute distances along the directions of the forces. By iterating these operations, the structure at which all force balances is obtained. This method is, in general, called structure relaxation. In the present paper, to reduce the number of iteration, the amounts of movement of atoms are determined by using the Broyden-Fletcher-Goldfarb-Shanno method [26-29], which is one of quasi Newton methods. The iteration was finished when the maximum magnitude of the force vector is smaller than 3.0×10^{-4} hartree/bohr. The ten kinds of amorphous structure are prepared varying the number of carbon atoms from 80 to 175, which are corresponding to the mass density from 1.59g/cc to 3.49g/cc.

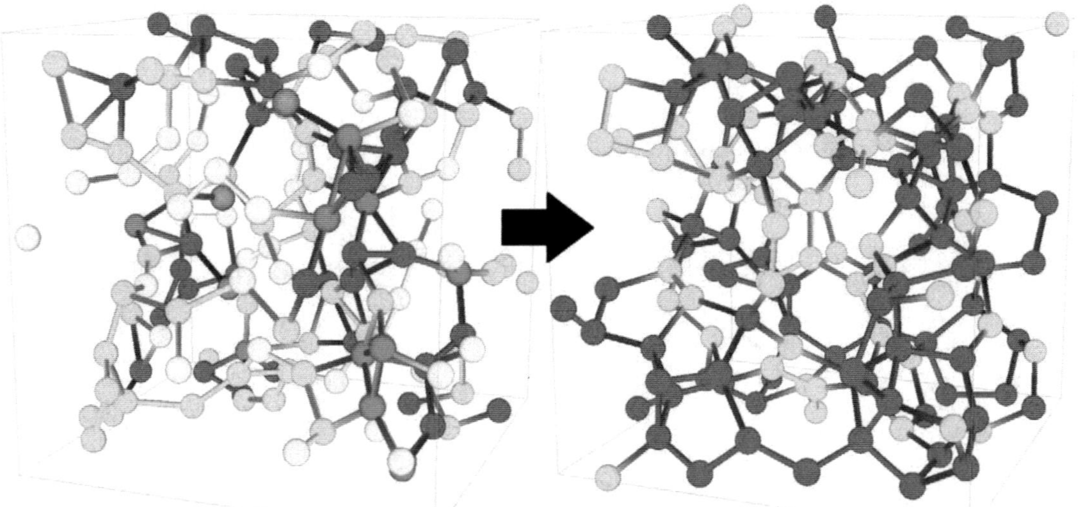

Figure 1. The structure relaxation from the left structure to the right structure. The color of atoms indicate the coordination numbers of atoms. The yellow, green and red spheres indicate sp^1, sp^2 and sp^3 atoms, respectively. The blue and purple spheres indicate lower and higher coordination atoms.

Figure 1 shows the structure change before and after the structure relaxation. From this figure, it is confirmed that all of carbon atoms which have higher or lower coordination numbers changes into the carbon atoms which have the coordination numbers from two to four.

The validation of the amorphous structure created by the above structure relaxation was to check the radial distribution function. All of the amorphous structures created in the present work have the radial distribution function like a liquid state.

By the above operation of structure relaxation, the forces acting on atoms balance, i.e. forces becomes zero. However, stress is applied to the whole of system since the fixed volume of simulation box under the periodic boundary condition. In some cases, the amorphous structures are narrowly kept by the bound by the fixed size of simulation box. Therefore, we should relax the amorphous structures by varying the volume of the simulation box, which is called volume relaxation here.

In the operation of the volume relaxation, the size of simulation and atomic positions are isotropically scaled. The atomic position after structure relaxation $\boldsymbol{r}$ and the side of initial simulation box L is scaled with the scaling factor s. Scaled atomic position becomes $\boldsymbol{r}' = s\boldsymbol{r}$ and the scaled side of initial simulation box becomes sL. The total energy $U(\boldsymbol{r}')$ is function of the scaling parameter s via the scaled atomic position. By iterative calculation of total energy varying s, we obtained the solution s_0 of an equation $\partial U(s\boldsymbol{r})/\partial s = 0$. As a result, atomic position and the side of simulation box after these volume relaxation are $\boldsymbol{r}'_0 = s_0\boldsymbol{r}$ and $L'_0 = s_0 L$, respectively. Since the volume of the simulation box was changed, the mass densities of the amorphous carbon structures after volume relaxation becomes 1.64 g/cc to 3.06 g/cc.

Note that isotropic stress is reduced by the volume relaxation. The derivative of total energy $\partial U(s\boldsymbol{r})/\partial s$ by the scaling factor s can be given by

$$\frac{\partial U(s\boldsymbol{r})}{\partial s} = -\frac{1}{s}\sum_i \boldsymbol{F}_i \cdot \boldsymbol{r}'_i \tag{1}$$

where $\boldsymbol{r}'_i$ is the position of i-th atom and the force acting on the i-th atom $\boldsymbol{F}_i = -\partial U(\boldsymbol{r}')/\partial \boldsymbol{r}'_i$. The right hand term in equation (1) corresponds to isotropic stress in the particle picture [30]. When the derivative of total energy $\partial U(s\boldsymbol{r}_0)/\partial s = 0$, then the isotropic stress becomes zero.

Journal of Physics: Conference Series **518** (2014) 012011 doi:10.1088/1742-6596/518/1/012011

2.3. Calculation of Bulk modulus and Young's modulus

For the ten kinds of amorphous structures by the above operations of structure relaxation and volume relaxation, Bulk modulus and Young's modules are evaluated by using the DFT calculation.

Bulk modulus can be calculated by the isotropic scaling operation same to the volume relaxation. Here, we redefine atomic positions and simulation box after the above volume relaxation as $\boldsymbol{r}$ and L_0. They are re-scaled by the scaling factor s, and then the bulk modulus is given by

$$K = \frac{1}{9{L_0}^3}\left[-\frac{2}{s^2}\frac{\partial U(s\boldsymbol{r})}{\partial s} + \frac{1}{s}\frac{\partial^2 U(s\boldsymbol{r})}{\partial s^2}\right] \tag{2}$$

In actual operation, the bulk modulus is obtained from same calculation to the volume relaxation.

Young's modulus is calculated by one dimensional scaling operation. In the present paper, we performed only the scaling in the x direction with the scaling factor s_x. The atomic positions are changed from $\boldsymbol{r} = (x, y, z)$ to $\boldsymbol{r}'' = (s_x x, y, z)$ and the side of the simulation box in x direction is changed from $L_{0,x}$ to $s_x L_{0,x}$. The Young's modulus is given by

$$E = \frac{1}{{L_0}^3}\frac{\partial^2 U(\boldsymbol{r}'')}{\partial {s_x}^2} \tag{3}$$

3. Results and discussion

By the operations of the structure relaxation and volume relaxation in section 2, the ten kinds of amorphous carbon structure, which has difference in mass density. The mass density of these amorphous structures are for the range from 1.64 g/cc to 3.06 g/cc.

Figure 2 shows the bulk modulus of amorphous carbon material as a function of mass density. The bulk modulus increases in proportion to the mass density. This proportionality coefficient is estimated as 133.9 GPa·cc/g. In addition, we calculated the bulk modulus of a perfect crystalline diamond. The evaluated value of bulk modulus of the perfect crystalline diamond by DFT was 450.6 GPa. The bulk modulus of the perfect crystalline diamond can be placed on the approximate straight line which generated by fitting the bulk modulus of amorphous carbon proportional to the mass density.

Figure 3 shows the Young's modulus of amorphous carbon material as a function of mass density. As well as the bulk modulus, the Young's modulus increases in proportion to the mass density. This proportionality coefficient is estimated as 434.3 GPa·cc/g. The Young's modulus of the perfect crystalline diamond was evaluated as 1116.2 GPa by DFT. The Young's modulus of the perfect crystalline diamond is also placed on the approximate straight line generated by fitting the bulk

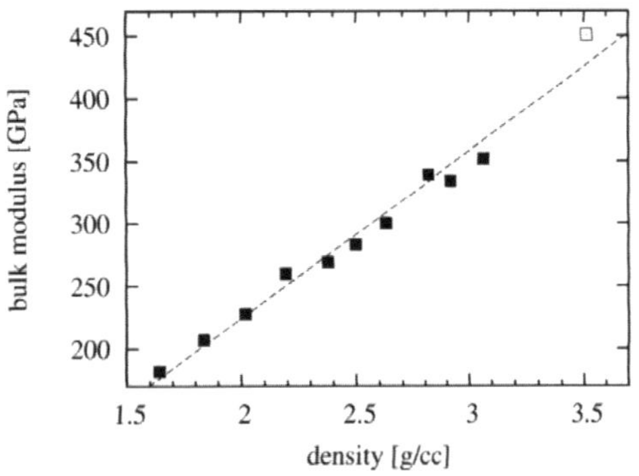

Figure 2. The bulk modulus of amorphous carbon material as a function of mass density. The white rectangle indicates the bulk modulus of diamond.

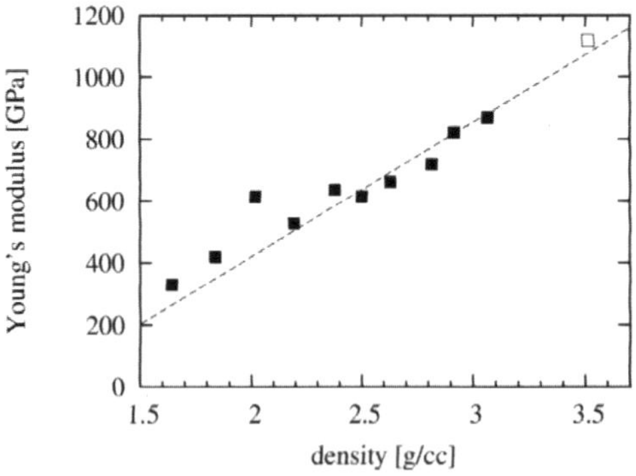

Figure 3. The Young's modulus of amorphous carbon material as a function of mass density. The white rectangle indicates the Young's modulus of diamond.

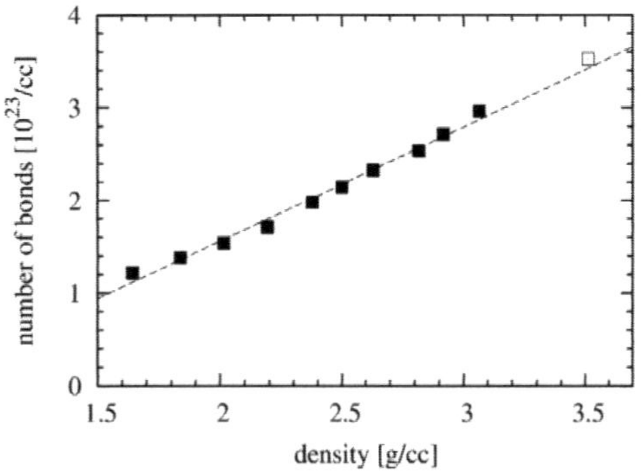

Figure 4. The number density of covalent bonds in amorphous carbon materials as a function of mass density. The white rectangle indicates diamond.

modulus of amorphous carbon proportional to the mass density.

The proportionality of the Young's modulus and mass density was also reported by experiment [5]. The proportionality coefficient estimated from experimental reported data is about 530 GPa·cc/g and it is agreement with our calculation results. We consider that the difference in the proportionality coefficient between the present calculation and experimental report is caused by the variability among sample structures of amorphous carbon in this calculation. Especially, for the small range of mass density, the amorphous structures include larger void spaces. The variety of the position and configuration of the large void spaces causes the variability of mechanical property.

To explain the dependences of the bulk modulus and the Young's modulus on mass density, we counted the covalent bonds of amorphous carbon materials. Figure 4 shows the number density of covalent bonds, which is the number of covalent bonds per unit volume, as a function of mass density. It is found that the number density of covalent bonds increases in proportional to mass density as well as the bulk modulus and Young's modulus. This fact implies that the bulk modulus and Young's modulus is simply proportion to the number of covalent bonds.

Further, sp^1, sp^2 and sp^3 atoms were counted in amorphous carbon materials, where the classification of carbon atoms into sp^1, sp^2 and sp^3 states is simply according to coordination number. Figure 5(a) shows sp^1, sp^2 and sp^3 fractions as functions of mass density. It is found that the sp^3 fraction increases in proportion to mass density. The present calculation result well agrees with the experimental measurement [9,10] on the fact that the linear increase of sp^3 fraction from 0.2 at 2.2 g/cc to 0.8 at 3.0g/cc.

However, the sp^3 fraction of the perfect crystalline diamond, which is 1.0, is not located on the approximate straight line of linear proportionality in figure 5(a). Therefore, we recommend the use of the number density of sp^3 carbon atom, which is the number of sp^3 carbon atoms per unit volume, in place of the sp^3 fraction. Figure 5(b) shows the number densities of sp^1, sp^2 and sp^3 atoms as functions of mass density. In the representation by the number densities of sp^3 atoms, the value of the perfect crystalline diamond is located on the approximation straight line of linear proportionality.

The representation by number density is superior in terms of sp^2 carbon atoms also. In the representation by fraction in figure 5(a), the sp^2 fraction decreases as mass density increases, while in the representation by number density in figure 5(b), the number density of sp^2 hardly changes for the range of 2.8g/cc or less. This fact implies that the decrease of sp^2 fraction is apparent decrease associated with by the increase of sp^3 atoms.

Thus, the total number density of covalent bonds and the number density of sp^3 atoms increase in proportion to mass density. This fact implies that the increases of the bulk modulus and Young's modulus are brought about the increase of sp^3 carbon atoms or simply covalent bonds. The bulk modulus and Young's modulus hardly had correlation with the number density of the sp^2 carbon atoms.

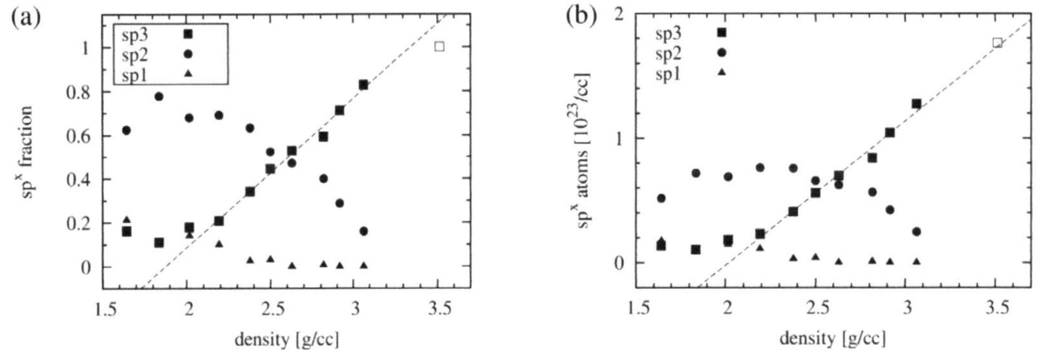

Figure 5. (a) The sp^1, sp^2 and sp^3 fraction as a function of mass density. (b) The number densities of sp^1, sp^2 and sp3 atoms as a function of mass density. The white rectangles indicate diamond.

Finally, we note the problem to advance the present analysis with the above numerical simulation. One problem is the variability among sample structures of amorphous carbon. Especially, the amorphous carbon has the small mass density, which has sp^2 rich structure, easily create larger void spaces. In order to more accurate evaluation of elastic modulus, the calculation of many sample structures is required. The other problem is that the space scale in the present simulation is small. If the size of sample amorphous structures are enough large, the variability among the sample structures becomes vanishingly small. Moreover, the real large scale amorphous carbon material, which means the material treated in real experiment, has certainly precipitated crystalline grains and long range order structure. These are effective to the mechanical property such has elastic modulus. To treat the precipitated crystalline grains and the long range order structure, at least sub-micro meter scale is required. Classical MD simulation can be used for the evaluation of mechanical property of the large scale amorphous structures, where the interaction potential model for classical MD should be improved to represent the realistic amorphous carbon material [18].

4. Conclusions

In this paper, we investigated the bulk modulus and Young's modulus of hydrogen-free amorphous carbon structures by using the first principle calculation on density functional theory. The amorphous carbon structures which have various mass densities for the range from 1.64 g/cc to 3.06 g/cc were reproduced by the operations of the structure relaxation from the random structures and the volume relaxation. For the reproduced amorphous, we showed that the bulk modulus and Young's modulus increases in proportion to the mass density. The number density of sp^3 carbon atoms in amorphous carbons also increases in proportion to the mass density, while that of sp^2 carbon atoms does not increase with the mass density. From this fact, we suggest that the increases of bulk modulus and Young's modulus in hydrogen-free amorphous carbon materials are produced by the increase of the sp^3 carbon atoms or the increase of covalent bonds simply.

As a new classification parameter, we proposed the number densities of sp^2 and sp^3 atoms in place of these fractions. The superiority of the representation by number density is evident from the two facts that the number density of sp^3 atoms can fit amorphous carbon and the perfect crystalline diamond by one straight line and that the remaining of the number density of sp^2 was shown. The number densities of sp^2 and sp^3 are obtained by simple transformation from the sp^2 and sp^3 fractions.

In future work, hydrogenated amorphous carbon will be analysed by similar method with DFT calculations. Not only elastic modulus, electronic property of amorphous carbon can be investigated by DFT. However, much more sample structures should be prepared to vanish the variability among the sample structures of amorphous carbon, especially for the range of small mass density. In association with this problem, the existence of the amorphous structures which have same sp^3/sp^2 ratio and but differs in bond orientation due to the difference of fabrication method was predicted [31]. MD is necessary for the simulation of various fabrication of amorphous carbon and the simulation of larger amorphous structure including the precipitated crystalline grains and the long range order structure. Therefore, we should use both MD and DFT simulations as the situation demands.

Acknowledgements
Our work was supported by Grants-in-Aid for Scientific Research (No. 23710135) from the Ministry of Education, Culture, Sports, Science, and Technology, Japan, and by the National Institute of Natural Science (NINS) Program for Cross-Disciplinary Study. Numerical simulations were carried out by using the Plasma Simulator at the National Institute for Fusion Science and IFERC-CSC at the Japan Atomic Energy Agency.

References
[1] Robertson J 1991 *Prog. Solid St. Chem.* **21** 199
[2] Robertson J 1992 *Surf. Coat.Tech.* **50** 185
[3] Robertson J 1992 *Diamond and Related Materials* **1** 397
[4] Lifshitz Y 1996 *Diamond and Related Materials* **5** 388
[5] Robertson J 2002 *Mater.Sci.Engineering* R **37** 129
[6] Lifshitz Y 1999 *Diamond and Related Materials* **8** 1659
[7] Demichelis F, Pirri C F, Tagliaferro A, Benedetto G, Boarino L, Spagnolo R, Dunlop E, Haupt J, and Gissler W 1993 *Diamond and Related Materials* **2** 890
[8] Savvides N and Bell T J 1993 *Thin Solid Films* **228** 289
[9] Robertson J 1993 *Diamond and Related Materials* **2** 984
[10] Fallon P J, Veerasamy V S, Davis C A, Robertson J, Amaratunga G, Milne W I and Koskinen J 1993 *Phys. Rev.* B **48** 4777
[11] Schultrich B, Scheibe H J, Drescher D and Ziegele H 1998 *Surf. Coatings Thechnol.* **98** 1097
[12] Beghi M G, Ferrari A C, Bottani C E, Libassi A, Tanner B K, Teo K B K and Robertson J 2002 *Diamond and Related Materials* **11** 1062
[13] Schultrich B, Scheibe H J, Grandremy G, Drescher D and Schneider D 1996 *Diamond Relat. Mater.* **5** 914
[14] Neyts E, Bogaerts A, Gijbels R, Benedikt J, Sanden M C M 2004 *Diamond and Related Materials* **13** 1873
[15] Marks N A, Cooper N C, McKenzie D R, McCulloch D G, Bath P, and Russo S P *Phys. Rev.* B **65** 075411
[16] Ito A M, Takayama A, Saito S, Ohno N, Kajita S and Nakamura H 2011 *Jpn. J. Appl. Phys.* **50** 01AB01
[17] Saito S, Ito A M, Takayama A and Nakamura H 2013 *Jpn. J. Appl. Phys.* **52** 01AL02
[18] Ito A M, Takayama A, Saito S and Nakamura H 2013 *Jpn. J. Appl. Phys.* **52** 01AL04
[19] Hohenberg P and Kohn W 1964 *Phys. Rev.* **136** B864
[20] Kohn W and Sham L J 1965 *Phys. Rev.* **140** A1133
[21] OpenMX web-site: http://www.openmx-square.org/
[22] Perdew J P, Burke K, and Ernzerhof M 1996 *Phys. Rev. Lett.* **77** 3865
[23] Morrison I, Bylander D M, Kleinman L 1993 *Phys. Rev.* B **47** 6728
[24] Ozaki T 2003 *Phys. Rev.* B **67** 155108
[25] Ozaki T and Kino H 2004 *Phys. Rev.* B **69** 195113
[26] Broyden C G 1970 *J. Inst. Math. Appl.* **6** 76
[27] Fletcher R 1970 *Comput. J.* **13** 317
[28] Goldrarb D 1970 *Math. Comp.* **24** 23
[29] Shanno D F 1970 *Math. Comp.* **24** 647
[30] Ito A and Nakamura H, 2008 *Theor. Appl. Mechanics Jpn.* **57** 159
[31] Saito S, Ito A M, Takayama A and Nakamura H 2013 *Jpn. J. Appl. Phys.* **51** 01AC05

Journal of Physics: Conference Series **518** (2014) 012012 doi:10.1088/1742-6596/518/1/012012

Multiple-height microstructure fabricated by deep reactive ion etching and soft resist masks combined with UV curing

R Sato, T Sawada, S Kumagai, and M Sasaki

Department of Advanced Science and Technology, Toyota Technological Institute, 2-12-1 Hisakata, Tenpaku-ku, Nagoya, 468-8511, Japan

E-mail: mnr-sasaki@toyota-ti.ac.jp

Abstract. Multiple-height microstructures are realized by deep reactive ion etching and UV-cured photoresist used in the embedded mask process. Although the UV-cured photoresist is a soft mask, its material property becomes stable against resist thinner and UV exposure. A layered resist pattern can be realized by stacking normal photoresist on the UV-cured photoresist. The normal photoresist can be selectively removed by the flush exposure and developing after the first Si etching. This technique is applied to two MEMS devices.

1. Introduction

3D structures are frequently required in MEMS devices. In many cases, the shape in its depth direction is determined by the projection of a top single mask pattern (so called 2.5D). The mask pattern is on the top surface, since the photolithography patterning is only effective on the planer surface. The challenge towards 3D is changing the depth profile in a controlled manner. A grey scale mask can give 3D resist structure, which can be transferred to Si substrate [1]. Multiple-height structures having digital steps are realized by the embedded mask process. This process usually uses masks of resist and hard materials (e.g., SiO_2, SiN, Al). To prepare SiO_2 [2] or SiN [3] films, the process temperature must be high. A high temperature should be avoided to maintain already fabricated structures. As for the metal film [4, 5], there is a risk of contamination of the plasma chamber. Soft resist mask is preferred since the plasma cleaning can be applied. Recently, UV curing of the photoresist has been found to realize layered resist masks [6]. This previous method relies on the ashing rate difference (only 2-times) relating with the glass transition temperature for the selectivity. The process margin for the demonstrated micromirror is tight [7].

In this study, a new fabrication technique for improving the selectivity is described. Two devices are demonstrated.

2. UV curing

UV curing is the treatment realized by heating the photoresist under vacuum and UV irradiation. UV curing makes the resist molecules link together to form larger molecules. UV cured photoresist becomes stable against thinner and UV exposure. The over-coating of the additional resist layer does not change the underlying pattern. UV exposure only patterns the top layer of the normal photoresist [6]. During the process, the process plasma irradiates UV to the photoresist mask. For removing the photoresist after the plasma process, the positive type photoresist is useful since it can be removed by developing. The above things are basically valid for g-line photoresist. The photoresist used is

Journal of Physics: Conference Series **518** (2014) 012012 doi:10.1088/1742-6596/518/1/012012

AZ1500 having thicknesses of about 3 μm. UV curing is at 100 °C, about 3 kPa for 40 min. The light (main wavelength 240-380 nm, ~1 W/cm²) of a high-pressure mercury lamp is irradiated.

3. Selective removal of normal resist against UV cured resist

Figure 1 shows one fabrication sequence. The 1st pattern is transferred to the photoresist and UV cured. Then the normal resist is coated and 2nd pattern is transferred. After the plasma processing (step 2), the top normal resist is removed by the flush exposure and the developing. The inset (b) in figure 1 shows the photo of the sample. The residue is observed and remains even at the over-dose condition (appropriate patterning condition is 70 mJ/cm²). This residue is considered to be the deposited film during the first Si etching using Bosch process. When ashing (corresponding to about 10 nm removal as for the photoresist) is introduced, the subsequent flush exposure and developing gives the result shown in inset (c). The residue is clearly removed and UV-cured photoresist remains stably. Then the 2nd Si etching can be applied.

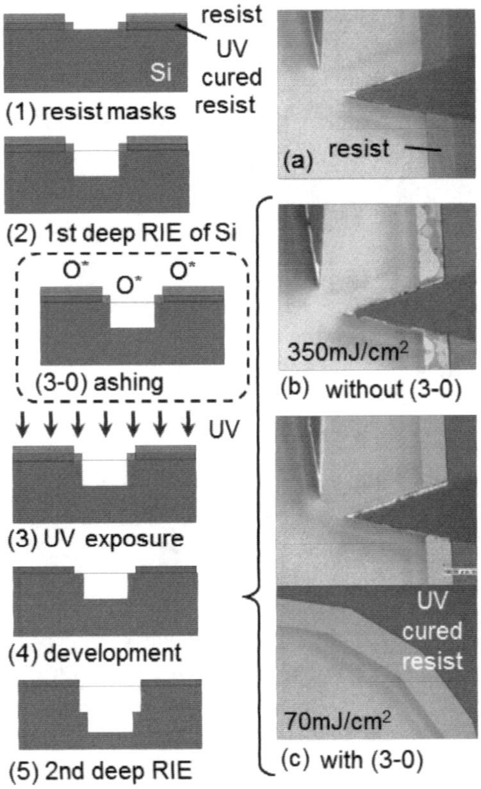

Figure 1. Fabrication sequence using two soft resist masks without and with ashing.

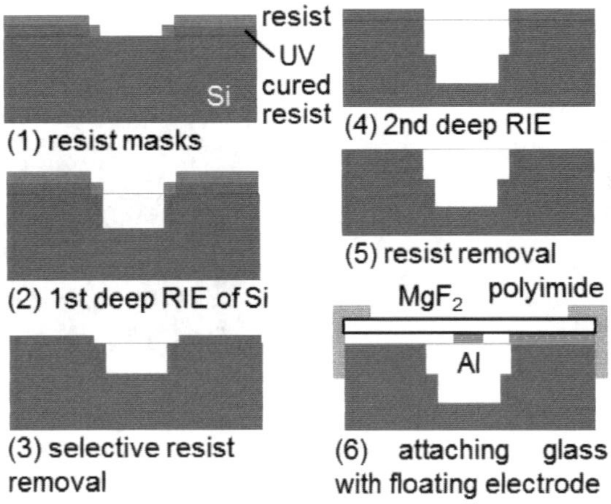

Figure 2. Fabrication sequence of plasma chip.

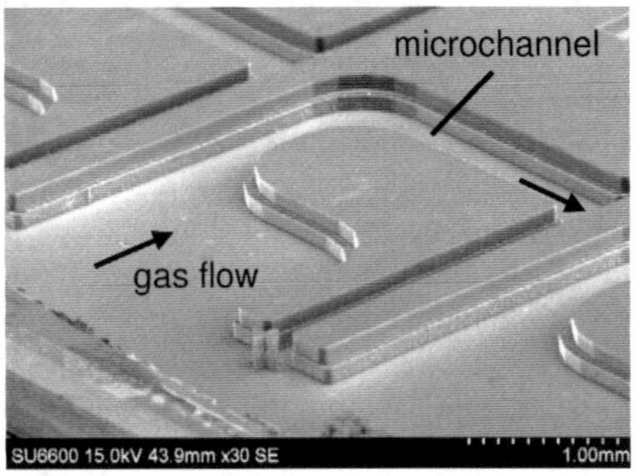

Figure 3. Fabricated gas channel chip with stepped sidewall.

4. Applications

4.1. Microchannel with stepped sidewall for MEMS plasma source

Small plasma source is fabricated [8]. Figure 2 shows the fabrication sequence of the device. Steps (1)-(5) are the processing of Si wafer using the technique proposed. Stepped sidewall is introduced because the effect for decreasing the ignition power is observed in the model experiment. Although the physical mechanism is not clear at present, the gas flow condition seems to relate with this. The stepped sidewall is fabricated by the embedded mask process. The first resist layer is patterned and UV cured. The next resist is spin-coated on that (step 1). The 1st vertical Si etching by 100 μm is conducted using the normal resist mask (step 2). Then, the normal resist is selectively removed (step 3). Additional Si etching is conducted by 100 μm (step 4). All resist mask is removed (step 5). Step 6 is attaching MgF_2 glass with patterned Al electrode, which enhances the plasma ignition. Figure 3 shows the fabricated chip with the microchannel.

Figure 4 shows MEMS chip set in the coil connected to 100 MHz power supply. Microplasma is generated at the apex of the triangle electrode. For the microplasma chip without stepped sidewall, the input power of 124 W is necessary for the ignition. The stepped sidewall decreases the power to 102 W (-18%), which is the significant difference.

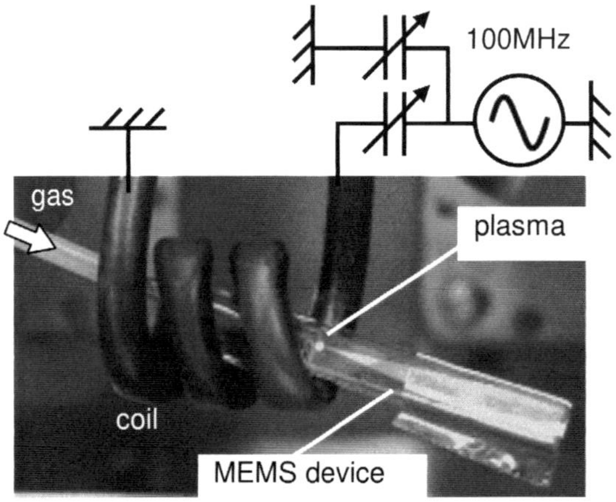

Figure 4. Operation of plasma chip using the microchannel with stepped sidewall.

4.2. IR emitter with local texturing for enhancing the emission

The microheater is fabricated for the IR emitter as shown in figure 5. The wavelength selectivity is realized since the almost IR is reflected back to the microheater and confined except the coupling with the surface plasmon polariton on the grating.

Figure 6 shows the fabrication sequence. SOI (silicon on insulator) wafer is used (step 1). The handle layer is Boron doped <0.02 Ω-cm and becomes the microheater. The first mask pattern is transferred on the device layer defining the slit and cover (step 2). The device Si layer is etched (step 3). The second mask pattern is transferred on the backside handle layer defining a part of the spiral heater (step 4). The center part is open. This mask is UV-cured. Then, the third mask pattern is transferred on the second mask pattern. This resist is not UV-cured keeping as the normal resist. These two patterns connect each other making the spiral heater structure (step 5). The handle Si layer is etched (step 6). The normal resist layer is removed with the flush exposure and developing (step 7). Using the underlying UV-cured resist mask remained, the additional Si etching is carried out. The main aim is for roughening Si surface at the center area of $3.15x3.15mm^2$. Since Si has the large refractive index (about 3.45 for IR) compared to that of air (about 1), IR is reflected back at the interface. Roughening surface (having the microstructure smaller than the wavelength) enhances the emission. Figure 7 shows the photo of Si surface (a) before and (b) after etching controlling the balance of the deposition gas and the etching gas. The surface roughness increases from 2 to 83 nm Ra. Experimentally 1.3-times larger emissivity is observed compared to the original flat surface at the wavelength of 650 nm. After removing the resist layer, the buried SiO_2 layer is removed releasing the heater structure (step 8). Figure 8 shows the fabricated microheater. The chip size is $6.5x6.5mm^2$. The center whitish region is the roughened area.

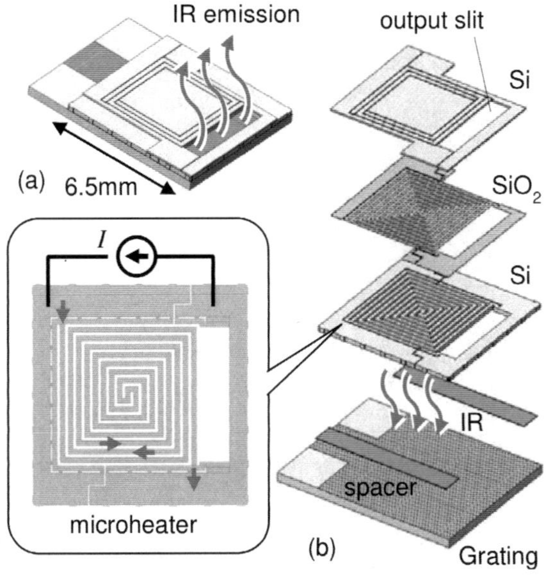

Figure 5. Schematic drawing of IR microemitter. (a) Assembled drawing. (b) Construction of elements.

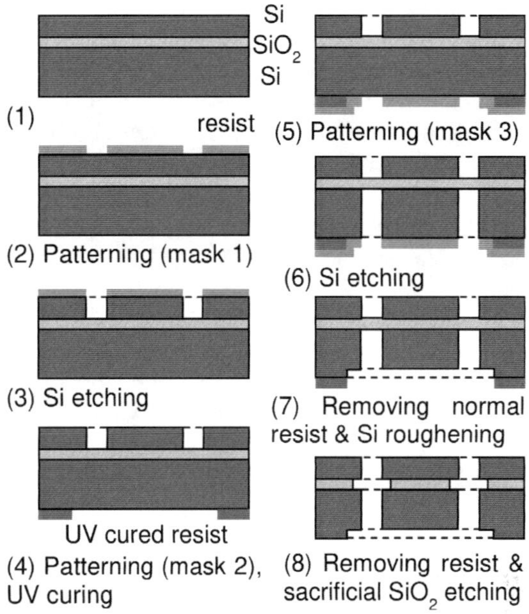

Figure 6. Fabrication sequence of microheater.

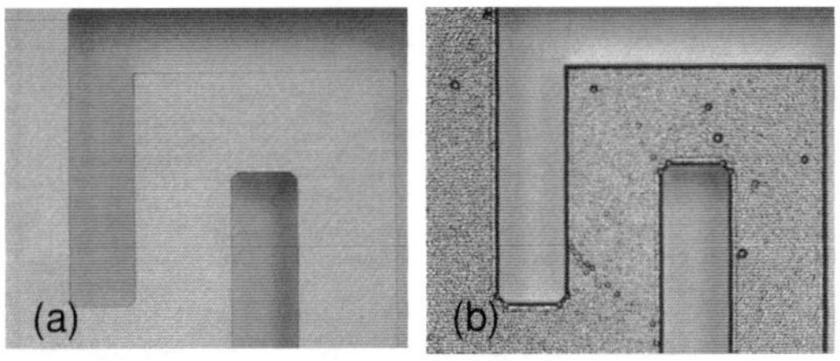

Figure 7. Surface morphology (a) before and (a) after Si etching for roughening the surface.

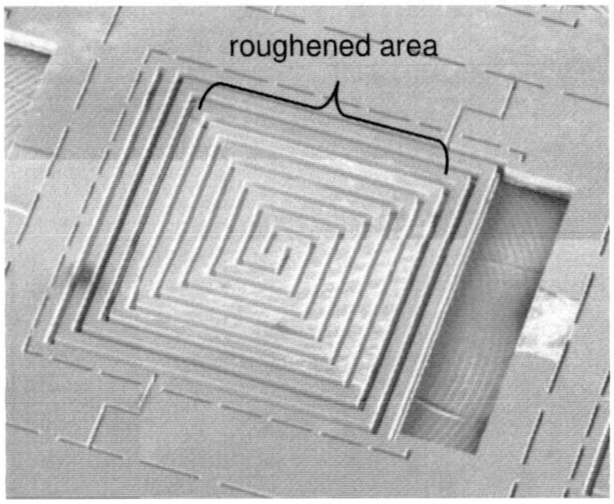

Figure 8. One fabricated microheater. The side with roughened area for enhancing IR emission.

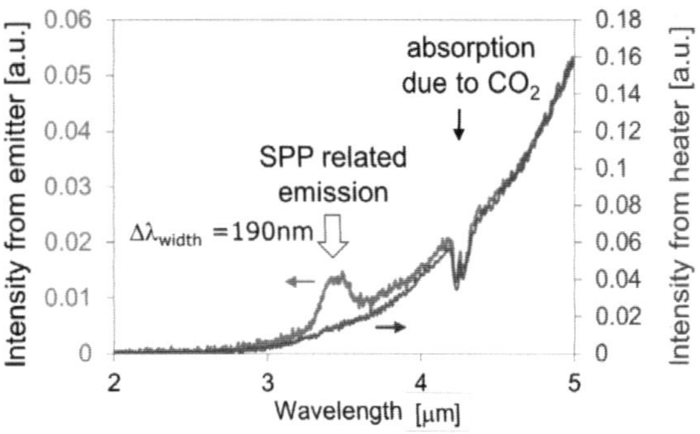

Figure 9. FT-IR spectra from the microemitter and the microheater reference (blackbody).

Figure 9 shows FT-IR (Thermo Scientific, Nicolet 6700) spectra. One curve is the emission from the output end. SPP related peak is at the wavelength of about 3.5 μm. Another blue curve is the reference directly observed at the microheater rough surface. The absorption at 4.2-4.3 μm can be attributed to CO_2 gas in the optical setup. The enhanced emissivity to the grating surface will contribute to the efficient transfer from the thermal energy to the IR wavelength designed.

5. Conclusion

Highly selective removal technique of normal resist against UV cured one is realized. The embedded mask process is combined using the normal resist stacked on UV-cured one. This demonstrates 2-step structures using only soft resist masks. The microchannel with stepped sidewall for MEMS plasma source and IR emitter with local texturing for CO_2 gas sensing are demonstrated showing the merit of the multiple-height structures.

This study is supported by MEXT Grant-in-Aid for Scientific Research on Innovative Areas (24110720), program for forming strategic research infrastructure from 2011, and Nanotechnology Platform Japan.

References

[1] Waits C M, Morgan B, Kastantin M, Ghodssi R 2005 *Sens. Actuators A* **119** 245.
[2] Sasaki M, Nakai F, Hane K, Yokomizo K, Hori K 2006 *J. Opt. A* **8** S391.
[3] Mita Y, Kubota M, Harada T, Marty F, Saadany B, Bourouina T, Shibata T 2006 *J. Micromech. Microeng.* **16** S135.
[4] Mita Y, Tixier A, Oshima S, Mita M, Gouy J-P, Fujita H, 2000 *IEEJ Trans. SM*, **E120** 357.
[5] Mita M, Mita Y, Toshiyoshi H, Fujita H, 2000 *IEEJ Trans. SM*, **E120** 493.
[6] Aonuma T, Kumagai S, Sasaki M 2012 *IEEJ Trans. SM*, **E132** 119.
[7] Kumagai S, Hikita A, Iwamoto T, Tomikawa T, Hori M, Sasaki M 2012 *Jpn. J. Appl. Phys.* **51** 01AB04.
[8] Sato R, Yasumatsu D, Kumagai S, Takeda K, Hori M, Sasaki M *Sens. Actuators A* in press 10.1016/j.sna.2013.09.018.

Fabrication of stable pn junction single-walled carbon nanotube thin films by position selective Cs plasma irradiation method

Y Abiko, T Kato, R Hatakeyama, and T Kaneko

Department of Electronic Engineering, Tohoku University,
6-6-05 Aoba, Aramaki, Aoba-ku, Sendai 980-8579, Japan

E-mail: kato12@ecei.tohoku.ac.jp

Abstract. Stable n-type thin film transistors (TFTs) are fabricated with Cs encapsulated single-walled carbon nanotubes (Cs@SWNTs). The transport property of SWNTs-TFTs clearly changes from p- to n-type characteristic after the Cs plasma irradiation. Based on the systematic investigations, it is revealed that there is an optimum ion energy for the effective Cs encapsulation, which is around 50 eV. Furthermore, it is also found that the n-type feature is stable even in water and high temperature (< 400 °C) conditions. The pn junction structure is also realized by position selective doping of Cs. This very stable pn junction TFT is important for the practical application of SWNTs-based thin film electronics.

1. Introduction

Thin film transistors (TFTs) are one of the most promising practical applications of single-walled carbon nanotubes (SWNTs) due to their flexible filament-like structure and high carrier mobility [1]. For the fabrication of the industrial electrical devices, it is an inevitable issue to utilize both p- and n-type transistors as basic components of the electrical circuits. Since oxygen and water molecules adsorbing on the surface of SWNTs are known to play a role as an electron acceptor against SWNTs, SWNTs-TFTs show the p-type semiconducting features. Up to now, there are several reports for the fabrication of n-type SWNTs-TFTs by functionalizing the outside surface of SWNTs [2-5]. However, the operation of n-type SWNTs-TFTs is limited only under the specific condition and the fabrication of stable n-type SWNTs-TFTs under the various environmental conditions has not been realized.

Here we report the successful fabrication of very stable n-type SWNTs-TFTs by encapsulating Cs atoms in SWNTs with a plasma ion irradiation method [6-10]. Through the systematical investigations, we found that there is a suitable energy range of Cs ions to be encapsulated. This finding is important to realize the high yield encapsulation of Cs atoms inside SWNTs. Since the graphitic network of carbon cells protects the inside Cs atoms from other reactive molecules existing outside of SWNTs, the n-type features are found to be very stable under the various environmental conditions such as air, water, and high temperatures.

2. Experiments

The Cs atom encapsulation is carried out by a plasma ion irradiation method, as shown in figure 1. Cs^+ plasmas are generated by a thermal contact ionization method [6-10]. The typical plasma parameters are as follows: electron density (n_e) $\simeq 10^9$ cm^{-3}, electron temperature (T_e) $\simeq$0.2 eV, and space potential

Journal of Physics: Conference Series **518** (2014) 012013 doi:10.1088/1742-6596/518/1/012013

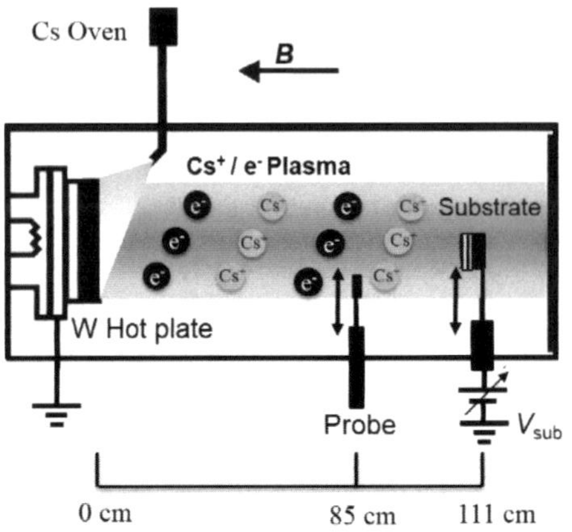

Figure 1. Schematic of experimental apparatus.

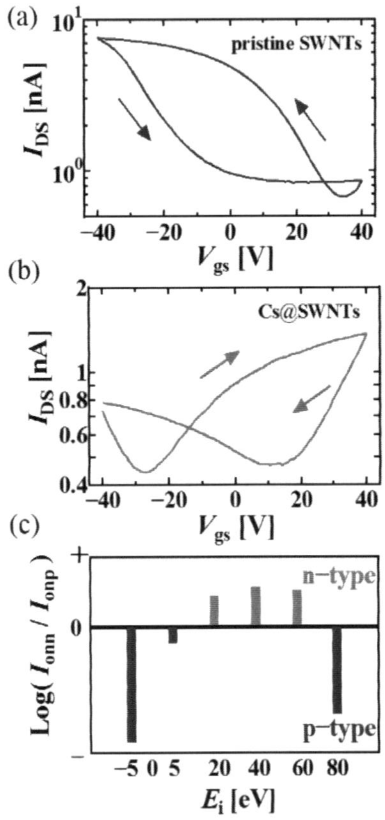

Figure 2. Typical I_{ds}-V_{gs} curves of (a) before (pristine SWNTs-TFTs) and (b) after (Cs@SWNTs-TFTs) Cs plasma irradiation. (c) I_{onn}/I_{onp} current ratio as a function of E_i.

(ϕ_s) ≃ -3 V. SWNTs are deposited on a SiO_2 (300 nm)/Si substrate. To promote the adsorption of SWNTs, the SiO_2 substrate surface is functionalized by 3-aminopropyltriethoxysilane (APTES) prior to the SWNTs deposition. The pairs of 40-nm-thick Au electrodes are fabricated as source-drain electrodes by a conventional photolithography technique. The length and width between source and drain are 50 μm and 100 μm, respectively. SWNTs-TFTs are put on a SUS plate, which is inserted into the plasma region. The energy of ions is controlled by applying dc bias voltages to the SUS plate. After the Cs irradiation, Cs irradiated SWNTs-TFTs are rinsed by purified water to remove the Cs atoms adsorbing on the outside of SWNTs. The transport properties of SWNTs-TFTs are measured by a vacuum probe station and semiconductor parameter analyzer.

3. Results and discussion

Figure 2 (a) and (b) show typical source-drain current (I_{ds}) vs. gate bias voltage (V_{gs}) curves of the same device before and after the Cs plasma irradiation, respectively. The source-drain voltage (V_{ds}) is fixed at 1 V. The threshold voltage of V_{gs} (V_{gsth}) clearly shifts to the negative V_{gs} direction after the Cs plasma irradiation. Furthermore, the currents in the positive V_{gs} region obviously increase after the Cs plasma irradiation. This indicates the transport property of SWNTs-TFTs drastically change from the p-type to n-type semiconducting features after the Cs plasma irradiation. Hysteresis was often observed in I_{ds}-V_{gs} curves even before (figure 2(a)) and after (figure 2(b)) the Cs irradiation, which may relate with the water molecule adsorbed along the interlayer between SWNTs and substrate. Suppression of hysteresis is important for the practical application of SWNTs-TFTs. Further progress is required to solve this problem. The mobility of pristine and Cs@SWNTs-TFT is 0.87 and 0.06 $cm^2V^{-1}s^{-1}$, respectively.

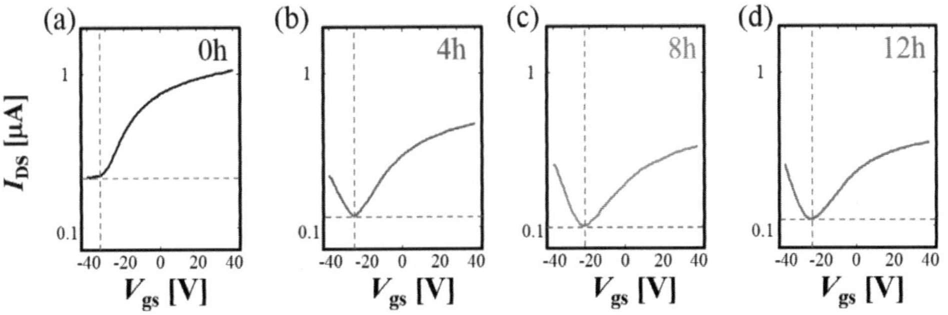

Figure 3. Typical I_{ds}-V_{ds} characteristic of Cs@SWNTs-TFT (a) before and (b-d) after H_2O soaking for different soaking time ((b) 4h, (c) 8h, and (d) 12h).

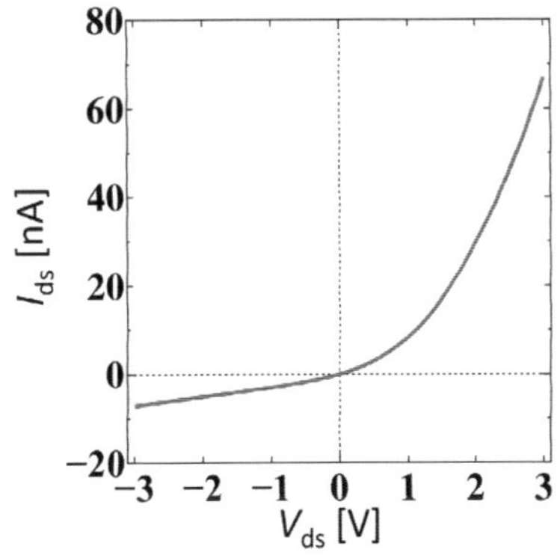

Figure 4. Typical I_{ds}-V_{ds} characteristic of the Cs@SWNTs/pristine SWNTs junction TFT.

In order to identify the optimum irradiation energy of Cs ions, the Cs ion irradiation is carried out under the various energy conditions. It should be noted that the energy of ions (E_i) is estimated from the potential difference between the space potential ϕ_s in plasmas and the substrate bias voltage V_{sub}. The plasma parameters are measured by a Langmuir probe. The saturated source-drain current (I_{ds}) at p (negative V_{gs}) and n (positive V_{gs}) region in $-40 \leq V_{gs} \leq 40$ range is defined as I_{onp} and I_{onn}, respectively. The saturated source-drain current (I_{ds}) ratio of n- to p-channel (I_{onn}/I_{onp}) is utilized as a guidepost to estimate the conducting type of SWNTs-TFTs (figure 2(c)). We defined the device as n-type characteristic when I_{onn} is higher than I_{onp} and V_{gsth} shift to negative V_{gs} direction compared with that of pristine device. Since pristine SWNTs thin film is intrinsic semiconductor, it is natural that p-type conduction is appeared even in Cs@SWNTs thin film transistor. Because the space potential in plasmas is -3 V, only electrons are irradiated to SWNTs-TFTs under the condition of V_{sub} = 0 V. In this case, the averaged I_{onn}/I_{onp} is $\simeq$ 0.2, which indicates SWNTs-TFTs work as the p-type semiconducting devices. With an increase in the ion energy, the clear transition from p- to n-type features is observed around E_i = 20 eV. The n-type transitions of SWNTs-TFTs are found up to E_i = 60 eV. Interestingly, the conducting type of SWNTs-TFTs does not change when the irradiation energy is over 80 eV. In order to identify the meaning of this threshold energy of Cs ion irradiation, the Raman measurement is carried out for the Cs irradiated SWNTs. We use the intensity ratio of G-band to Si-band (520 cm^{-1}) (I_G/I_{Si}) as a sign of the amount of SWNTs. It is to be noted that the I_G/I_{Si} is almost the same for each sample before the Cs irradiation. Since I_G/I_{Si} drastically decreases under the condition of E_i > 80 eV, SWNTs should be damaged in this high irradiation-energy range. Based on these results, it is found that there is an optimum irradiation energy range of Cs encapsulation, which is around E_i = 50 eV and the conducting type of SWNTs-TFTs does not change in the case of high energy (E_i > 80 eV) Cs ion irradiation due to the ion bombardment effects.

To confirm the stability of Cs encapsulated SWNTs (Cs@SWNTs), Cs@SWNTs-TFTs are soaked in the purified water for certain periods. Figure 3 shows soaking time dependence of I_{ds}-V_{gs} curve for Cs@SWNTs with exactly same device. The decreasing of I_{ds} and slight V_{gsth} shift to positive direction are observed between 0h (figure 3(a)) and 4h soaking (figure 3(b)), which should be caused by washing out of Cs atoms from SWNTs outside surface. Noticeably, the I_{ds}-V_{gs} does not show an obvious change after 8h (figure 3(c)) and the n-type feature is still maintained even after 12h soaking (figure 3(d)). This indicates that the origin of the transition from p- to n-type property should be caused by the encapsulated Cs atoms inside SWNTs, which is very stable even in the water. The stability of further longer scale such as week or month is under investigations. We also measure the stability of Cs@SWNTs-TFTs under the other conditions. The clear n-type feature is still observed in the air and after the high temperature (< 400 °C) annealing. This very stable n-type TFTs is important for the fabrication of SWNTs-based high performance thin-film electronic devices.

The pn junction structure is also fabricated by position selective doping of Cs. Prior to the Cs irradiation, the half of SWNTs channels is covered by polymer to prevent Cs ions doping. The basic device structure is same with pristine SWNTs-TFTs and Cs@SWNTs-TFTs. Figure 4 shows a typical I_{ds}-V_{ds} curve of Cs@SWNTs/pristine SWNTs junction TFT. The clear rectifying characteristic can be obtained, which indicates the pn junction structure is formed by our position selective doping method.

4.Conclusion

We have succeeded in the fabrication of very stable n-type SWNTs-TFTs with Cs@SWNTs. It is found that the transport property of SWNT-TFTs clearly changes from p- to n-type after the Cs plasma irradiation. Furthermore, the n-type property is very stable under the long time soak in water, air exposure, and after the high temperature annealing. The pn junction structure is also realized by position selective doping of Cs. This stable pn junction SWNTs-TFTs is very important for the industrial fabrication of high performance electrical circuits with SWNTs-TFTs.

References

[1] Wang C, Zhang J, Ryu K, Badmaev A, Arco L G D and Zhou C 2009 *Nano Lett.* **9** 4285

[2] Qi P, Vermesh O, Grecu M, Javey A, Wang Q and Dai H 2003 *Nano Lett.* **3** 347
[3] Nosho Y, Ohno Y, Kishimoto S and Mizutani T 2005 *Appl. Phys. Lett.* **86** 073105
[4] Li Z, Saini V, Dervishi E, Kunets V P, Zhang J, Xu Y, Biris A R, Sakamo J and Biris A S 2010 *Appl. Phys. Lett.* **96** 033110
[5] Biswas C, Lee S Y, Ly T H, Ghosh A, Dang Q N and Lee Y H 2011 *ACS Nano* **5** 9817
[6] Jeong G H, Farajian A A, Hatakeyama R, Hirata T, Yaguchi T, Tohji K, Mizuseki H and Kawazoe Y 2003 *Phys. Rev. B.* **68** 075410
[7] Izumida T, Hatakeyama R, Neo Y, Mimura H, Omote K and Kasama Y 2006 *Appl. Phys. Lett.* **89** 09312
[8] Kato T, Hatakeyama R, Shishido J, Oohara W and Tohji K 2009 *Appl. Phys. Lett.* **95** 083109
[9] Hatakeyama R, Li Y F, Kato T Y and Kaneko T 2010 *Appl. Phys. Lett.* **97** 013104
[10] Jeong G H, Hatakeyama R, Hirata T, Tohji K, Motomiya K, Sato N and Kawazoe Y 2001 *Appl. Phys. Lett.* **79** 4213

Journal of Physics: Conference Series **518** (2014) 012014 doi:10.1088/1742-6596/518/1/012014

Plasma etching of single fine particle trapped in Ar plasma by optical tweezers

T Ito[1], K Koga[1], D Yamashita[1], K Kamataki[2], N Itagaki[1,3], G Uchida[4] and M Shiratani[1]

[1] Graduate School of Information Science and Electrical Engineering, Kyushu University, 744 Motooka, Nishi-ku, Fukuoka, 819-0395, Japan
[2] Faculty of Arts and Science, Kyushu University, 744 Motooka, Nishi-ku, Fukuoka, 819-0395, Japan
[3] PRESTO, Japan Science and Technology Agency, Gobancho, Chiyoda-ku, Tokyo, 102-0076, Japan
[4] Joining and Welding Research Institute, Osaka University, 11-1 Mihogaoka, Ibaraki, Osaka, 567-0047, Japan

E-mail: siratani@ed.kyushu-u.ac.jp

Abstract. Physical and chemical interactions between plasmas and nano-featured surfaces are one important issue in the plasma processing. Here we optically trap single fine particle levitated at plasma/sheath boundary with an infrared laser to realize in-situ analysis of such interactions. We have measured time evolution of the diameter of the single fine particle in Ar plasma. The trapped particle was etched at an etching rate of 1 nm/min in Ar plasma. We also obtained a Raman peak at around 2950 cm^{-1} corresponding to C-H bonds in the single fine particle in Ar plasma. The results open a new possibility to observe directly interactions between plasma and single fine particle.

1. Introduction

Plasma processing plays a central role in nanotechnology and is widely employed for producing large scale integration, nano-films, garaphene, nanoparticles, and so on [1-28]. Spatial and temporal fluctuation of plasma parameters, such as electron density, electron energy distribution, kinetic energy of ions impinging on surfaces, and radical density, is an important cause of fluctuations of size, structure, and properties of nano-materials and nano-devices produced by plasma processing. However, relation between these two fluctuations has not been clarified yet. To clarify the relation, we have studied physical and chemical interactions between plasmas and fine particles suspended in plasmas [16-28]. Conventional approach of in-situ observation of such interactions is ensemble average observation, namely, interactions between plasmas and many fine particles are observed. Observation of single fine particle in plasmas offers an alterative way of studying physical and chemical interactions between plasmas and fine particles and quantitative evaluation of forces exerted on a fine particle. Here we report interactions between plasma and "single fine particle" trapped by using optical tweezers, namely, a single-beam gradient laser trap.

2. Experimental

Figure 1 shows a schematic diagram of the experimental apparatus. Experiments were carried out with a radio frequency low pressure plasma reactor, which was a stainless steel one of 38 mm in inner diameter and 31 mm in inner height and had two quartz windows: the top window was employed to observe fine particles in plasmas with a CCD camera and the bottom one was employed to irradiate Nd: YAG 1064-nm laser light to a fine particle for optically tweezing it [29-31]. A powered ring-electrode of 10 and 25 mm in inner and outer diameter was set at the bottom of the reactor and a grounded stainless steel mesh electrode of 25 mm in diameter was placed at 9 mm above the bottom of the reactor. Ar plasmas were generated at a total pressure of 40-100 Pa by applying 13.56MHz voltage. The discharge power was 1.9-2.9 W. For the pressure of 100 Pa and the discharge power of 2.9W, the peak-to-peak discharge voltage and self-bias DC voltage were 300 V and -55 V, respectively.

Particles injected into the plasmas were monodisperse methyl methacrylate-polymer (PMMA) spheres of 1.2 g/cm^3 and 10 μm in diameter. Particles injected into the plasmas were charged negatively and were suspended at the plasma/sheath boundary at the center of the powered electrode due to the balance between gravitational force and electrostatic force exerted on the particles. An optical trap is formed by tightly focusing a laser beam with an objective lens (Sigma, NIR Plan APO LMPAL-50-NIR) of high numerical aperture (NA=0.8). A dielectric particle near the focus experiences a force due to the transfer of momentum from the scattering of incident photons [16-18]. The resulting optical force is decomposed into two components: (1) a scattering force, in the direction of light propagation and (2) a gradient force, in the direction of the spatial light gradient. At the focal point of the irradiated laser light, one of the fine particles can be optically trapped in plasmas where electrostatic force, ion drag force, neutral drag force, and gravity were exerted on the fine particle. We observed the fine particle with a CCD camera (Sentech, STC-TC202USB-AT) with a spatial resolution of 1600×1200 pixels and 88 nm/pixel at a maximum frame rate of 15.32 fps.

We also carried out in-situ Raman spectroscopy of single fine particle in Ar plasma with an exposure time of 0.1 s. The Raman signal was accumulated over 30 times to improve the signal to noise ratio.

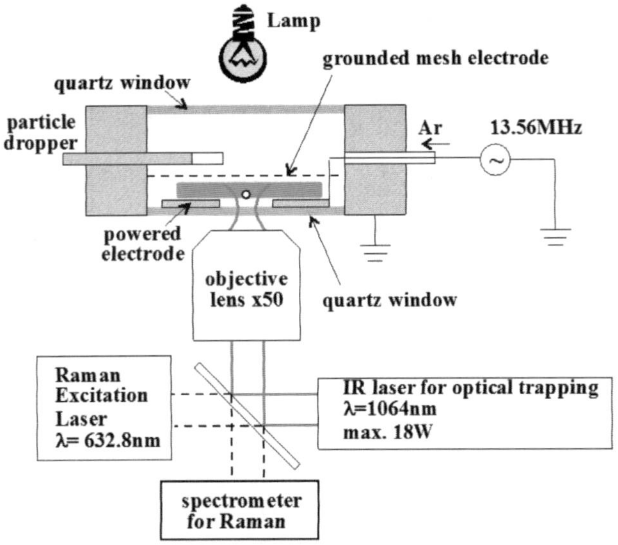

Figure 1. Schematic diagram of plasma reactor, optical tweezers, and in-situ Raman spectrometer.

3. Results and discussion

We succeeded in trapping and manipulating a single fine particle levitated at plasma/sheath boundary in plasmas by controlling precisely the focal point and power of the laser light. Figure 2 shows a typical image of the trapped fine particle obtained with the CCD camera. The particle was trapped around plasma/sheath boundary, at 0.5 mm above the powered electrode. The fine particle followed the shift of the focal point, and then it was released just after turning off the laser irradiation. Radiation pressure acts on the fine particle to the direction of the focal point due to the difference of refractive index between the fine particle and Ar gas. The electron temperature and electron density for 100 Pa, 2.9 W discharge plasma are 1.9 eV and 1.7×10^{9} cm^{-3}. The corresponding charge accumulated on a fine particle is -7300e=-1.7×10^{-15} C. Since the gravitational force is 6.1 (pN), the electric field needed for the counter balance to the gravitational force is 52 V/cm. A trapping laser power of 0.1 W can provide an optical trapping force of 100 pN, which is significantly large compared with the other forces exerted on the fine particle.

Figure 3 shows time evolution of the diameter of a fine particle trapped optically in Ar plasma. We observed a gradual decrease in the diameter of trapped fine particle due to interactions between the trapped particle surface and the Ar plasma. The diameter decreases linearly with time due to sputter etching at an etching rate of 1 nm/min. Since the ion flux to the fine particle is 6×10^{15} $cm^{-2}s^{-1}$, the apparent sputtering yield deduced from the etching rate is 0.8, being high compared with 0.1 in the previous reports [32]. This high sputtering yield may be attributed to a synergetic effect of simultaneous irradiation of ions and VUV photons from Ar plasma [19].

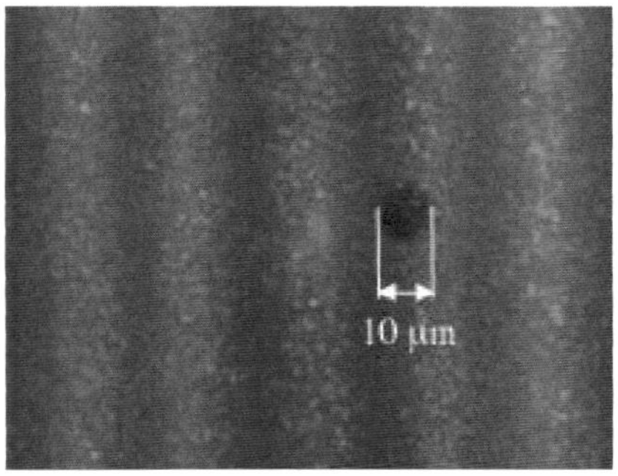

Figure 2. A typical image of single fine particle of 10 μm in diameter, trapped optically in Ar plasma.

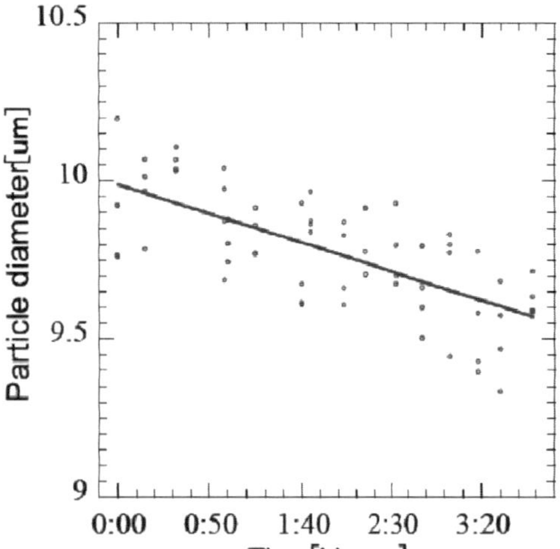

Figure 3. Time evolution of diameter of a fine particle trapped optically in Ar plasma.

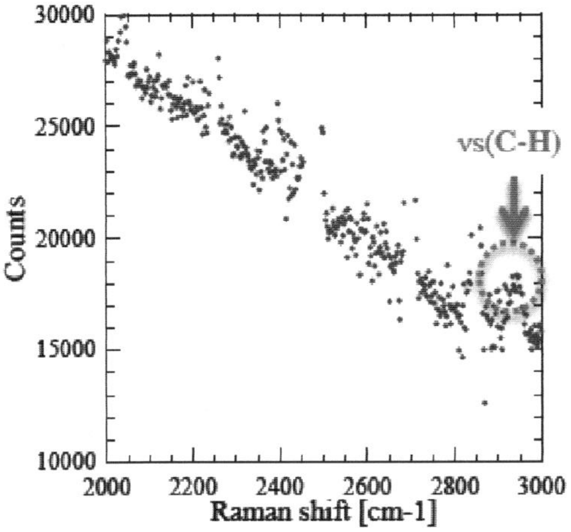

Figure 4. Raman spectrum of single fine particle in Ar plasma.

Figure 4 shows Raman spectrum of single acrylic fine particle in Ar plasma. There is a peak at around 2950 cm^{-1} corresponding to C-H bonds in the single fine particle.

The results in figures 3 and 4 open a new possibility to observe directly interactions between plasma and single fine particle.

4. Conclusions

Single fine particle was trapped in Ar plasma by using the optical tweezers with the IR laser. We demonstrated in-situ observation of interactions between plasmas and the trapped fine particle. The trapped fine particle was etched at an etching rate of 1 nm/min in Ar plasma. We also obtained a Raman peak at around 2950 cm^{-1} corresponding to C-H bonds in the single fine particle in Ar plasma.

Acknowledgments

This work was partly supported by the Grant-in-Aid for Scientific Research on Innovative Areas, the Ministry of Education, Culture, Sports, Science and Technology (MEXT) KAKENHI Grant Numbers 21110001 and 21110005.

References

[1] Hosokawa N, Matsuzaki R and Asamaki T 1974 *Jpn. J. Appl. Phys.* **13** [Suppl. 2, Part 1] 435
[2] Abe H, Yoneda M and Fujiwara N 2008 *Jpn. J. Appl. Phys.* **47** 1435
[3] Tsuda H, Takao Y, Eriguchi K and Ono K 2012 *Jpn. J. Appl. Phys.* **51** 08HC01
[4] Okada T, Kaneko T, Hatakeyama T and Tohji K 2006 *Chem. Phys. Lett.* **417** 288
[5] Komatsu S, Kurashima K, Kanda H, Okada K, Mitomo M, Moriyoshi Y, Shimuzu Y, Shiratani M,NakanoT and Samukawa S 2002 *Appl. Phys. Lett.* **81** 4547
[6] Komatsu S, Kurashima K, Shimizu Y, Moriyoshi Y, Shiratani M and Okada K 2004 *J. Physi. Chem. B* **108** 205
[7] Komatsu S, Okudo A, Kazami D, Golberg D, Li Y B, Moriyoshi Y, Shiratani M and Okada K 2004 *J. Physi. Chem. B* **108** 5182
[8] Komatsu S and Shiratani M 2014 *Jpn. J. Appl. Phys.* **53** 010202
[9] Saraswati T E, Tsumura S and Nagatsu M 2014 *Jpn. J. Appl. Phys.* **53** 010205
[10] Yang E and Nagatsu M 2014 *Jpn. J. Appl. Phys.* **53** 010206
[11] Ciolan M A, Motrescu I, Luca D and M. Nagatsu M 2014 *Jpn. J. Appl. Phys.* **53** 010207
[12] Kobayashi H, Chewchinda P, Inoue Y, Funakubo H, Hara M, Fujino M, Odawara O and Wada

H 2014 *Jpn. J. Appl. Phys.* **53** 010208
[13] Hashizume H, Ohta T, Takeda K, Ishikawa K, Hori M and Ito M 2014 *Jpn. J. Appl. Phys.* **53** 010209
[14] Oshima F, Stauss S, Inose Y and Terashima K 2014 *Jpn. J. Appl. Phys.* **53** 010214
[15] Bouchoule A, Plain A, Boufendi L, Blondeau J Ph and Laure C 1991 *J. Appl. Phys.* **70** 1991
[16] Watanabe Y, Shiratani M, Kubo Y, Ogawa I and S. Ogi S 1988 *Appl. Phys. Lett.* **53** 1263
[17] Kawasaki H, Fukuzawa T, Tsuruoka H, Yoshioka T, Shiratani M and Watanabe Y 1994 *Jpn. J. Appl. Phys.* **33** 4198
[18] Kawasaki H, Kida J, Sakamoto K, Fukuzawa T, Shiratani M and Watanabe Y 1998 *J. Appl. Phys.* **83** 5665
[19] Fukuzawa T, Kushima S, Matsuoka Y, Shiratani M and Watanabe Y 1999 *J. Appl. Phys.* **86** 3543
[20] Koga K, Kai M, Shiratani M, Watanabe Y and Shikatani N 2002 *Jpn. J. Appl. Phys.* **41** L168
[21] Nunomura S, Koga K, Shiratani M, Watanabe Y, Morisada Y, Matsuki N and Ikeda S 2005 *Jpn. J. Appl. Phys.* **44** L1509
[22] Shiratani M, Koga K, Iwashita S and Nunomura S 2008 *Faraday Discuss.* **137** 127
[23] Shiratani M, Koga K Iwashita S, Uchida G, Itagaki N and Kamataki K 2011 *J. Phys. D* **44** 174038
[24] Kamataki K, Miyata H, Koga K, Uchida G, Itagaki N and Shiratani M 2011 *Appl. Phys. Express* **4** 105001
[25] Kamataki K, Morita Y, Shiratani M, Koga K, Uchida G and Itagaki N 2012 *J. Inst.* **7** C04017
[26] Kamataki K, Koga K, Uchida G, Itagaki N, Yamashita D, Matsuzaki H and Shiratani M 2012 *Thin Solid Films* **523** 76
[27] Shiratani M, Koga K, Kamataki K, Iwashita S, Uchida G, Seo H and Itagaki N 2014 *Jpn. J. Appl. Phys.* **53** 010201
[28] Shiratani M, Morita Y, Kamataki K, Seo H, Uchida G, Itagaki N and Koga K 2014 *JPS Conf. Proc.* in press
[29] Ashkin A 1997 *Proc. Natl. Acad. Sci.* **94** 4853
[30] Ashkin A 2000 *IEEE J. Sel. Top. Quantum Electron.* **6** 841
[31] Neuman K C and Block S M 2004 *Rev. Sci. Instrum.* **75** 2787
[32] Yoshimura S, Tsukazaki Y, Kiuchi M, Sugimoto S and Hamaguchi S 2012 *J. Phys. D: Appl. Phys.* **45** 505201

Generation of line-shaped atmospheric pressure plasma on planar surface with diffuse coplanar surface barrier discharge

H Akamatsu

Kobe City College of Technology, 8-3 Gakuenhigashi-machi, Nishi-ku, Kobe, Hyogo 651-2194, Japan

akamatsu@kobe-kosen.ac.jp

Abstract. A line-shaped uniform atmospheric pressure plasma was generated by diffuse coplanar surface barrier discharge on a planar surface. In this work, the plate is assumed as a wall panel in operation rooms. In the wall panel sample, powered and grounded electrode lines were arranged alternately. Applying an negative high voltage pulse of V=-4 kV and f=13 kHz, a uniform atmospheric pressure plasma was observed between electrodes on the panel surface. The plasma had chemically active species of nitrogen molecule and nitrogen molecule ion which are effective for sterilization.

1. Introduction

Atmospheric pressure plasmas have received much attention as novel techniques for bio-application such as sterilization and bacteria inactivation [1-3]. In the application, reactive species originated from oxygen, nitrogen, and hydroxyl radical assume a key role [3].

In operation room of a hospital, wall panels are wiped off and cleaned by cleaning person after an operation in order to maintain cleanliness. This cleaning method has two problems. The one is that the labor force is needed for this work. The next is that an artificial error arises for the cleaning accuracy. In this study, a brand-new method for the cleaning of wall panels is proposed by using line-shaped atmospheric pressure plasma.

Figure 1 shows a schematic illustration of a wall panel in the proposing method. The wall panel, which is a dielectric material, has some line electrodes in the body. The adjacent electrodes are powered by pulsed high voltages, resulting in generation of line-shaped atmospheric pressure plasma on the wall surface by diffuse coplanar surface barrier discharge (DCSBD) [4-7]. In addition, the line-shaped plasma can be moved to the next electrodes by switching connection of the voltage and electrode lines. The switching of electrodes results in reduction of power supply capacity and cooling of the wall panel. The wall surface is exposed by the line-shaped atmospheric pressure plasma. Therefore the wall surface can be cleaned and sterilized automatically.

In this paper, a miniature-size wall panel sample, which has electrodes of DCSBD, was developed experimentally. Discharge property and optical emission spectrum of the generated plasma in a mixed gas of air and helium was evaluated.

2. Experimental procedure

2.1. Wall panel sample

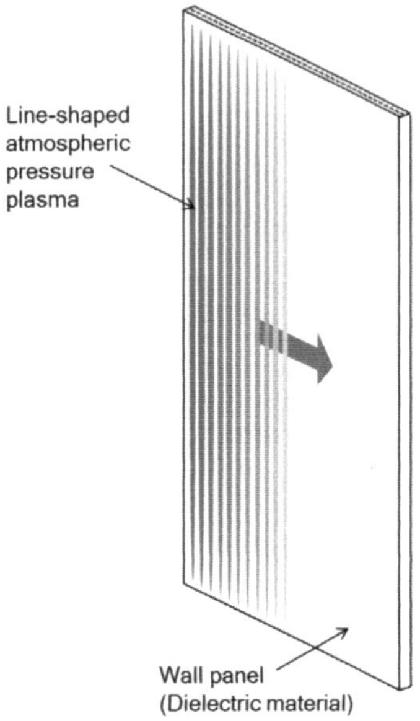

Figure 1. Scheme of cleaning and sterilization of wall panel in operation room by line-shaped atmospheric pressure plasma.

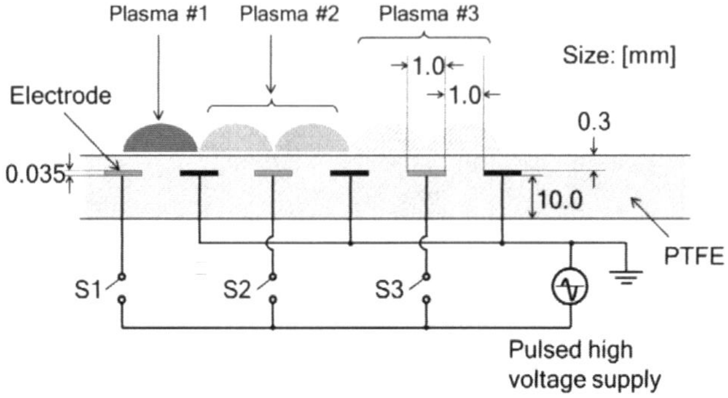

Figure 2. Schematic illustration of cross-section of wall panel sample.

Figure 2 shows a schematic cross-sectional view of a miniature size wall panel sample and electric circuit of DCSBD. The wall panel sample is made of polytetrafluoroethylene (PTFE) as a dielectric material. The wall panel sample has a size of length 197 mm x width 100 mm x thickness 10.3 mm. Some parallel stripline electrodes are embedded in the wall panel sample. Powered and grounded electrode lines are arranged alternatively. The electrode has a length of 157 mm, width of 1.0 mm, and thickness of 0.035 mm. The gap length between the powered and grounded electrodes is 1.0 mm. A distance between the electrode and the panel surface is 0.3 mm. The powered electrode lines are connected to a pulsed high voltage supply through reed relay switches S1-Sn. Turning on the switch S1 and turning off the switches S2 and S3, the plasma #1 is formed on the wall surface. Then the plasma #2 is formed by turning on the switch S2 and turning off the switches S1 and S3. Similarly, the plasma #3 is formed by turning on the switch S3 and turning off the switches S1 and S2. In this study, the only one switch of S1 is turned on experimentally, resulting in formation of the plasma #1.

2.2. Generation and measurement of diffuse coplanar surface barrier discharge plasma

Figure 3 shows an electrical circuit of diffuse coplanar surface barrier discharge generation and measurement. A wall panel sample was located in an acrylic acid case with the size of 200 mm x 200 mm x 50 mm. Helium gas, regulated by a flow meter (RK-1650, KOFLOC) was tentatively introduced into the case through a thin flex tube of ϕ6 mm in order to break down at lower applied voltage. Residual air was exhausted through a thin flex pipe of ϕ6 mm to the outside of the box. A pulsed negative high voltage generated by a collector resonance circuit was applied between a pair of electrode in the wall panel sample. The applied voltage was measured with a voltage divider (HV-P30, IWATSU) and a digital oscilloscope (TDS2024C, Textronix). A current with discharge was measured with a current probe (A621, Textronix) and the digital oscilloscope. Power consumption for the plasma generation was estimated by voltage-charge Lissajous plot. Electrical charge due to discharge was measurement as a voltage of a capacitance 0.1 μF.

A digital camera (DMC-GF5, Panasonic) was used for direct observation of diffuse coplanar surface barrier discharge plasma. Optical emission of diffuse coplanar surface barrier discharge plasma was evalulated by a multichannel spectrometer (SEC2000, BAS).

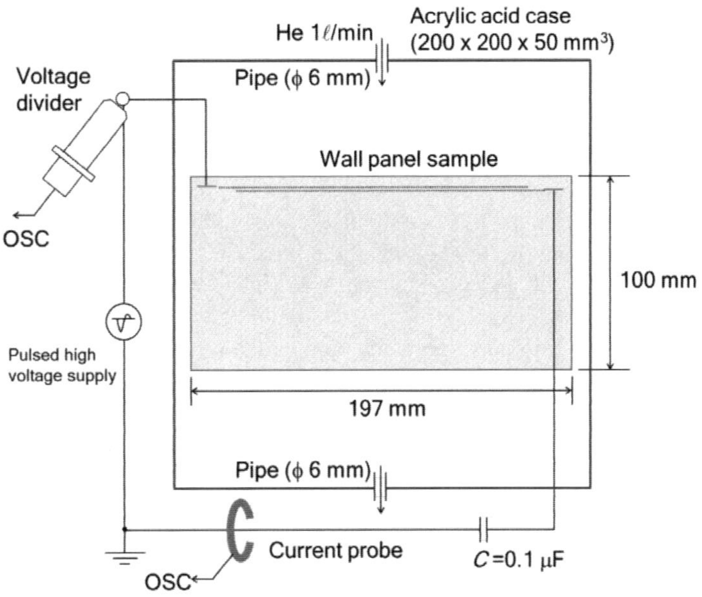

Figure 3. Electrical circuit of diffuse coplanar surface barrier discharge generation and measurement.

3. Result and discussion

3.1. Electrical property of diffuse coplanar surface barrier discharge

Figures 4 shows time evolutions of (a) applied voltage and (b) discharge current. The applied voltage has a peak value of -4 kV, pulse width (FWHM) of 15 μsec, and frequency of 13 kHz, respectively. The discharge current rises with the voltage. The peak value of the current is -0.3 A and the pulse width (FWHM) is 10 μsec. A positive current of 0.2 A and 18 μsec flows after the negative current.

Figure 5 shows a voltage-charge Lissajous plot of the diffuse coplanar surface barrier discharge plasma generation. Voltage-charge Lissajous plots of dielectric barrier discharge, generally, have a

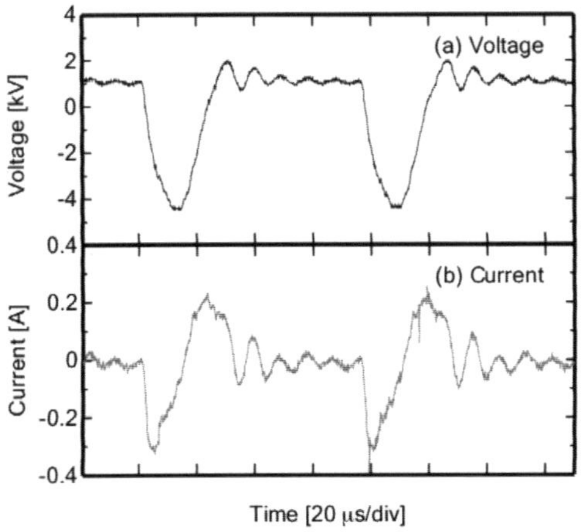

Figure 4. Time evolutions of applied voltage and discharge current.

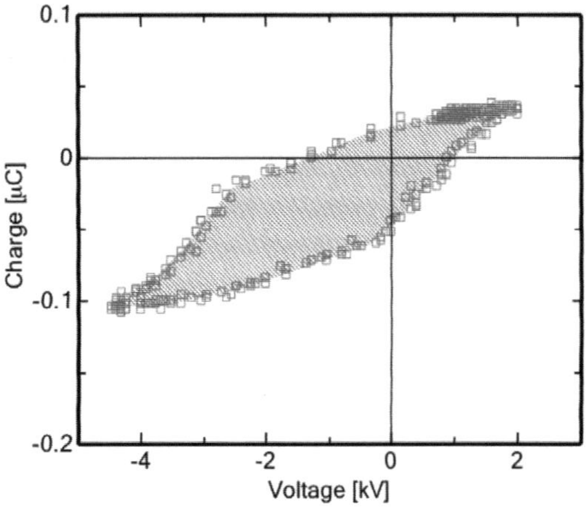

Figure 5. Voltage-charge Lissajous plot of the diffuse coplanar surface barrier discharge plasma.

parallelogram shape. The large gradient of the parallelogram indicates an equivalent capacitance of dielectric material. Another gradient indicates a total capacitance of discharge area and dielectric material. In figure 4, the total capacitance of the pair electrode system was estimated to be 156 pF. In addition, the area of the voltage-charge plot gives energy consumption for the plasma generation. The energy of 31.2 mJ was estimated as a consumed energy per a voltage cycle.

3.2. Observation of diffuse coplanar surface barrier discharge plasma

Figure 6 shows direct observation of the panel surface (a) without and (b) with applying the high voltage. The photographs indicate the upper end of the panel in figure 3. As shown in figure 6 (b), a line-shaped uniform plasma can be observed on the wall panel surface. The width of the plasma was estimated to be 4 mm.

Figure 7 shows an optical emission spectrum of the diffuse coplanar surface barrier discharge plasma on the wall panel sample. In the spectrum, some peaks originated from nitrogen and helium are observed. The peaks of nitrogen molecules $N_2(C^3\Pi_u$-$B^3\Pi_g)$ are located at 315.8 nm, 337.1 nm, 353.6 nm, 357.6 nm, 371.0 nm, 375.4 nm, and 380.4 nm. The peaks of N_2(C-B) are observed at 399.7 nm and 405.8 nm. In addition, peaks of nitrogen molecule ion $N_2^+(B^2\Sigma_u$-$X^2\Sigma_g)$ are located at 391.4 nm and 427.8 nm. The excited nitrogen molecules was generated the diffuse coplanar surface barrier discharge in the mixed helium and air. In figure 7, chemically active species of originated from oxygen and hydroxyl radical are not observed.

The excited nitrogen molecules radiate ultraviolet ray [3]. The UV emission are effective for self-sterilization of the wall panel sample.

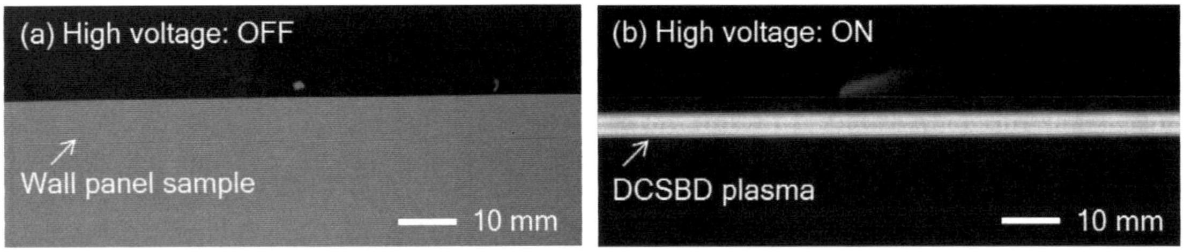

Figure 6. Direct observation of diffuse coplanar surface barrier discharge plasma on the wall panel surface.

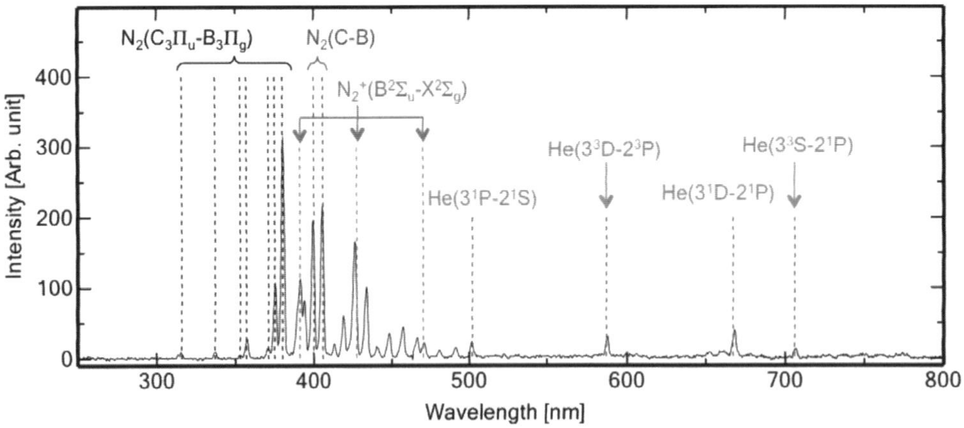

Figure 7. Optical emission spectrum of diffuse coplanar surface barrier discharge plasma.

4. Conclusion

In this paper, atmospheric pressure plasma has been experimentally generated with diffuse coplanar surface barrier discharge. When the wall panel sample was powered by pulsed voltage of -4 kV and 13 kHz, line-shaped uniform plasma was formed on a wall panel sample surface. The diffuse coplanar surface barrier discharge plasma has excited nitrogen molecules and ions. The nitrogen molecules are effective for self-sterilization of wall panel sample surface since the molecules radiate ultraviolet rays.

Acknowledgments

This work was supported by JSPS KAKENHI Grant-in-Aid for Young Scientists (B) No. 24760241.

References

[1] Goree J, Bin Liu, Drake D, Stoffeis E 2006 *IEEE Trans. Plasma Sci.* **34** 1317
[2] Lee K, Paek K-h, Ju W-T, Lee Y 2006 *J. Microbiol.* **44** 269
[3] Eto H, Ono Y, Ogino A, Nagatsu M 2008 *Appl. Phys. Lett.* **93** 221502
[4] Bonova L, Bucek A, Plecenik T, Zahoranova A, Cernak M 2008 *Chem. Listy* **102**, s1452
[5] Hoder T, Sira M, Kozlov K V, Wagner H–E 2011 *J. Phys. D: Appl. Phys.* **41**, 034212
[6] Cernak M, Kovacik D, Rahel' J, St'ahel P, Zahoranova A, Kubincova J, Toth A, Cernakova 'L 2011 *Plasma Phys. Control. Fusion* **53**, 124031
[7] Kobayashi H, Kakiuchi H, Yasutake K, Suzuki T, Noborisaka M, Negishi N 2013 *J. Phys: Conference Series* **441** 012024.

Journal of Physics: Conference Series **518** (2014) 012016 doi:10.1088/1742-6596/518/1/012016

Control of the area irradiated by the sheet-type plasma jet in atmospheric pressure

T Kawasaki[1], K Kawano[1], H Mizoguchi[1], Y Yano[1], K Yamashita[1], M Sakai[1], G Uchida[2], K Koga[3] and M Shiratani[3]

[1] Department of Mechanical and Electrical Engineering, Faculty of Engineering, Nippon Bunri University, 1727 Ichigi, Oita 870-0397, Japan
[2] Joining and Welding Research Institute, Osaka University, 11-1 Mihogaoka, Ibaraki, Osaka 567-0047, Japan
[3] Department of Electronics, Faculty of Information Science and Electrical Engineering, Kyushu University, 744 Motooka, Nishi-ku, Fukuoka 819-0395, Japan

E-mail: kawasaki@nbu.ac.jp

Abstract. The sterilization effect has been investigated using the sheet-type plasma jet, which was generated between asymmetric electrodes with dielectric plates in gas flow released into the atmospheric air. In this paper, it is indicated there is a possibility that the plasma jet irradiation area can be controlled only by supplied gases without changing a generator structure. The irradiation area control was evaluated from both the sterilization area size and the oxidizing substances distribution. The oxidizing substance distribution was obtained using the chemical reagent prepared in our laboratory. The width of the sheet-type plasma jet was able to be controlled by N_2 addition into He gas. As a result, the width of the sterilization area was able to be controlled within the range of 2 to 12 mm at a constant height without changing the generator structure. On the other hand, the evaluation from the oxidizing substances distribution indicated that the irradiation area cannot be controlled in one direction.

1. Introduction

Atmospheric pressure non-thermal plasma technologies have been actively studied and developed for environmental applications [1-4]. Recently, plasma chemical reactions and its applications including a liquid phase in addition to a gas phase have attracted special interest all over the world not only for environmental applications but also for biomedical applications [5-7]. In this case, an atmospheric pressure non-thermal plasma jet has attracted attention of many researchers and has been commonly used by reason of special features [8]. Various types of plasma jet have been developed and expected to be applied in a wide range of fields, such as nanomaterial processing, sterilization, microbe inactivation and biomedical applications [9-17].

A commonly-used plasma jet generator is convenient for selective plasma irradiation to a narrow area confined because a glass tube with an inner diameter of a few millimeters is generally used as a body. It means that it is difficult to obtain the uniform irradiation to a wide range area of a target. Therefore, a plasma jet array has been developed as a countermeasure [18]. On the other hand, in the application of a plasma jet as new technologies, the elucidation of plasma chemical reactions including a liquid phase is a matter of great importance. Especially, it is said that reactive oxygen species (ROS) and reactive nitrogen species (RNS) play important roles [19]. The spatiotemporal distribution of reactive species has been actively studied by optical and chemical measurements [7].

In this study, the atmospheric pressure sheet-type plasma jet has been studied for a wide range sterilization method [20]. This time it became clear there is a possibility that a plasma jet irradiation area can be controlled without changing a generator structure, and that are reported in this paper. The irradiation area control was evaluated from both the sterilization area (clear zone size) size and the oxidizing substances distribution.

2. Experimental setup

Figure 1 shows a schematic diagram of the experimental setup including the atmospheric pressure sheet-type plasma jet generator. The generator consists of two glass plates (thickness 1.2 mm) with a gap distance of 1.2 mm, high voltage (H.V.) electrode made onto one glass plate, and grounded electrode (12 mm × 12 mm). The height and width of the H.V. electrode were fixed at 13 mm and 5 mm, respectively. The irradiation distance between the tip of the glass plate and the surface of the target placed onto the grounded electrode was fixed at 5 mm. A sinusoidal 1 kHz H.V. was applied to the generator using a function generator (Iwatsu, SG-4105) and an amplifier (Trek, 20/20C). The mixture gas of He and N_2 was supplied into the gap between the glass plates at a flow rate of 2.0 L/min. Applied voltage and discharge current waveforms were recorded using an oscilloscope (Lecroy, LT374M) via a voltage monitor in the amplifier and a current monitor (Pearson Electronics, 2877), respectively. Figure 2 shows the applied voltage and current waveforms obtained at 7 $kV_{o\text{-}p}$ in He gas. It is the typical waveform of a dielectric barrier discharge with many current pulses generating only around the polarity change. Discharge power was measured from V-Q Lissajous figures obtained using a 220 nF capacitor connected to the grounded electrode in series. A digital camera (Canon,

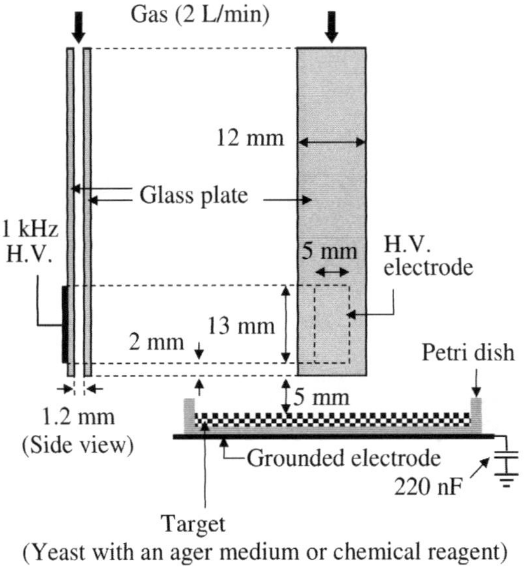

Figure 1. Schematic diagram of the sheet-type plasma jet generator.

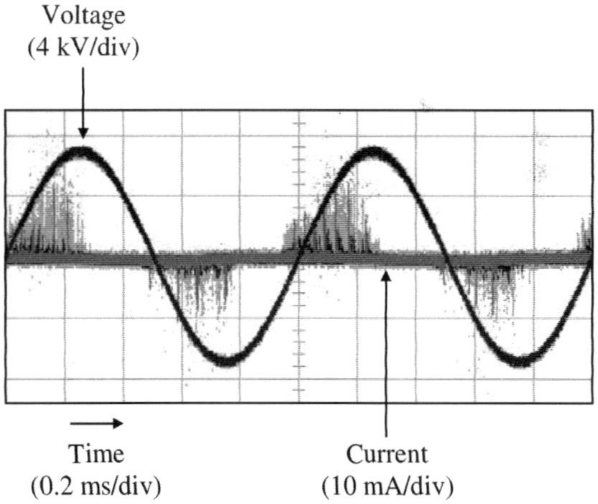

Figure 2. Applied voltage and current waveforms at 7 $kV_{o\text{-}p}$ in the case of supplied gas He.

PowerShot G11 or G1X) was used for taking the discharge photographs at an exposure time of 2 s.

Two kinds of targets were used in this paper. The yeast with an ager medium and the chemical reagent were prepared in our laboratory for the evaluation of an irradiation area control from the clear zone size and oxidizing substances distribution, respectively. *Saccharomyces selevisiae* (yeast) was used in order to obtain the clear zone size. Yeast was grown on a potato dextrose (PD) medium at

30°C for 16 hours with shaking. The PD solid medium was made by the addition of agar at a final concentration of 1.5%. The culture liquid of 1 mL was spread on Petri dishes (polystyrene, ϕ 90 mm) and it was made to dry for 30 minutes. The resulting thickness of the culture was 7 mm. After that, yeast with an agar medium in the Petri dish was placed onto the grounded electrode, and the plasma jet was irradiated to it for 30 s. Static culture was carried out at 30°C after the plasma jet irradiation. The width and height (mm) of the clear zones were obtained by the image processing using ImageJ software [21] in order to evaluate the irradiation are control. The oxidizing substances distribution was visualized by iodine-starch reaction. The chemical reagent (oxidizing substances visualizing reagent) which makes it possible was prepared in our laboratory. The plasma jet was irradiated to the chemical reagent in the Petri dish for 30 s, and the color reaction was occurred on its surface. After that, the absorbance measurement was performed in order to obtain the numerical value of the color reaction for the evaluation of the irradiation area control. The measuring apparatus for absorbance consists of a tungsten-halogen lamp (B&W Tek, BPS101), a collimating lens (Ocean Optics, 74-UV), a XY stage (Yashima Optical, YAX-4), a spectrometer (Ocean Optics, USB4000) and optical fiber cables. All of numerical data shown in this paper are the average value of more than three measurements.

3. Results

3.1. Evaluation from the clear zone size

Figure 3 shows the influence of the N_2 addition into He gas on the plasma jet generation at a constant voltage of 8 $kV_{o\text{-}p}$ and the resulting clear zones for an irradiation time of 30 s. The outward appearance of the plasma jet was slightly changed by the addition of N_2 with a concentration of 0.5% into He gas. Between the N_2 concentration of 0.5% and 8%, the plasma jet became weaker from the square edge of the high voltage electrode, and then it was significantly changed from a sheet-type to a spot-type when the N_2 concentration was increased to 12%. Accordingly, the shape and size of the clear zones were gradually changed while keeping a sufficient sterilizing property. It was reported that the generation of dielectric barrier discharge in atmospheric pressure is restrained at the electrode square edge in nitrogen [22]. In this study, it is considered that this phenomenon was remarkable with increasing the N_2 concentration in He gas, and the plasma jet was finally changed from a sheet-type to a spot-type. As a result, the discharge volume was decreased and its power was also decreased.

Figure 4 (a) and (b) show the average width and height of the clear zones as a function of the N_2 concentration at applied voltages from 4 to 8 $kV_{o\text{-}p}$, respectively. The width of the clear zones was gradually decreased with increasing the N_2 concentration in the all applied voltages, as shown in figure

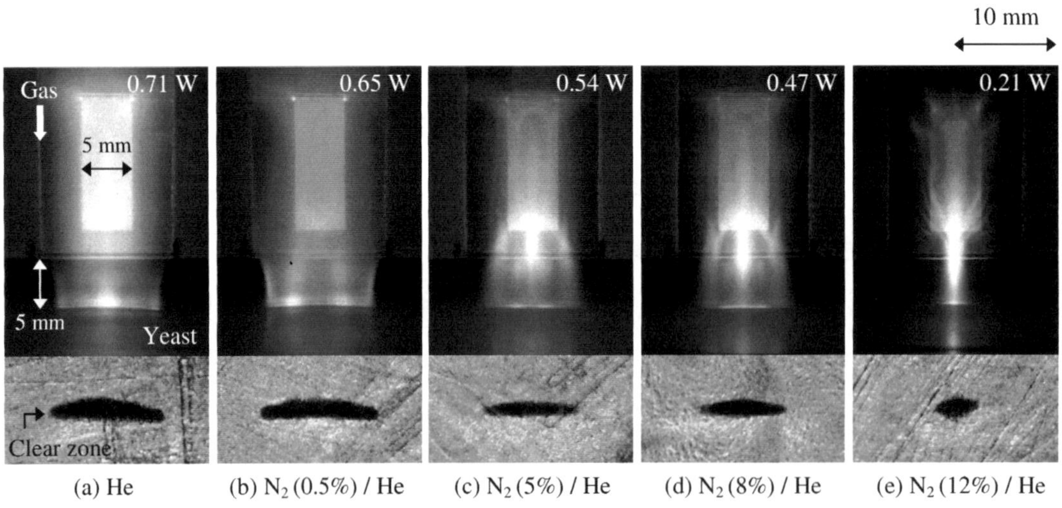

Figure 3. Influence of the N_2 addition into He gas on the plasma jet generation at 8 $kV_{o\text{-}p}$ and the resulting clear zone for an irradiation time of 30 s.

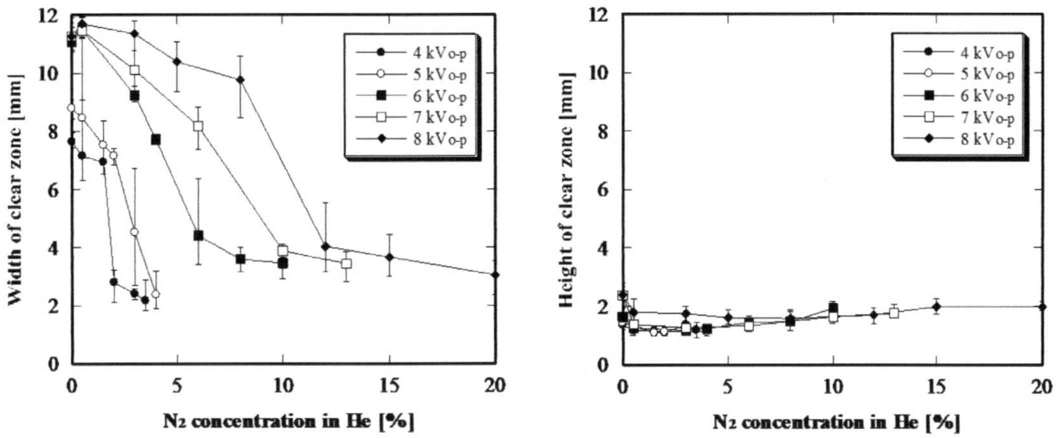

(a) Width of clear zone (b) Height of clear zone

Figure 4. Influence of the N_2 concentration in He on the clear zone size at voltages in the range from 4 to 8 $kV_{o\text{-}p}$.

4 (a). However, the decrease was stopped around the width of 2 - 4 mm at the N_2 concentration dependent on the applied voltage. These results indicate that the width of the clear zone can be controlled without changing a generator structure. The maximum width of 12 mm is equal to the gas outlet width of the plasma jet generator. Therefore, it may be possible to enlarge the width by improving the generator structure. On the other hand, the height of the clear zones was not affected by the N_2 addition, as shown in figure 4 (b). It means that the width can be controlled at a constant height.

3.2. Evaluation from the oxidizing substance distribution

Figure 5 shows the plasma jet irradiated to the chemical reagent and the resulting oxidizing substances distribution in the cases of (a) He only and N_2 (10%) / He. The applied voltage and the irradiation time were 7 $kV_{o\text{-}p}$ and 30 s, respectively. The shape of the plasma jet was also changed when the chemical reagent was used as the target. The stable spot-type plasma jet was able to be obtained under the conditions of applied voltage 7 $kV_{o\text{-}p}$ and N_2 concentration 10%. The oxidizing substances distribution

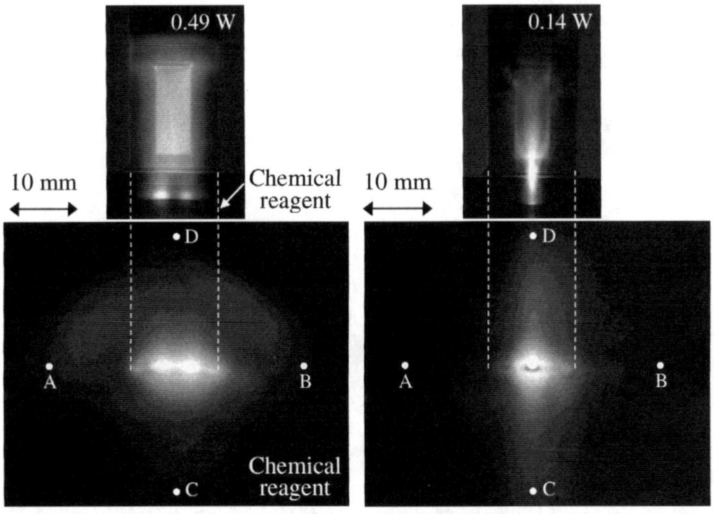

(a) He (b) N_2 (10%) / He

Figure 5. Evaluation of the irradiation area control from the oxidizing substances distribution using the chemical reagent in the cases of (a) He only and (b) N_2 (10%) / He. The applied voltage and irradiation time were 7 $kV_{o\text{-}p}$ and 30 s, respectively. The black background and white part in below figures indicate the unreacted reagent and color reaction by oxidizing substances, respectively. Points A-D are used in the next figure.

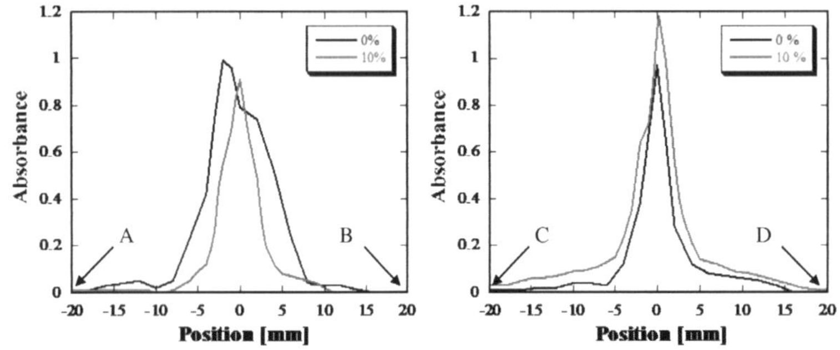

(a) Along the line AB (b) Along the line CD

Figure 6. Oxidizing substances distribution obtained by an absorbance measurement along the line AB and the line CD. A, B, C and D were shown in figure 5.

was able to be easily visualized using the chemical reagent prepared in our laboratory. This result indicates that the oxidizing substances were widely irradiated not only just under the generator but also around it. Figure 6 shows the numerical data of the oxidizing substances distribution shown in figure 5 by an absorbance measurement. The measurement was performed along the line AB (a) and CD (b) shown in figure 5. The width of the irradiation area was able to be also controlled judging from the oxidizing substances distribution just like the sterilization, as shown figure 6 (a). On the other hand, the height of the irradiation area was not able to be controlled, as shown figure 6 (b), although it was constant in the evaluation from the sterilization, as shown figure 4 (b).

Figure 7 shows the irradiation time dependency of the oxidizing substances distribution visualization, and its numerical data of the width (a) along the line AB and height (b) along the line CD were obtained by an absorbance measurement, as shown in figure 8. These results indicate that the width of the oxidizing substances distribution was kept within plus/minus 10 mm even when the irradiation time was increased, as shown in figure 8 (a). Only absorbance was increased in that range. On the other hand, figure 8 (b) indicates that its height was slightly increased around plus/minus 20 mm with increasing the irradiation time, and the absorbance was significantly increased in that range,

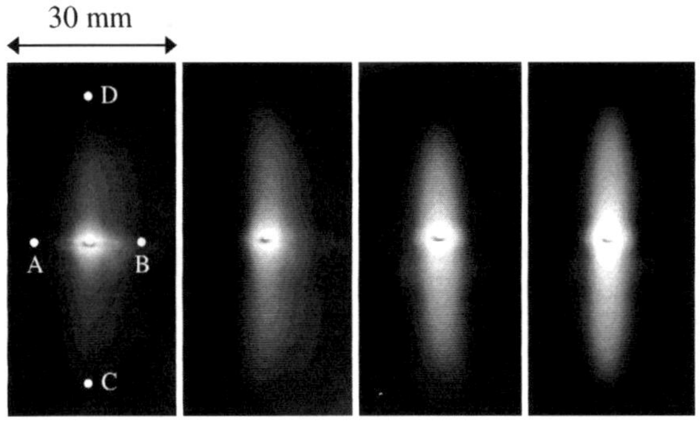

(a) 30 s (b) 60 s (c) 120 s (d) 180 s

Figure 7. Influence of an irradiation time on the oxidizing substances distribution in the case of N_2 (10%) / He. Points A-D are used in the next figure. The black background and white part in below figures indicates the unreacted reagent and color reaction by oxidizing substances, respectively

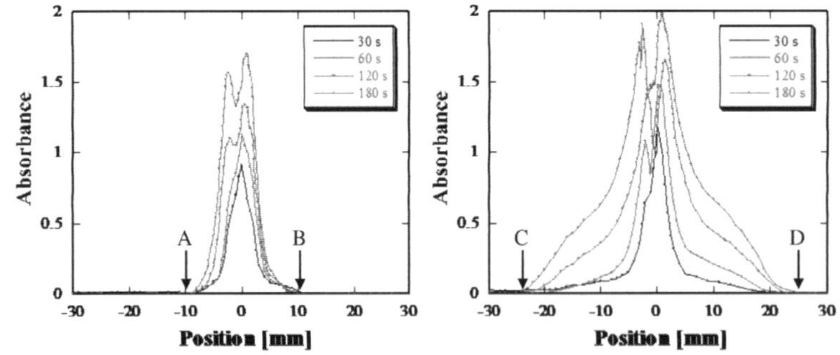

(a) Along the line AB (b) Along the line CD

Figure 8. Oxidizing substances distribution obtained by an absorbance measurement along the line AB and the line CD. A, B, C and D were shown in figure 7.

especially for more than 120 s. It became clear that the irradiation area was not able to be controlled in one direction when it is evaluated from the oxidizing substances distribution. The irradiation control of oxidizing substances is an important subject of study for the safe use of the plasma jet technologies.

4. Conclusions

The irradiation control method using the sheet-type plasma jet generator was proposed, and evaluated from both the sterilization area and the oxidizing substances distribution. The shape of plasma jet was able to be changed from a sheet-type to a spot-type by the N_2 addition into He gas without changing the generator structure. The width of an irradiation area was able to be controlled at a constant height in the evaluation from sterilization area. It was able to be also controlled just like the sterilization judging from the oxidizing substances distribution. On the other hand, the height of the irradiation area was not controlled with the strong time dependence, although it was constant in the evaluation from the sterilization. The irradiation control of oxidizing substances is an important subject of study for the safe use of the plasma jet technologies.

Acknowledgements

This work was partly supported by JSPS KAKENHI Grant Number 25820113 (Young Scientists (B)) and 21110001 (Scientific Research on Innovative Areas).

References

[1] Chang J S, Lawless P A and Yamamoto T 1991 *IEEE Trans. Plasma Sci.* **19** 1152
[2] Urashima K, Chang J S and Ito T 1997 *IEEE Trans. Ind. Appl.* **33** 879
[3] Kanazawa S, Sumi T, Sato N, Ohkubo T, Nomoto Y, Kocik M, Mizeraczyk J and Chang J S 2005 *IEEE Trans. Ind. Appl.* **41** 200
[4] Takashima K, Suzuki T, Nomura Y, Hinata Y, Hayashi H, Kurita H and Mizuno A 2013 *Int. J. Environ. Sci. Tech.* **7** 142
[5] Bruggeman P and Leys C 2009 *J. Phys. D: Appl. Phys.* **42** 053001
[6] Takeuchi N, Oishi R, Kitagawa Y and Yasuoka K 2011 *IEEE Trans. Plasma Sci.* **39** 3358
[7] Kanazawa S, Furuki T, Nakaji T, Akamine S and Ichiki R 2012 *Int. J. Environ. Sci. Tech.* **6** 166
[8] Schutze A, Jeong J Y, Babayan S E, Park J, Selwyn G S and Hicks R F 1998 *IEEE Trans. Plasma Sci.* **26** 1685
[9] Deng X T, Shi J J, Shama G and Kong M G 2005 *Appl. Phys. Lett.* **87** 153901
[10] Deng X T, Shi J J, Chen H L and Kong M G 2007 *Appl. Phys. Lett.* **90** 013903
[11] Rupf S, Lehmann A, Hannig H, Schafer B, Schubert A, Feldmann U and Schindler A 2010 *J. Med. Microbiol.* **59** 206
[12] Korachi M, Gurol C, and Aslan N 2010 *J. Electrostat.* **68** 508
[13] Ikawa S, Kitano K and Hamaguchi S 2010 *Plasma Process. Polym.* **7** 33
[14] Yamazaki H, Ohshima T, Tsubota H, Yamaguchi H, Jayawardena J A and Nishimura Y 2011 *Dent. Mater. J.* **30** 384
[15] Tanaka H, Mizuno M, Ishikawa K, Nakamura K, Kajiyama H, Kano H, Kikkawa F and Hori M 2011 *Plasma Medicine* **1** 265.
[16] Cunningham S, Nakajima T, Kurita H, Yasuda H, Takashima K and Mizuno A 2012 *Int. J. Environ. Sci. Tech.* **6** 54
[17] Iseki S, Nakamura K, Hayashi M, Tanaka H, Kondo H, Kajiyama H, Kano H, Kikkawa F and Hori M 2012 *Appl. Phys. Lett.* **100** 113702
[18] Kim J Y, Shim J B and Kim S O 2011 *IEEE Trans. Plasma Sci.* **39** 2300
[19] Dobrynin D, Fridman A and Starikovskiy A Y 2012 *IEEE Trans. Plasma Sci.* **40** 2163
[20] Kawasaki T, Hoketsu S, Morisaki M, Sato K, Baba H, Umeda S, Nakagawa Y, Wakamatsu S, and Sakai M 2013 *Int. J. Environ. Sci. Tech.* **7** 26
[21] Abramoff M D, Magalhaes P J and Ram S J 2004 *Biophotonics International* **11** 36
[22] Abolmasov S N, Shirafuji T and Tachibana K 2005 *IEEE Trans. Plasma Sci.* **33** 941

Journal of Physics: Conference Series **518** (2014) 012017 doi:10.1088/1742-6596/518/1/012017

Growth enhancement effects of radish sprouts: atmospheric pressure plasma irradiation vs. heat shock

T Sarinont[1], T Amano[2], S Kitazaki[3], K Koga[1], G Uchida[4], M Shiratani[1] and N Hayashi[3]

[1] Graduate School of Information Science and Electrical Engineering, Kyushu University 744 Motooka, Nishi-ku, Fukuoka, 819-0395, Japan
[2] Department of Electrical Engineering, School of Engineering, Kyushu University 744 Motooka, Nishi-ku, Fukuoka, 819-0395, Japan
[3] Interdisciplinary Graduate School of Engineering Science, Kyushu University 6-1 Kasuga-kouen, Fukuoka, 816-8580, Japan
[4] Joining and Welding Research Institute, Osaka University, 11-1 Mihogaoka, Ibaraki, Osaka, 567-0047, Japan

E-mail : t.sarinont@plasma.ed.kyushu-u.ac.jp

Abstract. We compare growth enhancement effects due to atmospheric air dielectric barrier discharge plasma irradiation and heat shock to seeds of radish sprouts (*Raphanus sativus L.*). Interactions between radicals and seeds in a short duration of 3 min. lead to the growth enhancement of radish sprouts in a long term of 7 days and the maximum average length is 3.7 times as long as that of control. The growth enhancement effects become gradually weak with time, and hence the ratio of the average length for plasma irradiation to that for control decreases from 3.7 for the first day to 1.3 for 7 day. The average length for heat shock of 60°C for 10 min. and 100°C for 3 min. is longer than that for control, and the maximum average length is 1.3 times as long as that of control. Heat shock has little contribution to the growth enhancement due to plasma irradiation, because the maximum temperature due to plasma irradiation is less than 60°C.

1. Introduction

The demand for food will continue to increase towards 2050 as a result of population growth by an additional 2.7 billion people [1]. Increases in food production per hectare of land have not kept pace with increases in population, leading to the global food crisis. Especially, the current world food crisis is the result of the combined effects of competition for cropland from the growth in biofuels, low cereal stocks, high oil prices, speculation in food markets and extreme weather events [1]. One possible solution to the global food crisis is to improve agricultural productivity by some means. Most improvement methods of agricultural productivity are irradiations of γ-ray, ^{60}Co-ray, laser, electric and magnetic field. Such irradiations often induce damage of cells of plant seeds [2-4]. Atmospheric pressure plasmas become prevalent in various applications, ranging from material processing to biomedicine [5-20]. Atmospheric pressure non-thermal plasmas have recently been employed to agricultural crops and the accelerate seed germination in order to improve agricultural productivity [5]. For interactions with living tissues, the choice of an adequate plasma source is crucial. It was operated under atmospheric pressure, electrically and chemically safe, and may not cause any thermal damage to the living object. Recently, we have found that atmospheric pressure dielectric barrier discharge (DBD) plasma irradiation to seeds of plants enhance their growth [16-20]. The growth promotion is attributed not to an increase in size of each cell but to an increase in the number of cells. Plasma irradiation may raise the temperature of seeds. In many species, plants germination and growth are significantly affected by the temperature treatment level [19-20]. For instance, short exposure of seeds to high temperatures generally stimulates germination, whereas prolonged exposure reduces seed germination [21-26]. Here, we focus on the effects of temperature rise due to plasma irradiation to seeds of radish sprouts (*Raphanus sativus L.*) on the plant growth.

2. Experimental

Experiments were carried out using a atmospheric pressure scalable DBD device described elsewhere [16-20]. The device consisted of 20 electrodes of a stainless rod of 1 mm in outer diameter and 60 mm in length covered with a ceramic tube of 2 mm in outer diameter as shown in Figure 1. The electrodes were arranged parallel at spacing of 0.2 mm. DBD plasma was generated in air between the electrodes by supplying 10 kHz AC high voltage (Logy Electric, LHV-09K). The temperature and humidity of air were 24-26°C and 57-61%, respectively. The discharge voltage and current were measured with a high-voltage probe (Tektronix, P6015A) and a Rogowski coil (URD, CTL-28-S90-05Z-1R1), respectively. The peak-to-peak discharge voltage and current were 9.2 kV and 0.2 A, respectively. The corresponding discharge power density was 1.49 W/cm^2, which was deduced from voltage/charge Lissajous plots. There was little air flow around the DBD device, and therefore, spatial profiles of radical densities were governed by diffusion and gas phase reactions.

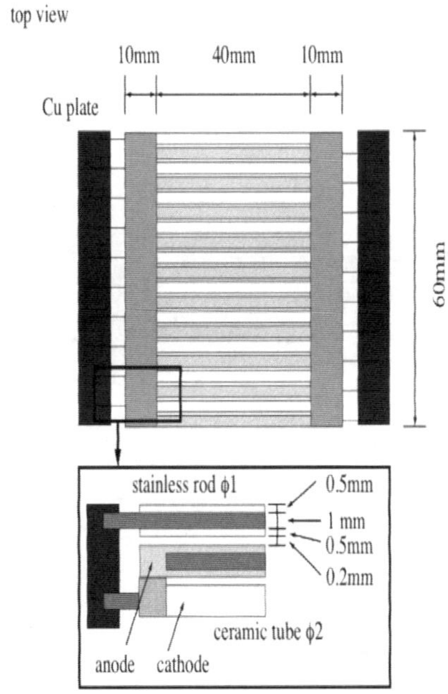

Figure 1. Schematic representation of the scalable atmospheric pressure DBD device.

10 seeds of radish sprouts (*Raphanus sativus L.*) were arranged horizontally at intervals of 5 mm at 3 mm below the electrodes DBD plasma was irradiated to the seeds for 3 min. Seeds with and without plasma irradiation were cultivated for 7 days in an incubator at 22°C and 60% relative humidity in the dark with pure water feed. Their total length from primary root to stark was measured with an image analysis system and statistical significance of the total length was evaluated every day.

We also examined effects of heat shock. To apply heat shock, 10 seeds in a quartz petri dish were put on hot plate. The hot plate temperature was set at 60°C or 100°C for 3 min. or 10 min. Then the seeds were cooled down at room temperature or -20°C for 10 min. Seeds with and without heat shock were cultivated for 3 days in an incubator at 22°C and 60% relative humidity in the dark with pure water feed. Their total length was measured every day. Time evolution of the surface temperature of seeds was measured with an IR-camera (NEC, TH7800N).

3. Results and discussion

First, we examined effects atmospheric pressure DBD plasma irradiation to seeds on growth of plants. Figure 2(a) shows growth curves for radish sprouts with and without plasma irradiation. The symbols and bars in figure 2(a) show the average length and standard error, respectively. The length with plasma irradiation is longer than that without plasma irradiation for 7 days. Interactions between seeds and radicals produced by the plasmas in a short duration of 3 min. lead to the growth enhancement of

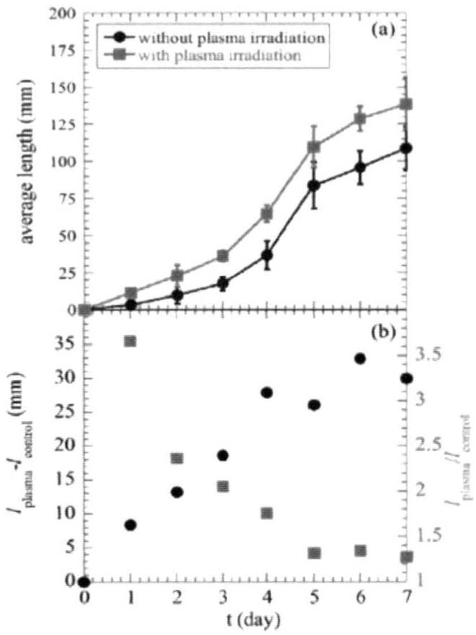

Figure 2. (a) Growth curve of plants and (b) time evolution of a ratio between the length for plasma irradiation and that for control together with a difference between the length for plasma irradiation and that for control.

radish sprouts in a long term of 7 days. Figure 2(b) shows time evolution of a ratio $[l_{plasma}]/[l_{control}]$ of the average length for plasma irradiation to that for control together with a difference $[l_{plasma}\text{-}l_{control}]$ between the length for plasma irradiation and that for control. The difference increases with time during 7 days, namely, plasma irradiation promotes plant growth for 7 days. The ratio decreases from

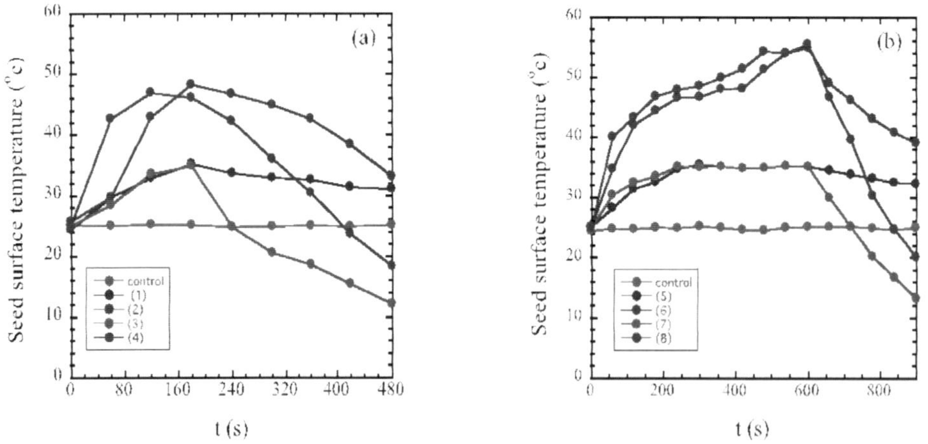

Figure 3. Time evolution of surface temperature of seeds on heat treatments for (a) 3min. and (b) 10 min.

3.7 at the first day to 1.3 at 7 day. Therefore, the response of plants to the plasma irradiation becomes gradually weak with time.

Plasma irradiation to seeds raises the temperature of seeds up to 30°C. Such heat shock due to plasma irradiation may contribute to the growth enhancement in figure 2. To clarify effects of heat shock to the seeds, dry heat treatments to seeds in a quartz petri dish were carried out with a hot plate. 8 kinds of heat treatments were applied to seeds: (1) They are heated on hot plate of 60°C for 3 min. then they are cooled down to room temperature, (2) They are heated on hot plate of 100°C for 3 min. then they are cooled down to room temperature, (3) They are heated on hot plate of 60°C for 3 min. then they are cooled down to -20°C, (4) They are heated on hotplate of 100°C for 3 min. then they are cooled down to -20°C, (5) They are heated on hot plate of 60°C for 10 min. then they are cooled down to room temperature, (6) They are heated on hot plate of 100°C for 10 min. then they are cooled down to room temperature, (7) They are heated on hot plate of 60°C for 10 min. then they are cooled down to -20°C, (8) They are heated on hot plate of 100°C for 10 min. then they are cooled down to -20°C.

Figures 3(a) and 5(b) show time evolution of the seed surface temperature. The seed surface temperature gradually increases with time. For the heat treatment of 3 min., the highest seed surface

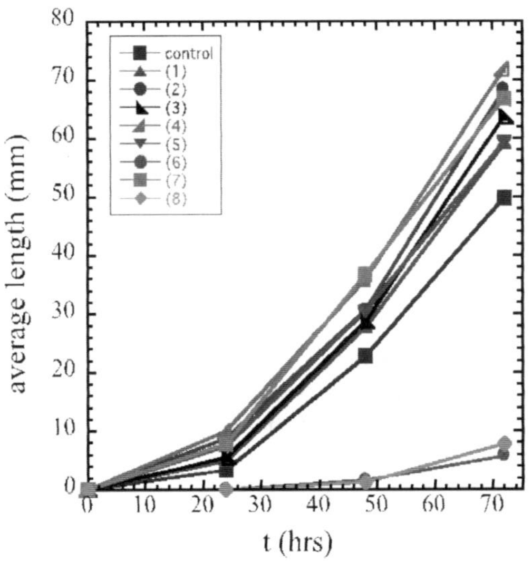

Figure 4. Growth curve of plants for control and .(1) heated on 60°C plate for 3 min. then cooled down to room temperature, (2) heated on 100°C plate for 3 min. then cooled down to room temperature, (3) heated on 60°C plate for 3 min. then cooled down to -20°C, (4) heated on100°C plate for 3 min. then cooled down to -20°C, (5) heated on 60°C plate for 10 min. then cooled down to room temperature, (6) heated on 100°C plate for 10 min. then cooled down to room temperature, (7) heated on 60°C plate for 10 min. then cooled down to -20°C, (8) heated on 100°C plate for 10 min. then cooled down to -20°C.

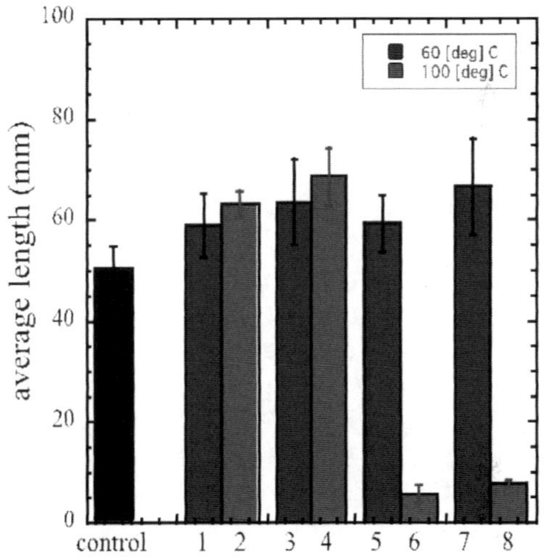

Figure 5. Average length of *Raphanus sativus L.* after 3days cultivation for control and .(1) heated on 60°C plate for 3 min. then cooled down to room temperature, (2) heated on 100°C plate for 3 min. then cooled down to room temperature, (3) heated on 60°C plate for 3 min. then cooled down to -20°C, (4) heated on100°C plate for 3 min. then cooled down to -20°C, (5) heated on 60°C plate for 10 min. then cooled down to room temperature, (6) heated on 100°C plate for 10 min. then cooled down to room temperature, (7) heated on 60°C plate for 10 min. then cooled down to -20°C, (8) heated on 100°C plate for 10 min. then cooled down to -20°C.

temperature is 35°C for the hot plate temperature of 60°C and 48°C for the hot plate temperature of 100°C. For the heat treatment of 10 min., the highest seed surface temperature is 35°C for the hot plate temperature of 60°C and 55°C for the hot plate temperature of 100°C.

Figure 4 shows time evolution of the average length of *Raphanus sativus L.* and figure 5 shows the average length after 3 days cultivation. Temperature had a significant influence on the germination of seeds of plant since first day after cultivation as shown on Figure 4. The heat shock treatments of (1)-(5), and (7) leads to growth enhancement and the length ratio of heat treatment plants to control after 3 days cultivation ranges from 1.2 to 1.4, which are much below the ratio of 2 for plasma irradiation. The heat treatments of (6) and (8) suppress significantly the growth. Both the temperature and the duration of treatments have influence on germination and growth of plants. The low temperature heat treatments tend to enhance germination, whereas high temperature short heat treatments enhance growth of plants. However, high temperature long heat treatments suppress germination, which are similar to the results in [25].

Based on the results in figures 2-5, the temperature rise of seeds due to plasma irradiation has little effects on the plant growth, because the maximum surface temperature of seed during plasma irradiation is 30 °C. Moreover our previous results show that irradiation of photons and ions to seeds has little effects on plant growth. Therefore, interactions between radicals such as OH, H_2O_2, O and O^- and seeds may bring about growth enhancement of plants in a long term of 7 days. One possible mechanism is that plasma irradiation leads to secrete some hormones in seeds, intracellular chemical signals and messengers induce the growth enhancement.

4. Conclusions

We applied plasma irradiation and heat shock to seeds of radish sprouts to study their effects of growth of radish sprouts. We obtained the following conclusions:

1) Plasma irradiation for 3 min. enhances plant growth in a long term of 7 days.
2) The response of plants to the plasma irradiation becomes gradually weak with time, and the ratio of plant length with plasma irradiation to control decreases from 3.7 at the first day to 1.3 at 7 day.
3) Temperature rise during plasma irradiation has little effect on the growth enhancement.

The remaining issues of clarification of the growth enhancement mechanisms are identification of plasma generated species responsible for the growth enhancement, and stimulated signal pathways in the plants for the growth enhancement.

Acknowledgement

This work was supported by JSPS KAKENHI grant number 2434014.

References

[1] Nellemann C, MacDevette M, Manders T, Eickhout B, Svihus B, Prins A G and Kaltenborn B P 2009 *The environmental food crisis – The environment's role in averting future food crises* ISBN: 978-82-7701-054-0
[2] Nastuta A, Topala I, Grigoras C, Pohoata V and Popa G 2011 *Jpn. J. Appl. Phys.* **44** 105204
[3] Pimentel D, Hurd L E, Bellotti A C, Forster M J, Oka I N, Shores O D and Whitman R J 1973 *Science* **182** 4111
[4] Victor J, Thannickal V and Fanberg B 2000 *Am. J. Physiol.* **279** L1005
[5] Lee J, Shon C H, Kim Y S, Kim S, Kim G C and Kong M G 2009 *New J. Phys.* **11** 115026
[6] Fridman A, Chirokov A and Gutsol A 2005 *Jpn. J. Appl. Phys.* **38** R1
[7] Chirokov A, Gutsol A and Fridman A 2005 *Pure Appl. Chem.* **77** 487
[8] Ito M, Ohta T and Hori M 2012 *J. Korean Phys. Soc.* **60** 6
[9] Zhou Z 2011 *Agr. Sci.* **2** 23
[10] Kong M G, Kroesen G, Nosenko T, Shimizu T, Van Dijk J and Zimmermann J 2009 *New J. Phys.* **11** 115012
[11] Laroussi M and Leipold F 2004 *Int. J. Mass Spectrom.* **233** 81

[12] Stoffels E, Keift I E and Sladek R E J 2003 *Jpn. J. Appl. Phys.* **36** 2908
[13] Dobrynin D, Fridman G, Friedman G and Fridman A 2009 *New J. Phys.* **11** 115020
[14] Hayashi N, Tsutsui S, Tomari T and Guan W 2008 *IEEE Trans. on Plasma Science* **36** 1302
[15] Hayashi N and Yagyu Y 2008 *Trans. of the Material Research Society of Japan* **33** 791
[16] Kitazaki S, Yamashita D, Matsuzaki H, Uchida G, Koga K and Shiratani M 2010 *Proc. IEEE TENCON* 1960
[17] Kitazaki S, Koga K, Shiratani M and Hayashi N 2012 *MRS Proc.* 1469
[18] Kitazaki S, Koga K, Shiratani M and Hayashi N 2012 *Jpn. J. Appl. Phys.* **51** 11PJ02
[19] Kitazaki S, Sarinot T, Koga K, Shiratani M and Hayashi N 2014 *Current Appl. Phys.* in press
[20] Sarinont T, Kazunori Koga, Satoshi Kitazaki, Giichirou Uchida, Nobuya Hayashi and Masaharu Shiratani 2014 *JPS Conf. Proceedings* in press
[21] Wehmeyer N, Hernandez L D Finkelstein R R and Vierling E 1996 *Plant Physiol.* **112** 747
[22] Gashaw M and Michelson A 2002 *Plant Ecol.* **159** 83
[23] Schoffl F, Prandl R and Reindl A 1998 *Plant Physiol.* **117** 1135
[24] Gashaw M and Michelsen A 2001 *Plant Ecol.* **159** 83
[25] Thomas B P, Morris E C and Auld D T 2003 *Austral Ecol.* **28** 674
[26] Boehm K A, Saunders A, Werner J and Lis T J 2003 *Mol. Cell. Biol.* **23** 7628

Molecular-structure variation of organic materials irradiated with atmospheric pressure plasma

K Takenaka, A Miyazaki and Y Setsuhara

Joining and Welding Research Institute, Osaka University, 11-1 Mihogaoka, Ibaraki, Osaka 567-0047, Japan

E-mail: k_takenaka@jwri.osaka-u.ac.jp

Abstract. The effect of atmospheric pressure He plasma on the molecular structure of polyethylene terephthalate (PET) has been investigated. The plasma composition was analyzed using optical emission spectroscopy. In addition to strong He emission lines, lines due to O and N radicals were also detected. The change in the molecular structure of the PET film surface was investigated using Fourier transform infrared spectroscopy and X-ray photoelectron spectroscopy. It was found that plasma irradiation led to oxidation and degradation of the surface due to chemical and physical effects of the active species. The results demonstrate the feasibility of observing the interaction of plasma with organic material on a local scale.

1. Introduction

Atmospheric pressure plasma is attractive for a variety of processes used in the fabrication of microelectronic devices, such as chemical vapor deposition (CVD), etching and ashing. It is a dense plasma that contains more activated species (ions, radials) than conventional low-pressure plasma, and so can speed up plasma processes. Its application in the fields of medicine and biology has also expanded remarkably [1-11]. For medical applications, it is essential to perform fundamental studies at the molecular or atomic level in order to clarify the complex interactions of plasma, which contains ions, radicals, electrons and photons, with biomolecules such as DNA, amino acids and proteins. In previous studies, we have investigated physical and chemical interactions of plasma with the surface of soft materials, including biomolecules, using X-ray photoelectron spectroscopy (XPS). The plasmas studied were generated using low-inductance antenna (LIA) modules at low pressures of 0.13-13 Pa [12-25]. The present study extends this work by considering the interactions of atmospheric pressure He plasma generated in air with polyethylene terephthalate (PET) films. The chemical composition of the plasma was analyzed using optical emission spectroscopy (OES). The effect of the plasma on the PET films was investigated using XPS and Fourier transform infrared (FTIR) spectroscopy.

2. Experimental

The atmospheric pressure He plasma was generated using a micro-hollow cathode source, as shown in figure 1. The cathode had a hole diameter of 0.3 mm and was operated at a DC voltage of 1000 V. He gas was supplied at a flow rate of 1 slm. The OES analysis was carried out using a high-resolution fiber-optic spectrometer (Ocean Optics HR-4000). PET films with a thickness of 100 μm were exposed to the plasma. The change in the chemical bonding state of the nanoscale

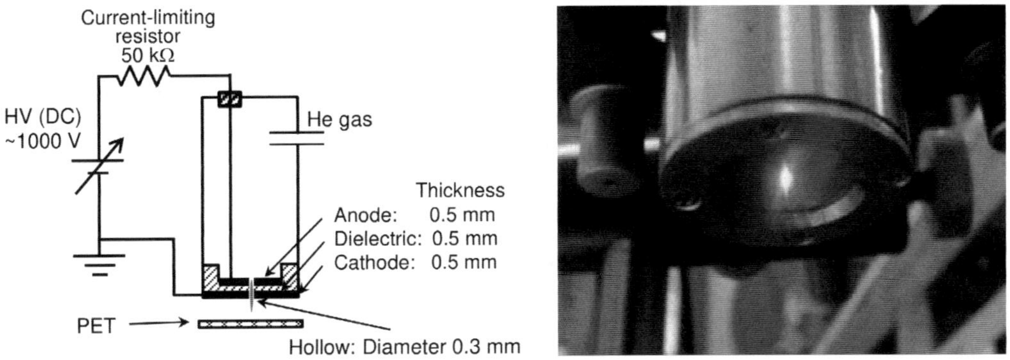

Figure 1. Schematic diagram and photograph of micro-hollow cathode plasma source.

surface layer exposed to the plasma was investigated using XPS (AXIS-165x, Shimadzu) with non-monochromatized Mg Kα radiation (photon energy 1253.6 eV), and an FTIR system (Nicolet iS50, Thermo Scientific) equipped with an infrared microscope (Continuum, Thermo Scientific) with an attenuated total reflection (ATR) accessory.

3. Results

3.1 Optical emission spectroscopic analysis in atmospheric pressure He plasma

Figure 2 shows a typical OES spectrum of the atmospheric pressure He plasma. In addition to strong He lines, emissions due to O and N radicals and N_2 molecules can be observed. Furthermore, a broad peak attributed to NO_x is observed from 400 nm to 700 nm. The radicals are considered to be produced by excitation and dissociation of N_2 and O_2 when the plasma comes into contact with air.

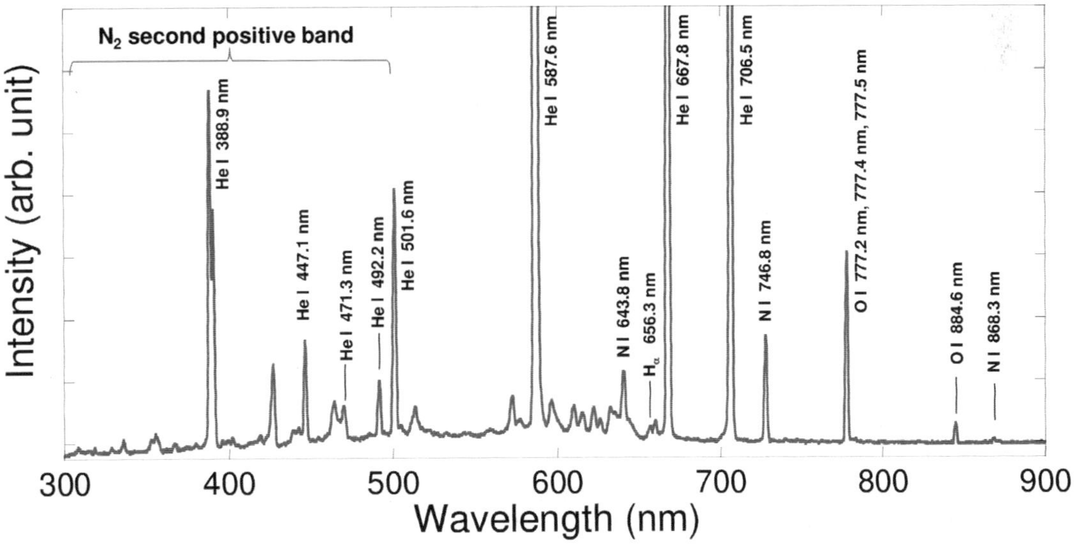

Figure 2. Optical emission spectrum of atmospheric pressure He plasma.

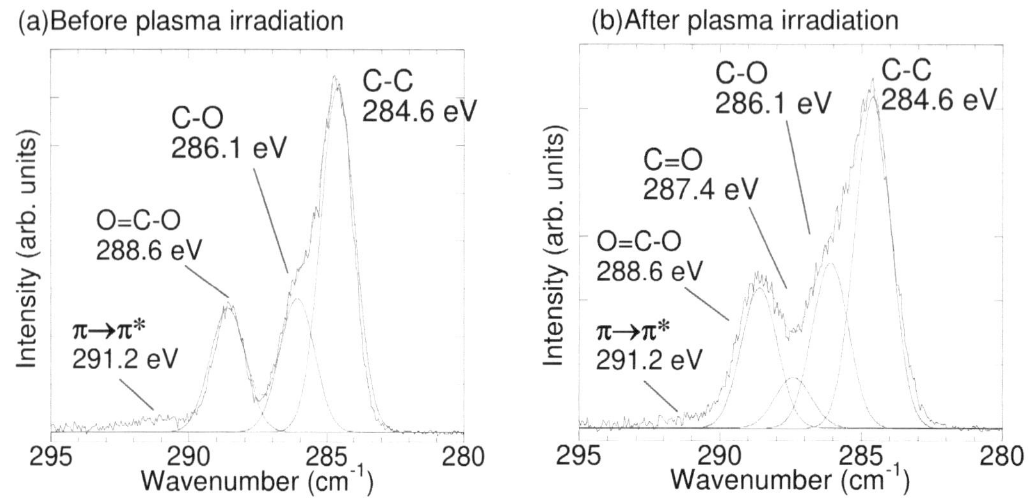

Figure 3. XPS C 1s spectra of PET films (a) before plasma irradiation and (b) after plasma irradiation.

3.2 XPS analysis of nano-surface on PET films

Figure 3 shows typical XPS C1s spectra of a PET film before and after irradiation with atmospheric-pressure He plasma. The spectra are normalized using the intensity of the C-C bond peak. The spectrum of the unirradiated PET sample in figure 3 (a) can be decomposed into four main components: a component at 284.6 eV due to C-C bonds, a component at 286.1 eV due to C-O bonds, a component at 288.6 eV due to O=C-O bonds and a component at 291.2 eV due to the π→π* shake-up satellite of the phenyl group [26]. As seen in figure 3 (b), following plasma exposure, the intensity of the peaks due to O=C-O and C-O bonds increased, and a new peak due to C=O bonds (287.4 eV) appeared. This indicates that a nanoscale surface region of the PET is oxidized by the reactive oxygen species, and/or the molecular structure is degraded.

3.3 FT-IR analysis of PET films

Figure 4 shows FTIR spectra (normalized at 1250 cm^{-1}) of the PET surface before and after plasma treatment, in the narrow wavenumber range 700-2000 cm^{-1}. The main absorption bands appear at 1712 cm^{-1} (C=O stretching), 1250 cm^{-1} (ring O=C-O, glycol O-C-H bending), 1123 and 1113 cm^{-1} (CH_2 twisting or ring mode, ring O=C-O), and 725 cm^{-1} (ring C-H) [27-31]. A broad peak also appears at about 900-1000 cm^{-1}, and is attributed to C-C and C=C groups that was responsible for the soot structure [32,33]. Following plasma irradiation, the spectrum in figure 4 (b) contains peaks attributed to a part of oxidized and degraded PET surface. Following plasma irradiation, the intensity of the O=C-O peak is seen to decrease. Furthermore, oxidation of the PET surface is confirmed from the FTIR analysis. For low-pressure (about 1 Pa) plasma irradiation of PET, the intensity of the C=O peak (1712 cm^{-1}) tends to decrease relative to that for the O=C-O and O-C-H peaks in the glycol segment (1250 cm^{-1}) with increasing plasma density. This indicates that more C=O bond breaking is taking place due to plasma irradiation. For the atmospheric pressure He plasma, the FTIR results suggest that O=C bonds are broken by chemical and/or physical effects that occur in the deep-surface region.

The changes in the PET structure determined by the FTIR analysis are not consistent with the XPS results, which indicated oxidation occurring. his is because XPS is a surface-sensitive technique and FTIR is not. The plasma contains not only radicals that modify the top surface but also a significant number of photons that can penetrate deep into the PET, and the resulting modification effects can be

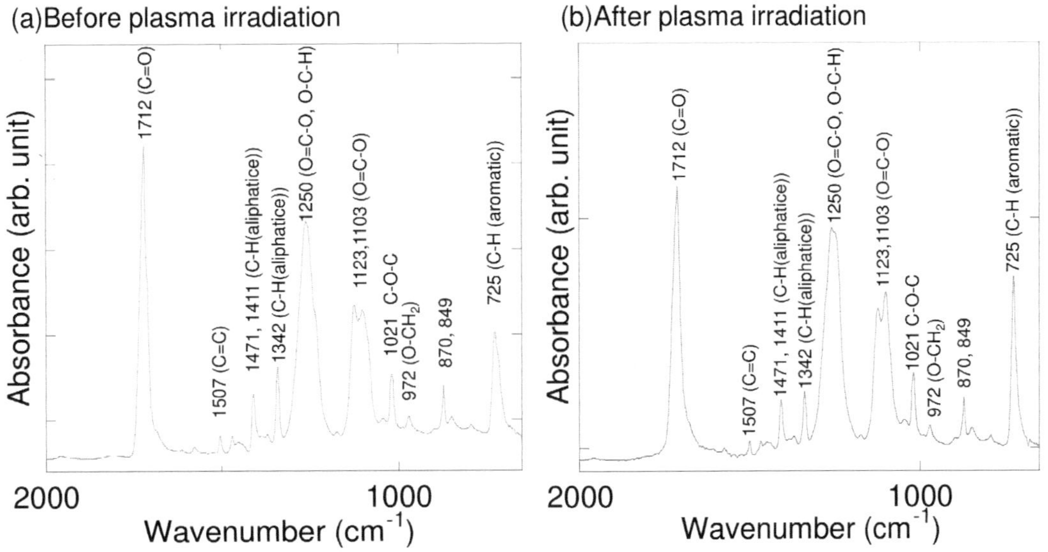

Figure 4. FTIR-ATR spectra of PET films (a) before plasma irradiation and (b) after plasma irradiation.

detected by FTIR. Therefore, the difference between the XPS and FTIR results is due to their being sensitive to different regions.

3.3 FT-IR mapping analysis of PET films

To investigate the variation of the chemical structure in the plasma irradiated area of the PET film, FTIR mapping was carried out. Figure 5(a) shows a plan-view optical micrograph of the PET surface,

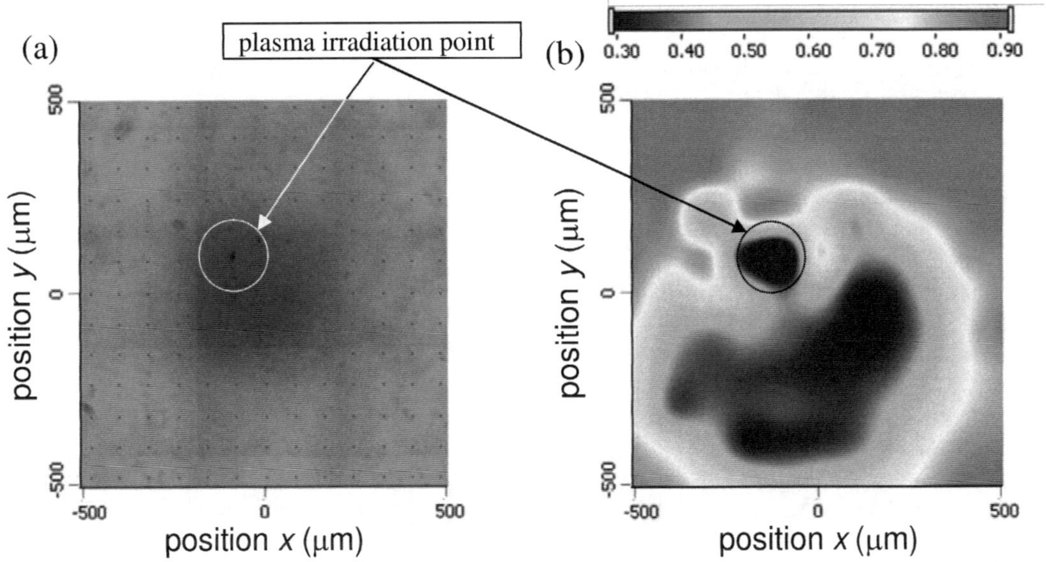

Figure 5.Plan-view (a) optical micrograph of PET surface and (b) FTIR map of intensity ratio of C=O peak (1712 cm^{-1}) to O=C-O peak (1250 cm^{-1}).

with the plasma irradiation point indicated. Figure 5(b) shows a FTIR map of the intensity ratio of the C=O peak (1712 cm^{-1}) to the O=C-O peak (1250 cm^{-1}). A large change in the intensity ratio is seen near the irradiation point, and the chemical structure gradually changes with distance from this point. The latter is considered to be due to a gas flow effect, which causes radial transport of the plasma across the PET surface. The resulting surface modification gives rise to the pattern shown in shown in figure 5 (b). These results demonstrate the feasibility of observing the interaction of plasma with organic material on a local scale. A more detailed study of the interaction of atmospheric plasma He plasma and organic material will soon be carried out.

4. Summary

Plasma interactions with organic materials have been investigated by using atmospheric-pressure He plasma. Optical emission spectrum in the atmospheric-pressure He plasma has been measured. The spectrum shows considerable emissions of He lines, and the emission of O and N radicals attributed to air has been confirmed in the atmospheric-pressure He plasma. Variation in molecular structure of PET film surface irradiated with the plasma has been observed via FT-IR and XPS. These results via FT-IR and XPS indicate that the PET surface irradiated with the atmospheric-pressure He plasma was oxidized and/or degraded by chemical and/or physical effect due to irradiation of active species. These FT-IR mapping results exhibit the feasibility of observation of plasma/organic materials interaction in a small area due to local irradiation of atmospheric-pressure He plasma.

Acknowledgments

This work was partly supported by The Grant-in-Aid for Scientific Research on Innovative Areas "Plasma Medical Innovation" (24108003) from The Ministry of Education, Culture, Sports, Science and Technology (MEXT) and The Grant-in-Aid for Young Scientists (B) (24760597) from the Japan Society for the Promotion of Science (JSPS).

References

[1] Moisan M, Barbeau J, Moreau S, Pelletier J, Tabrizian M and Yahia L H 2001 *Int. J. Pharm.* **226**, 1
[2] Laroussi M 2005 *Plasma Process. Polym.* **2** 391
[3] Boudam M K, Moisan M, Saoudi B, Popovici C, Gherardi N and Massines F 2006 *J. Phys. D: Appl. Phys.,* **29** 3494
[4] Weltmann K -D, Brandenburg R, Woedtke T, Ehlbeck J, Foest R, Stieber M and Kindel E 2008 *J. Phys D: Appl. Phys.,* **41** 194008
[5] Raiser J and Zenker M 2006 *J. Phys. D: Appl. Phys.,* **39** 3520
[6] Stalder K R and Woloszko J 2007 *Contrib. Plasma Phys.,* **47** 64
[7] Yonson S, Coulombe S, Lveill V, Leask R L 2006 *J. Phys. D: Appl. Phys.,* **39** 3508
[8] Coulombe S, Lveill V, Yonson S and Leask R L 2006 *Pure Appl. Chem.,* **78** 1147
[9] Stoffels E, 2007 *Contrib. Plasma Phys.,* **47** 40
[10] Fridman G, Friedman G, Gutsol A, Shekhter A B, Vasilets V N and Fridman A 2008 *Plasma Process. Polym.,* **5** 503
[11] Morfill1 G E, Kong M G and Zimmermann J L 2009 *New J. Phys.* **11** 115011
[12] Setsuhara Y, Cho K, Takenaka K, Shiratani M, Sekine M and Hori M, 2011 *Thin Solid Films* **519** 6721
[13] Cho K, Setsuhara Y, Takenaka K, Shiratani M, Sekine M and Hori M, 2011 *Thin Solid Films* **519** 6810
[14] Takenaka K, Setsuhara Y, Cho K, Shiratani M, Sekine M and Hori M, 2010 *Jpn. J. Appl. Phys.* **49** 08JA02
[15] Setsuhara Y, Cho K, Takenaka K, Shiratani M, Sekine M and Hori M, 2010 *Surf. Coatings Technol.* **205** S355.

[16] Setsuhara Y, Cho K, Takenaka K, Shiratani M, Sekine M and Hori M, 2011 *Surf. Coatings Technol.* **205** S484
[17] Setsuhara Y, Cho K, Takenaka K, Shiratani M, Sekine M and Hori M, Ikenaga E, and Zaima S 2010 *Thin Solid Films* **518** 3561
[18] Setsuhara Y, Cho K, Takenaka K, Shiratani M, Sekine M and Hori M, 2010 *Thin Solid Films* **518** 6320
[19] Cho K, Takenaka K, Setsuhara Y, Shiratani M, Sekine M, Hori M, Ikenaga E, Kondo H, Nakatsuka O and Zaima S 2010 *Ceramic Transactions* **219** 183
[20] Setsuhara Y, Cho K, Takenaka K, Ebe A, Shiratani M, Sekine M, Hori M, Ikenaga E, Kondo H, Nakatsuka O and Zaima S 2009 *Thin Solid Films* **518** 1006
[21] Setsuhara Y, Cho K, Shiratani M, Sekine M, Hori M, Ikenaga E and Zaima S 2009 *Thin Solid Films* **518** 3555
[22] Setsuhara Y, Cho K, Shiratani M, Sekine M and Hori M, 2009 *Thin Solid Films* **518** 6492
[23] Cho K, Setsuhara Y, Takenaka K, Shiratani M, Sekine M and Hori M, 2012 *Jpn. J. Appl. Phys.* **51** 01AJ02
[24] Takenaka K, Cho K, Setsuhara Y, Shiratani M, Sekine M and Hori M 2013 *J. Phys. Conference Series* **441** 012001
[25] Setsuhara Y, Cho K, Shiratani M, Sekine M and Hori M 2013 *Current Applied Physics* **13** S59
[26] Miyayama T, Sanada N, Iida S-I, Hammond J S and Suzuki M 2008 *Appl. Surf. Sci.* **255** 951
[27] Lee C-T, Wu C-H and Lin M-S 2004 *Polym. Degrad. Stab.* **83** 435
[28] Jaššo M, Krump H, Hudeca I, St'ahel' P, Kováčik D and Šíra M, 2006 *Surf. Coatings Technol.* **201** 57
[29] Patel M R, Patel J V and Sinha V K, 2005 *Polym. Degrad. Stab.* **90** 111
[30] Fávaro S L, Rubira A F, Muniz E C and Radovanovic E 2007 *Polym. Degrad. Stab.* **92** 1219
[31] Kormunda M, Pavlik J 2010 *Polym. Degrad. Stab.* **95** 1783
[32] Kirchner U, Scheer V and Vogt R 2000 *J. Phys. Chem. A* **104** 8908
[33] Armand X, Herlin N, Voicu I and Cauchetier M 1997 *J. Phys. Chem. Solids* **58** 1853

Investigation of plasma-organic materials interaction in aqueous solution with atmospheric pressure plasmas

K Takenaka, A Miyazaki and Y Setsuhara

Joining and Welding Research Institute, Osaka University, 11-1 Mihogaoka, Ibaraki, Osaka 567-0047, Japan

E-mail: k_takenaka@jwri.osaka-u.ac.jp

Abstract. An investigation was carried out into the interaction of an atmospheric pressure plasma with an organic material in an aqueous solution. The degradation of methylene blue (MB) by plasma exposure through the gas/liquid interface was studied. In the optical emission spectrum of the plasma, in addition to strong He emission lines, emissions due to O and OH radicals formed by dissociation of water were present. The change in the absorbance of a MB aqueous solution during plasma exposure was measured using UV-Vis spectroscopy. The absorption peak intensity decreased with plasma exposure time and complete decolorization occurred after 10 min. Since plasma exposure was found to decrease the pH of water, the effect of changes in pH on MB degradation in the absence of a plasma was investigated using solutions with different pH values. However, varying the pH was found to have no influence on MB degradation. The results indicated that MB degradation occurred due to interactions between MB and radicals across the plasma/liquid interface.

1. Introduction

There has been a remarkable increase in the number of medical and biological applications of plasmas [1-11]. In particular, high-pressure non-equilibrium plasmas, including atmospheric pressure plasma, are attractive in the field of medicine because thermal damage to cells is negligible if the plasma parameters and experimental setup are appropriately tuned. However, in order to fully understand the complex interactions of plasmas with biomolecules such as DNA, amino acids and proteins, which are the building blocks of cells and organs, it is essential to perform fundamental studies at the molecular or atomic level. In particular, it is important to understand plasma interactions with organic material through the plasma/liquid interface, because in many cases the active species interact with biomolecules through water. In the present study, the interaction of an atmospheric pressure He plasma with a methylene blue (MB) aqueous solution was investigated in order to determine the degradation effects of plasma exposure on such organic molecules.

2. Experimental

As shown in Figure 1, the atmospheric pressure He plasma was produced using a micro-hollow cathode plasma source, which operates at a low DC voltage of a few hundred volts, making it easy and efficient to apply to solutions. In the present experiment, a voltage of 800 V was used. He gas was supplied at a flow rate of 400 sccm. The plasma was used to irradiate a 10 mg/l aqueous solution of MB. The distance between the plasma source and the surface of the solution was 1 mm.

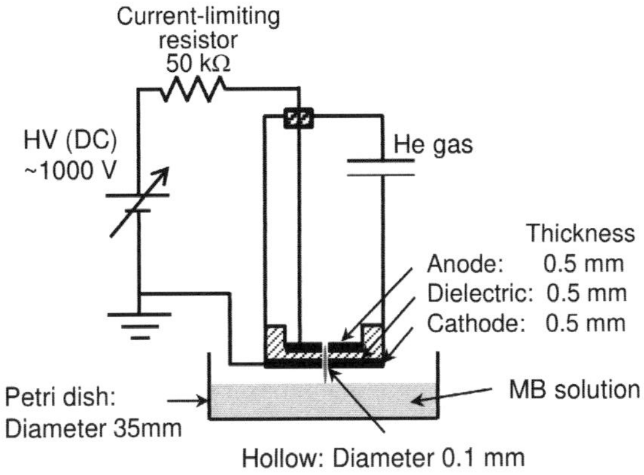

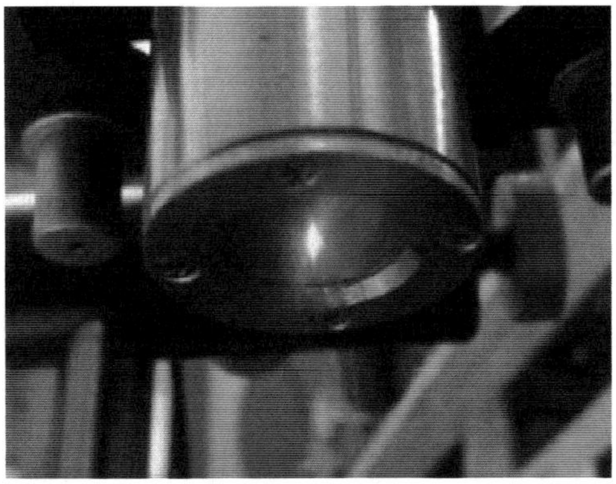

Figure 1. Schematic diagram and photograph of micro-hollow cathode plasma

Optical emission spectroscopy (OES) was carried out on the plasma using a high-resolution fiber-optic spectrometer (HR-4000, Ocean Optics). The absorbance of the MB solution was measured by using an ultraviolet-visible (UV-vis) spectrophotometer (UV-2450, Shimadzu). The pH of the aqueous solutions was measured using a pH meter (HM-25G, DKK-TOA Corporation) with pH electrodes (ELP-036, DKK-TOA Corporation).

3. Results

Figure 2 shows a typical OES spectrum for the atmospheric pressure He plasma. In addition to strong He emission lines, emissions due to O and OH radicals formed by dissociation of water are seen. Furthermore, a broad peak attributed to NOx is observed from 400 nm to 700 nm. The hydroxyl radical (OH) has a higher oxidation reduction potential (ORP = 2.85 V) than ozone (2.07 V), hydrogen peroxide (1.78 V) or hypochlorous acid (1.63 V), and is environmentally friendly because its lifetime is only around 10^{-6} sec [12,13]. Therefore, the use of an atmospheric pressure

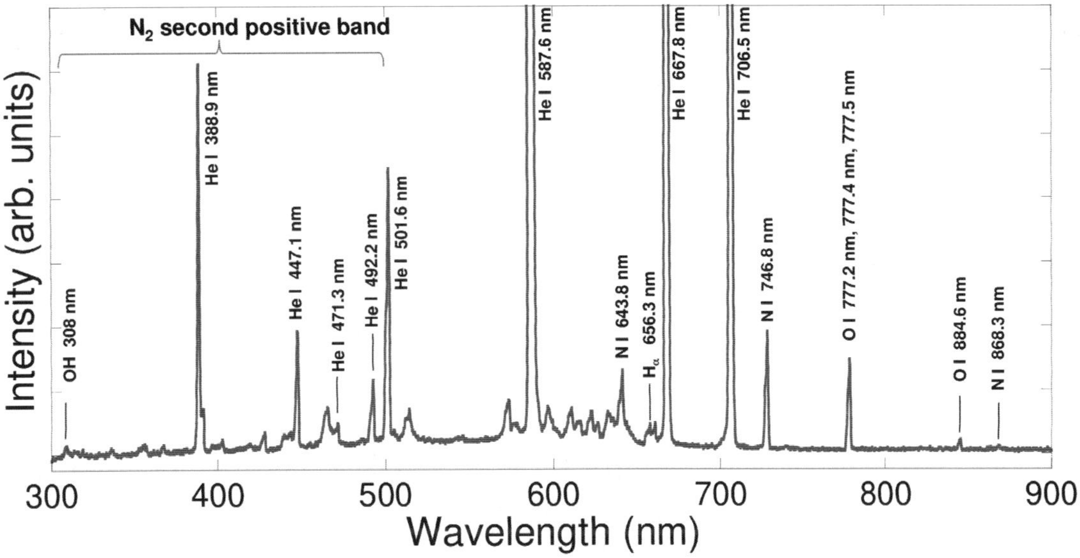

Figure 2. Optical emission spectra of atmospheric pressure He plasma.

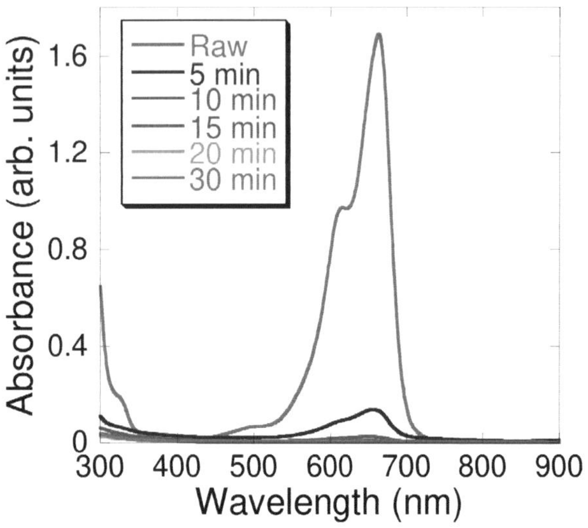

Figure 3. Time evolution of absorbance spectrum of MB solution during plasma treatment.

He plasma is a promising method for fluid sterilization, removal of discoloration, and oxidative dissolution.

Figure 3 shows the time evolution of the UV-Vis absorbance spectrum of the MB aqueous solution during exposure to the plasma. The absorbance peak intensity, which reflects the concentration of MB molecules, decreased with increasing plasma exposure time. Following plasma exposure for 10 min, the solution became completely colorless. Figure 4 shows the change in the estimated MB concentration with plasma exposure time. A large reduction is seen during the first 5 min.

One of the most notable chemical effects that plasmas have on solutions is acidification. To investigate whether this has some influence on the reduction of the MB concentration, deionized water was first exposed to the same plasma for varying lengths of time, and its pH was measured. Figure 5 shows the change in pH with plasma exposure time. The initial pH was about 6.4, and it decreased during the first 10 min of exposure and then became almost constant. The reason for the reduction in pH is considered to be the formation of nitric acid (HNO_3) by dissolution of the NO generated in the plasma,

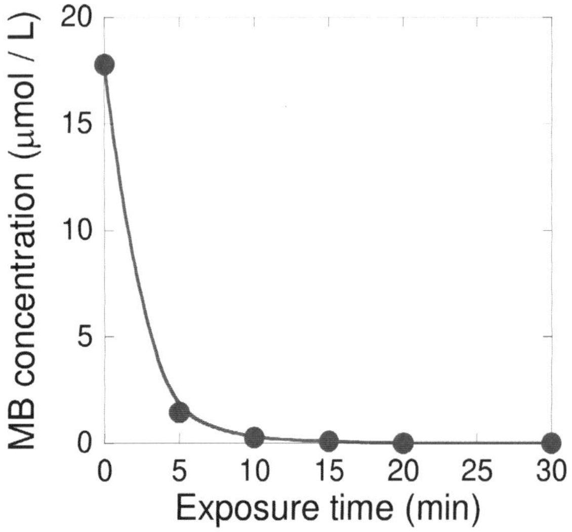

Figure 4. Variation of concentration of MB solution with plasma exposure time.

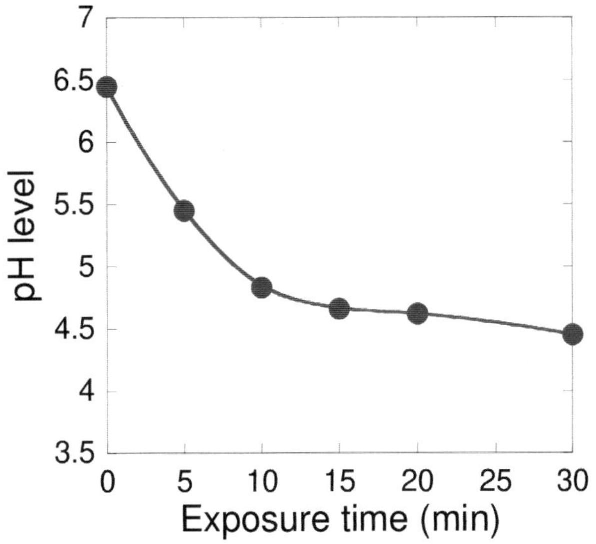

Figure 5. Variation of pH level of solution irradiated with atmospheric pressure He plasma.

as shown in the OES spectrum in Figure 2 [15]. To investigate whether the formation of HNO_3 had some influence on the degradation of the MB, aqueous MB solutions with pH values of 2.4 and 5.6 were prepared and their absorbance spectrum was measured (not shown here). It was found that the solution pH had no effect on MB degradation. Therefore, it can be concluded that the decrease in the MB concentration during plasma exposure is due to radicals entering the solution from the plasma.

4. Summary

An investigation has been carried out into the effect of an atmospheric pressure He plasma on the degradation of an aqueous solution of methylene blue (MB). In the optical emission spectrum of the plasma, in addition to strong He emission peaks, emissions due to O and OH radicals produced by dissociation of water were present. When the MB solution was exposed to the plasma, UV-Vis spectroscopy determined that the MB concentration decreased with exposure time, with the solution becoming completely colorless after 10 min. To determine whether this was the effect of changing pH, the absorbance was measured for MB solutions with different pH values. However, the pH was found to have no influence on MB degradation. The results indicated that MB degradation occurred due to interactions between MB and radicals across the plasma/liquid interface.

Acknowledgments

This work was partly supported by The Grant-in-Aid for Scientific Research on Innovative Areas "Plasma Medical Innovation" (24108003) from The Ministry of Education, Culture, Sports, Science and Technology (MEXT) and The Grant-in-Aid for Young Scientists (B) (24760597) from the Japan Society for the Promotion of Science (JSPS).

References

[1] Moisan M, Barbeau J, Moreau S, Pelletier J, Tabrizian M and Yahia L H 2001 *Int. J. Pharm.*,**226**, 1

[2] Laroussi M 2005 *Plasma Process. Polym.*, **2** 391

[3] Boudam M K, Moisan M, Saoudi B, Popovici C, Gherardi N and Massines F 2006 *J. Phys. D: Appl. Phys.*, **29** 3494

[4] Weltmann K -D, Brandenburg R, Woedtke T, Ehlbeck J, Foest R, Stieber M and Kindel E 2008 *J. Phys D: Appl. Phys.*, **41** 194008

[5] Raiser J and Zenker M 2006 *J. Phys. D: Appl. Phys.*, **39** 3520

[6] Stalder K R and Woloszko J 2007 *Contrib. Plasma Phys.,* **47** 64
[7] Yonson S, Coulombe S, Lveill V, Leask R L 2006 *J. Phys. D: Appl. Phys.,* **39** 3508
[8] Coulombe S, Lveill V, Yonson S and Leask R L 2006 *Pure Appl. Chem.,* **78** 1147
[9] Stoffels E, 2007 *Contrib. Plasma Phys.*, **47** 40
[10] Fridman G, Friedman G, Gutsol A, Shekhter A B, Vasilets V N and Fridman A 2008 *Plasma Process. Polym.,* **5** 503
[11] Morfill G E, Kong M G and Zimmermann J L 2009 *New J. Phys.,* **11** 115011
[12] Blanco J, Malato S, Fernández-Ibañez P, Alarcón D, Gernjak W and Maldonado M I 2009 *Renewable and Sustainable Energy Rev.*, **13** 1437
[13] Tochikubo F, Uchida S and Watanabe T 2004 *Jpn. J. Appl. Phys.,* **43** 315
[14] Bruggeman P, Ribel E, Maslani A, Degroote J, Malesevic A, Rego R, Vierendeels J, and Leys C 2008 *Plasma Source. Sci. Technol.,* **17** 025012
[15] Moussa D, Abdelmalek F, Benstaali B, Addou A, Hnatiuc E, and Brisset J L 2005 *EPJ Appl. Phys.,* **29** 189

26th International Symposium on Plasma Sciences for Materials (SPSM26) IOP Publishing
Journal of Physics: Conference Series **518** (2014) 012020 doi:10.1088/1742-6596/518/1/012020

Formation of carbon nanoparticle using Ar+CH_4 high pressure nanosecond discharges

K Koga[1, 2], X Dong[1], S Iwashita[2], U Czarnetzki[2] and M Shiratani[1]

[1] Graduate School of Information Science and Electrical Engineering, Kyushu University, 744 Motooka, Nishi-ku, Fukuoka 819-0395, Japan
[2] Institute for Experimental Physics V, Ruhr-University Bochum, Germany

E-mail: koga@ed.kyushu-u.ac.jp

Abstract. We have studied formation of carbon nanoparticles using Ar+CH_4 high pressure nanosecond discharge non-thermal plasmas. Transition pressure from uniform glow discharges to filamentary ones was clarified to obtain conditions under which uniform glow discharges are sustained. We have produced nanoparticles using the glow discharges, and then we have collected nanoparticles on the grounded electrode by the filtered vacuum collection method. Size distribution analysis reveals that the CH_4 concentration is an important parameter in controlling nanoparticle growth. We have also studied film deposition on the powered electrode and the grounded electrode. The deposition rate on the powered electrode is 7 times higher than that on the grounded electrode. Optical emission observations suggest that radical generation rate near the powered electrode is twice higher than that near the grounded electrode, leading to high deposition rate on the powered electrode.

1. Introduction

Nanoparticles attract a great deal of attention because of their size-dependent properties that differ significantly from properties of bulk materials. Their applications expand into a large number of fields such as environmental, chemical and biomedical engineering [1-4]. So far various kinds of the nanoparticle synthesis methods have been developed [5]. Solution process is one of the succeeded methods but nanoparticles are easily agglomerated in solution without using dispersants, resulting in difficulties of their size control. Moreover, unwanted surface modification such as oxidation occurs with ease in solution. Plasma is a practical method to overcome these problems. Because nanoparticles are charged negatively in plasmas, they tend to grow without their agglomeration. In addition, surface modification of nanoparticles can be controlled by using plasmas. For instance, surface modification of Si nanoparticles has been realized using twin plasmas [6]. Formation kinetics of nanoparticles in low pressure non-thermal plasmas and in high pressure thermal plasmas and their applications have been extensively studied [7-26], while nanoparticle synthesis using high pressure non-thermal plasmas has been attracted relatively little attention. For the high pressure thermal plasmas such as arc plasmas, a large amount of radicals are produced by high density plasmas at high gas pressure and hence productivity of nanoparticles is high. Such high pressure discharges lead to high gas temperature which is one of the major drawbacks for applications to direct nanoparticle deposition to substrates of low melting materials such as organic ones [27]. Although the low pressure non-thermal plasmas are sustained at low gas temperature, they offer rather low productivity of nanoparticles. To overcome difficulties for the low pressure plasma and high pressure thermal plasmas, high pressure non-thermal

Journal of Physics: Conference Series **518** (2014) 012020 doi:10.1088/1742-6596/518/1/012020

plasmas are promising because they can generate high density radicals at high pressure and low gas temperature [28, 29].

In this study, we have studied nanoparticle formation and film deposition using Ar+CH_4 high pressure nanosecond discharges to develop a large area process for a high yield nanoparticle synthesis, because nanosecond discharges can generate high pressure non-thermal plasmas.

2. Experimental

Experiments were performed using a nanosecond discharge reactor made of stainless steel. The inner diameter and length of the reactor were 58 mm and 77 mm, respectively. Two pieces of stainless steel electrode of 15 mm in diameter were placed at the center of the reactor as shown in figure 1 [30, 31]. The gap between the powered electrode and the grounded electrode was 0.2 or 0.5 mm. Ar diluted CH_4 was employed as ingredient gas, because nanoparticle growth in low pressure CH_4 plasmas is quite slow due low gas phase reaction rates of radicals in the plasmas [32] and hence explosive nanoparticle formation in high pressure non-thermal plasmas is expected to be suppressed by the combination of CH_4 and nanosecond discharges. Ar gas dilution allows to generate discharges at higher pressure compared to pure CH_4. The CH_4 concentration was 0.5-100 %. The total gas flow rate was 100 sccm. The total gas pressure was set at 6300 Pa for nanoparticle synthesis. Short voltage pulses were generated using a dc power source equipped with a high speed switch (Behlke HTS-81) and supplied to the upper electrode to generate discharges. Figure 2 shows the typical waveform of discharge voltage and discharge current. The maximum peak discharge voltage was -670 V and the maximum discharge current was -8.8 A. The discharge duration was 280 ns. The repetition frequency was 5 kHz. Although the electrodes were covered by carbon films during CH_4 containing discharges, the films had little influence on the shape and peak value of the waveform of discharge voltage and current. We employed nanosecond discharges, because longer duration discharges such as μs discharges tend to be filamentary discharges due to increase of conductivity, and current through the discharge and therefore eventual filamentation.

Nanoparticles deposited on the electrodes were collected by the filtered vacuum collection method, which has been widely employed for dust particle collection in nuclear fusion reactors [33-36]. For the

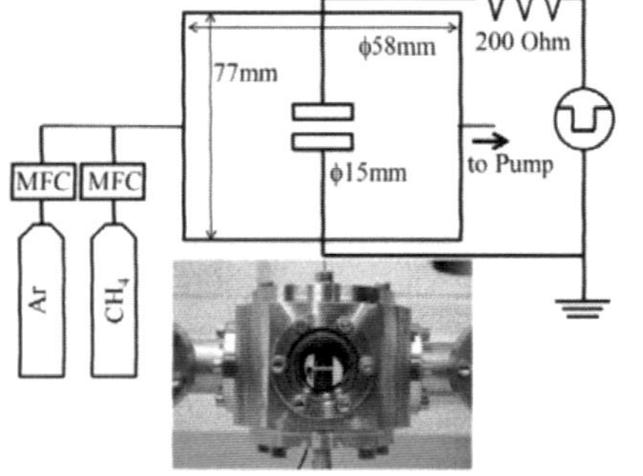

Figure 1. Schematic of nanosecond discharge reactor. MFC denotes mass flow controller.

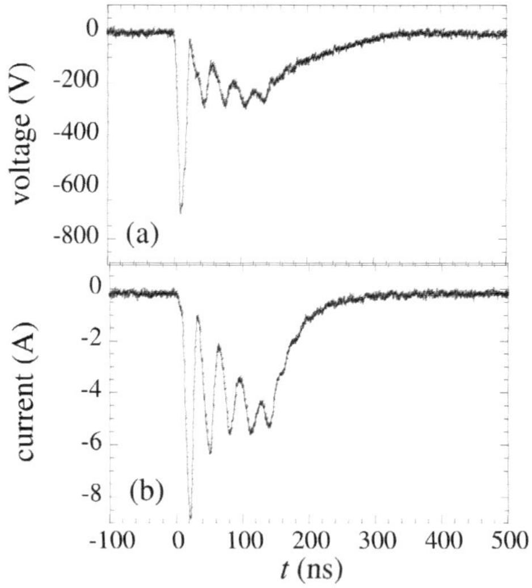

Figure 2. Typical waveform of (a) discharge voltage and (b) discharge current. Conditions: CH_4 (5%) + Ar (95%), 100 sccm, 6300 Pa.

method, a polycarbonate membrane filter (ADVANTEC K010A) with many holes of 100 nm in diameter was put into a filter holder. The filter holder was set at 0.5 mm above the grounded electrode, then particles on the electrode were sucked for 10 minutes with a rotary pump. The size and shape of the nanoparticles collected on the membrane filter were obtained with a scanning electron microscope (Carl Zeiss Leo 1530). The lower detection limit of size for the SEM measurements was 40 nm.

3. Results and discussion

Our nanosecond discharges are classified into two kinds: uniform glow discharges as shown in figure 3(a), and filamentary discharges as shown in figure 3(b). Uniform glow discharges are generated at lower pressure, whereas filamentary discharges are generated at higher pressure. This is because the larger expansion of the plasma channels and their overlapping tend to take place at the higher pressure. The filamentary discharges are streamers [37]. Glow discharges should be sustained for nanoparticle synthesis to realize production of size controlled nanoparticles in large area. We have studied pressure of transition from glow discharges to filamentary ones to obtain conditions under which the glow discharges are continuously generated. For the measurements, we employed two kinds of electrode surface: metal surface and carbon film surface. Figure 4 shows CH_4 concentration dependence of the transition pressure at which the discharges change from glow discharges to filamentary ones. For carbon film surface electrodes, uniform glow nanosecond discharges are sustained below 1800 Pa in pure CH_4 and below 10000 Pa in pure Ar. Uniform glow discharges tend to be sustained at the lower pressure. They change into filamentary discharges above a transition pressure. The transition pressure increases with decreasing the CH_4 concentration. The transition pressure for the metal surface is higher than that for the carbon film surface. The secondary electron emission coefficient of carbon films deposited on the stainless steel electrodes is 40 % of that of stainless steel [38] and this difference is not important for the discharge mode change. Non-uniformity of electron emission from electrode surface may bring about non-uniformity of current density, namely, filamentary discharges. Carbon films deposited on the electrode surface have low thermal conductivity, low electrical conductivity, and rough microstructure of the surface compared to the bare metal electrode surface. Low thermal

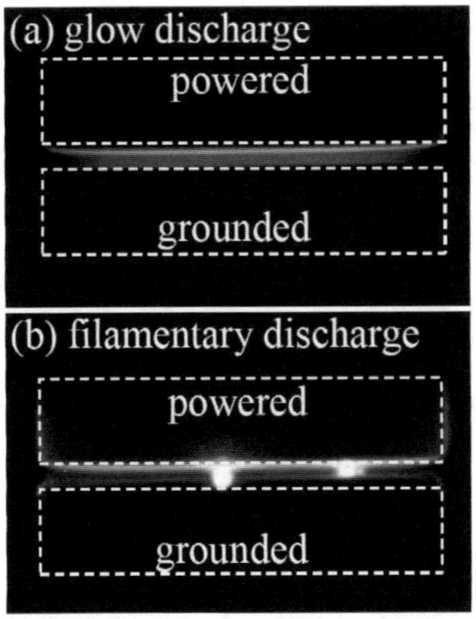

Figure 3. Photo images of (a) glow discharge and (b) filamentary discharge. Conditions: electrode gap 0.5 mm, CH_4 concentration 5 %, shutter speed 1/100 s.

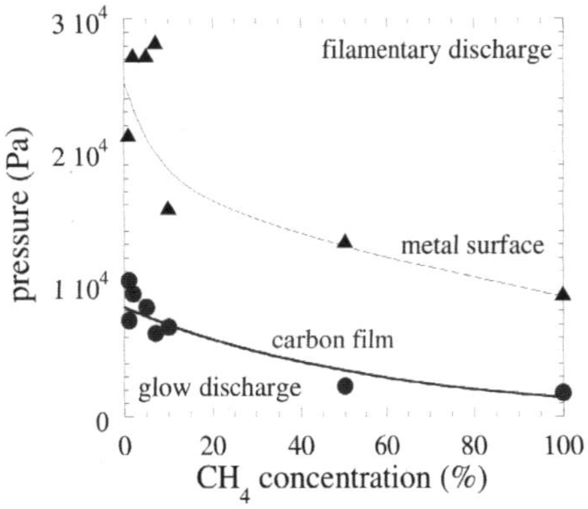

Figure 4. CH_4 concentration dpendence of pressure of transition from glow discharges to filamentary discharges. Solid circles and triangles indicate the transition pressure below which glow sicharges are sustained for carbon film surface electrodes and metal surface ones, respectively. Electrode gap is 0.2 mm.

conductivity tends to enhance localized thermionic emission, low conductivity tends to lead to localized accumulation of charge on the surface, and rough microstructure tends to enhance localized field electron emission. Due to these tendencies, the transition pressure for the metal surface is higher than that for the carbon film surface.

Based on the results mentioned above, we generated uniform glow discharges under the CH_4 concentration of 1% and the total pressure of 6300 Pa. Carbon films deposited on the powered electrode are much thicker than that on the grounded electrode. The deposition rate on the powered electrode and the grounded electrode are 0.751 nm/s and 0.107 nm/s, respectively. Since the repetition frequency of the discharges is 5 kHz, the deposition rate per each discharge on the powered electrode and the grounded electrode are 1.50×10^{-4} nm/shot and 2.15×10^{-5} nm/shot, respectively. As shown in figure 3(a), the optical emission intensity near the powered electrode is about twice of that near the grounded electrode, indicating higher generation rate of radicals near the powered electrode, leading to the higher deposition rate on the powered electrode. Based on the assumptions that (1) deposited films are hydrogenated amorphous carbon whose mass density is 1.4 g/cm^3 [39] and (2) all carbon containing radicals in gas phase deposit either on the powered electrode or on the grounded electrode, we deduce the lower limit density of carbon containing radicals in gas phase from the deposition rate. The lower limit density is 6.0×10^{13} cm^{-3}, corresponding to 0.40 % of the initial CH_4 concentration. Gas residence time between two electrodes is 1.9×10^{-2} s, whereas the characteristic time of diffusion toward the electrodes is 2.1×10^{-5} s. Therefore, the diffusion is the dominant transport process of carbon containing radicals for the film deposition.

We collected nanoparticles accumulated on the grounded electrode by the filtered vacuum collection method. The nanoparticles collected on the membrene filter were observed with SEM as

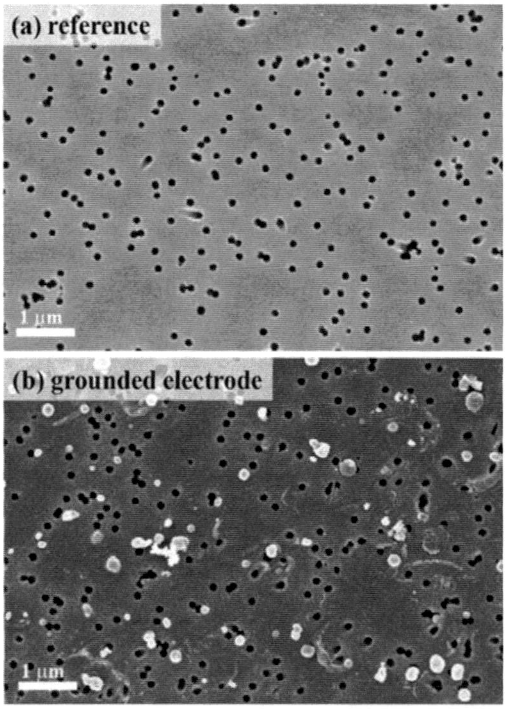

Figure 5. SEM images of (a) membrane filter and (b) nanoparticles collected on the filter. Scale bar indicates 1 μm. Conditions: CH_4 concentration 5%, total flow rate 100 sccm, total pressure 6300 Pa, frequency 5 kHz, and discharge time 30 minutes.

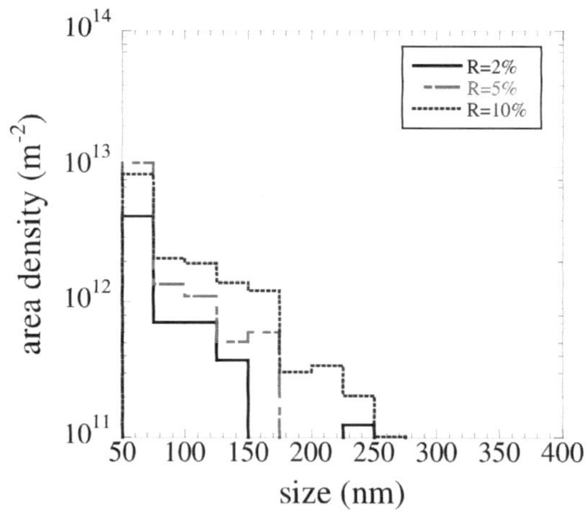

Figure 6. Size distribution of collected nanoparticles as a parameter of CH_4 concentration R.

shown in figure 5(b). Collected nanoparticles are classified into three kinds: spherical nanoparticles, agglomerated nanoparticles and flakes. The areal density ratio for there three types is 175:3:4, respectively. Flakes are generated by peeling off deposited films due to internal stress of the films. Although the detail formation kinetics of spherical nanoparticles and agglomerated ones is unclear at present, they are probably generated and grow due to gas phase reactions. Figure 6 shows size distribution of collected nanoparticles as a parameter of CH_4 concentration R. For R=2 %, most nanoparticles are below 150 nm in size. The area density of nanoparticles and their size dispersion tend to be higher and larger for the higher CH_4 concentration. These results indicate that CH_4 concentration is one of the key parameters in controlling nanoparticle growth.

4. Conclusions

We studied formation of carbon nanoparticles by Ar+CH_4 high pressure nanosecond discharge non-thermal plasmas. Following conclusions are obtained in this study.

(1) For carbon film surface electrodes, uniform glow nanosecond discharges are sustained below 1800 Pa in pure CH_4 and below 10000 Pa in pure Ar. They change into filamentary discharges above a transition pressure. The transition pressure increases with decreasing the CH_4 concentration.

(2) The deposition rates on the powered electrode and the ground electrode center are 0.751 nm/s (1.50×10^{-4} nm/shot) and 0.107 nm/s (2.15×10^{-5} nm/shot), respectively for the CH_4 concentration of 1% and the pressure of 6300 Pa. The spatial profile of optical emission intensity suggests that radical generation rate near the powered electrode is much higher than that near the grounded electrode, leading to the unbalance of deposition rate on these two electrodes.

(3) Nanoparticles are produced by the nanosecond discharges. For the CH_4 concentration of 2%, most nanoparticles are below 150 nm in size. The area density of nanoparticles and their size dispersion tend to be higher and larger for the higher CH_4 concentration. These results indicate that CH_4 concentration is one of the key parameters in controlling nanoparticle growth.

Acknowledgements

This work was partly supported by Kyushu Industrial Technology Center, and the Grant-in-Aid for Scientific Research on Innovative Areas, MEXT KAKENHI Grant Numbers 21110001 and 21110005.

References

[1] Joachim C, Gimzewski J K and Airam A 2000 *Nature* **408** 541
[2] Majumder D D, Banerjee R, Ulrichs C, Mewis I and Goswami A 2007 *IETE Tech. Rev.* **24** 9
[3] Rutkevych P P, Ostrikov K and Xu S 2007 *Phys. Plasmas* **14** 043502
[4] Gill V, Guduru P R and Sheldon B W 2008 *Int. J. Solids Structures* **45** 943
[5] Koch C C 2002 *Nanostructured Materials* (Norwich: Noyes Publications)
[6] Uchida G, Yamamoto K, Kawashima Y, Sato M, Nakahara K, Kamataki K, Itagaki N, Koga K, Kondo M and Shiratani M 2011 *Phys. Status Solidi C* **8** 3017
[7] Bouchoule A, Plain A, Boufendi L, Blondeau J Ph and Laure C, 1991 *J. Appl. Phys.* **70** 1991
[8] Boufendi L, Plain A, Blondeau J Ph, Bouchoule A, Laure C and Toogood M 1992 *Appl. Phys. Lett.* **60** 169
[9] Watanabe Y, Shiratani M, Kubo Y, Ogawa I and Ogi S 1998 *Appl. Phys. Lett.* **53** 1263
[10] Shiratani M, Maeda S, Koga K and Watanabe Y, 2000 *Jpn. J. Appl. Phys.* **39** 287
[11] Koga K, Matsuoka Y, Tanaka K, Shiratani M and Watanabe Y 2000 *Appl. Phys. Lett.* **77** 196
[12] Watanabe Y, Shiratani M and Koga K 2001 *Plasma Sources Sci. Technol.* **11** A229
[13] Koga K, Kai M, Shiratani M, Watanabe Y and Shikatani N 2002 *Jpn. J. Appl. Phys.* **41** L168
[14] Sharpe J P, Rohde V, Sagara A, Suzuki H, Komori A and Motojima O 2003 *J. Nucl. Mater.* **313** 455
[15] Nunomura S, Koga K, Shiratani M, Watanabe Y, Morisada Y, Matsuki N and Ikeda S 2005 *Jpn. J. Appl. Phys.* **44** L1509

[16] Kakeya T, Koga K, Shiratani M, Watanabe Y and Kondo M 2006 *Thin Solid Films* **506-507** 288
[17] Gonzalez-Aguilar J, Moreno M and Fulcheri 2007 *J. Phys. D* **40** 2361
[18] Shiratani M, Koga K, Iwashita S and Nunomura S 2008 *Faraday Discuss.* **137** 127
[19] Koga K, Iwashita S, Kiridoshi S, Shiratani M, Ashikawa N, Nishimura K, Sagara A and Komori A 2009 *Plasma Fusion Res.* **4** 34
[20] Shiratani M, Koga K, Iwashita S, Uchida G, Itagaki N and Kamataki K 2011 *J. Phys. D* **44** 174038
[21] Kamataki K, Miyata H, Koga K, Uchida G, Itagaki N and Shiratani M 2011 *Appl. Phys. Express* **4** 105001
[22] Kamataki K, Morita Y, Shiratani M, Koga K, Uchida G and Itagaki N 2012 *J. Inst.* **7** C04017
[23] Kamataki K, Koga K, Uchida G, Itagaki N, Yamashita D, Matsuzaki H and Shiratani M 2012 *Thin Solid Films* **523** 76
[24] Kim Y, Hatozaki K, Hashimoto Y, Uchida G, Kamataki K, Itagaki N, Seo H, Koga K and Shiratani M 2013 *Jpn. J. Appl. Phys.* **52** 11NA07
[25] Shiratani M, Koga K, Kamataki K, Iwashita S, Uchida G, Seo H and Itagaki N 2014 *Jpn. J. Appl. Phys.* **53** 010201
[26] Shiratani M, Morita Y, Kamataki K, Seo H, Uchida G, Itagaki N and Koga K 2014 *J. Phys. Conf. Series* in press
[27] Alexandrov S E and Hitchman M L 2005 *Chem. Vap. Deposition* **11** 457
[28] Napartovich A P 2001 *Plasmas and Polymers* **6** 1
[29] Fridman A, Chirokov A and Gustol A 2005 *J. Phys. D: Appl. Phys.* **38** R1
[30] Mueller S, Luggenhoelscher D and Czarnetzki U 2011 *J. Phys. D: Appl. Phys.* **44** 165202
[31] Donko Z, Schulze J, Muller S and Czarnetzki U 2011 *Appl. Phys. Lett.* **98** 251502
[32] Tachibana K, Nishida M, Harima H and Urano Y 1984 *J. Phys. D: Appl. Phys.* **17** 1727
[33] Sharpe J P, Sagara A, Rohde V, the ASDEX-upgrade experiment team, Suzuki H and Komori A 2003 *J. Nucl. Mater.* **313-316** 455
[34] Sharpe J P, Masaki K, Sagara A, Humrickhouse P W, Skinner C H, the NSTX team, Tanabe T, Miya N and the JT-60U team 2005 *J. Nucl. Mater.* **337-339** 1000
[35] Koga K, Iwashita S, Kiridoshi S, Shiratani M, Ashikawa N, Nishimura K, Sagara A, Komori A and the LHD experimental group 2009 *J. Plasma Fusion Res.* **4** 34
[36] Iwashita S, Nishiyama K, Uchida G, Seo H, Itagaki H, Koga K and Shiratani M 2013 *Fusion Eng. and Design* **88** 28
[37] Kogelshatz U 2002 IEEE Trans. Plasma Sci. **30** 1400.
[38] Ruzic D, Moore R, Manos D, and Cohen S, 1982 *J. Vac. Sci. Technol.* **20** 1313
[39] Robertson J 2002 *Mater. Sci & Eng.* **37** 129

Journal of Physics: Conference Series **518** (2014) 012021 doi:10.1088/1742-6596/518/1/012021

Time-resolved optical emission spectroscopy on three-dimensionally integrated micro solution plasma in He/H_2O mixture

Y Himeno, Y Ogura and T Shirafuji

Department of Physical Electronics and Informatics, Osaka City University, 3-3-138 Sugimoto, Sumiyoshi-ku, Osaka 558-8585, Japan

E-mail: m12tb0s46@ex.media.osaka-cu.ac.jp

Abstract. We have performed time-averaged and time-resolved optical emission spectroscopy (OES) on three-dimensionally integrated micro solution plasma (3D IMSP) in He/H_2O mixture. The results of time-resolved OES on 3D IMSP in He/H_2O mixture have shown that duration of the optical emission of OH($A^2\Sigma^+ \rightarrow X^2\Pi$ around 309 nm) is shorter than that in Ar/H_2O mixture. Possible causes of this difference are discussed by means of difference in the energy of metastable states of Ar and He.

1. Introduction

Nonthermal atmospheric pressure plasmas in contact with liquid is widely studied for variety of plasma application, such as nano-material synthesis, surface modification, water treatment, sterilization, recycling of rare materials, and decomposition of toxic compound [1-5]. Various plasma generation methods are used for these purposes, which include glow discharge [6], dielectric barrier discharge [7-9], radio-frequency discharge [10, 11] and microwave discharge [12, 13].

In our previous works, we have developed three-dimensionally integrated micro solution plasma (3D IMSP), in which a large amount of tiny plasma (microplasma [14]) is generated in a porous dielectric material filled with a gas/liquid mixed medium. 3D IMSP can be applied to gold nanoparticle synthesis [15, 16] and to decomposition of organic substances in water [17, 18]. Moreover, 3D IMSP have shown superior performance in the case of decomposition of methylene blue molecules in comparison to conventional solution plasma [19 - 21].

Recently, we have reported the results of time-resolved optical emission spectroscopy (OES) on 3D IMSP in Ar/H_2O mixture [22], in which we have observed that relatively high optical emission intensity of OH($A^2\Sigma^+ \rightarrow X^2\Pi$ around 309 nm) is kept even after the main peak pulse discharge has finished. In this paper, we report the results of time-resolved OES on 3D IMSP in He/H_2O mixture in order to compare to that in Ar/H_2O mixture.

2. Experimental set up

Figure 1 shows schematic illustration of the experimental set up, which is the same to that in our previous on 3D IMSP in Ar/H_2O mixture [22]. Table 1 is the experimental conditions for time-resolved OES in this and previous work. These conditions are the same to that in our previous report, expect that the feed gas is He [22]. The details are described in elsewhere [22]. Here after, 3D IMSP in

Ar/H_2O mixture and that in He/H_2O mixture are referred to as simply Ar/H_2O and He/H_2O, respectively.

3. Results and discussion

3.1 Time-averaged wide-range OES

Before starting the time-resolved OES, we performed OES in the wide range of 280 to 800 nm in order to see which emissive species can be detected in the case of He/H_2O. Figure 2 shows the result. A series of atomic He emission lines (e.g. 388, 501, 587, 667, 706 nm etc.) is observed, because He is used for generating gas bubbles in porous dielectric material. Furthermore, OH ($A^2\Sigma^+ \rightarrow X^2\Pi$ around 309 nm), Hβ ($4d \rightarrow 2p$ at 486 nm), Hα ($3d \rightarrow 2p$ at 656 nm) and O ($3p^5\mathrm{P} \rightarrow 3s^5S^0$ at 777 nm) emission lines are observed. Hereafter, OH ($A^2\Sigma^+$) and OH ($X^2\Pi$) are represented as OH (A) and OH (X), respectively.

A series of peaks observed at 315, 337, 353, 357, 375, 380 nm etc. corresponds to the emission of second positive-system band (SPB; $C^3\Pi \rightarrow B^3\Pi$) of N_2. One of the causes of the N_2 SPB is the discharge of air, in the isolated pores that is incorporated in the porous dielectric material during its fabrication. Another is the parasitic air discharge between outer surface of the glass tube and grounded electrode surrounding it. Since these discharges are isolated from liquid medium, they do not contribute to treatment of liquid.

In comparison to our previous report on the wide-range OES in case of Ar/H_2O [22], the result of He/H_2O exhibit stronger optical emission intensity of Hα relative to that of OH(A). This can be

Table 1. Experimental conditions for time-resolved OES on 3D IMSP in He/H_2O mixture (this work) and Ar/H_2O mixture (previous work).

Liquid medium	Deionized Water
Electrical conductivity (μS/cm)	0.77
pH	5.94
Volume of liquid (mL)	150
Temperature (°C)	25
Voltage amplitude (kV)	5
Frequency (kHz)	20
Pulse width (μs)	2.0
He or Ar flow rate (L/min)	1.1

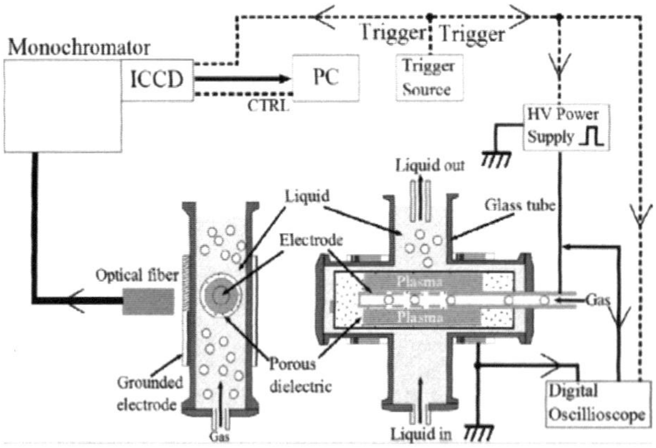

Figure 1. Schematic illustration of the experimental set up for time-resolved optical emission spectroscopy on 3D IMSP.

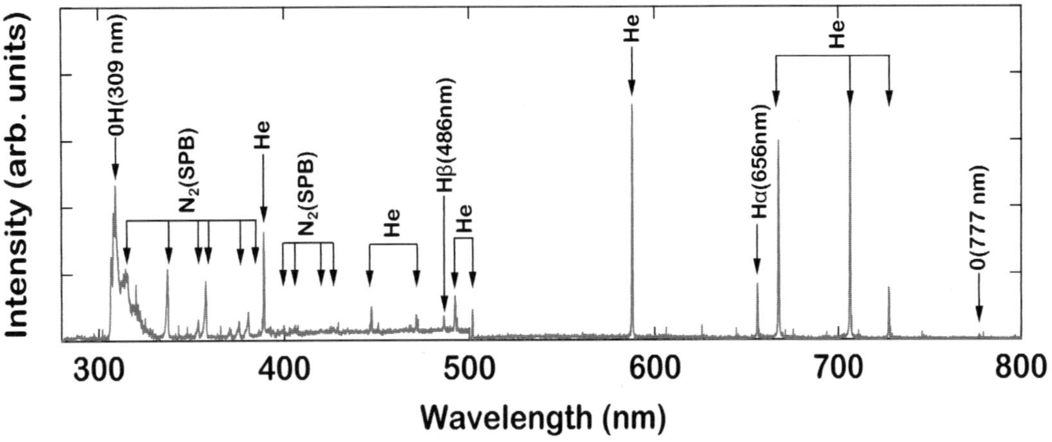

Figure 2. Wide-range OES performed on 3D IMSP in He/H_2O mixture.

explained by difference in energy level of metastable states of Ar and He. In order to emit Hα lines, H atoms should be once excited to the energy level of 3*d* (12.1 eV) or 4*d* (12.7 eV). The energy 19.8 eV of metastable state of He is high enough to excite H atoms to 3*d* or 4*d* levels, while the energy 11.5 eV of metastable of Ar is not high enough.

3.2 Time-resolved OES

After OH(A) emission has been confirmed in wide-range OES, we have investigated temporal behavior of OH(A) because it is believed to be the important reactant in water purification processes [4, 23]. Figure 3 (a) shows time-resolved OES spectra in the range of 300-325 nm in the case of He/H_2O, which correspond to OH(A) emission band and are acquired after triggering pulse voltage of the power source. In the case of Ar/H_2O, OH(A) emission spectra are observed for 500 ns after triggering the discharge [22]. In case of He/H_2O, on the other hand, OH(A) emission spectra are not observed after 300 ns. In order to discuss in details, temporal variation of OH(A) intensity is compared the other parameters of voltage, current and emission intensity of He (388 nm).

Figure 3-(b) shows temporal variation of voltage, current, emission intensity of He (388 nm) and OH(A) (309 nm). After a trigger signal is send to the power source, voltage increases up to 5 kV and current peak is observed during voltage rise time during 0 to 200 ns. After that, weak current is observed during 200 to 500 ns. Intensity of optical emission of He is observed only during the voltage rise time during 0 to 200 ns, and is not observed during 200 to 500 ns. These characteristics are different from those in the case of Ar/H_2O mixture [22].

Regarding the intensity of optical emission of OH(A), the main peak is observed during 0 to 200 ns, and weak emission of OH(A) is observed during 200 to 300 ns. These characteristics are slightly different from those in the case of Ar/H_2O [22]. The first difference is the fact that intensity of OH(A) after the main peak in the case of He/H_2O is relatively lower than that in the case of Ar/H_2O. The second difference is the fact that duration (200-300 ns) of the weak emission is shorter than that (200-500 ns) in the case of Ar/H_2O.

These differences are considered to be attributed to the differences in the contribution of metastable state of Ar($^3P_0$ and $^3P_2$, (Ar_m, hereafter)) and He(2^3S and 2^1S, (He_m, hereafter). In the plasma containing H_2O vapor with Ar or He, OH(A) can be generated mainly through following reactions [24 - 26];

$$H_2O + e(> 9.1\text{ eV}) \rightarrow H(n = 1) + OH(A) + e \quad (1)$$
$$H_2O + Ar_m \rightarrow OH(A) + H + Ar \quad (2)$$

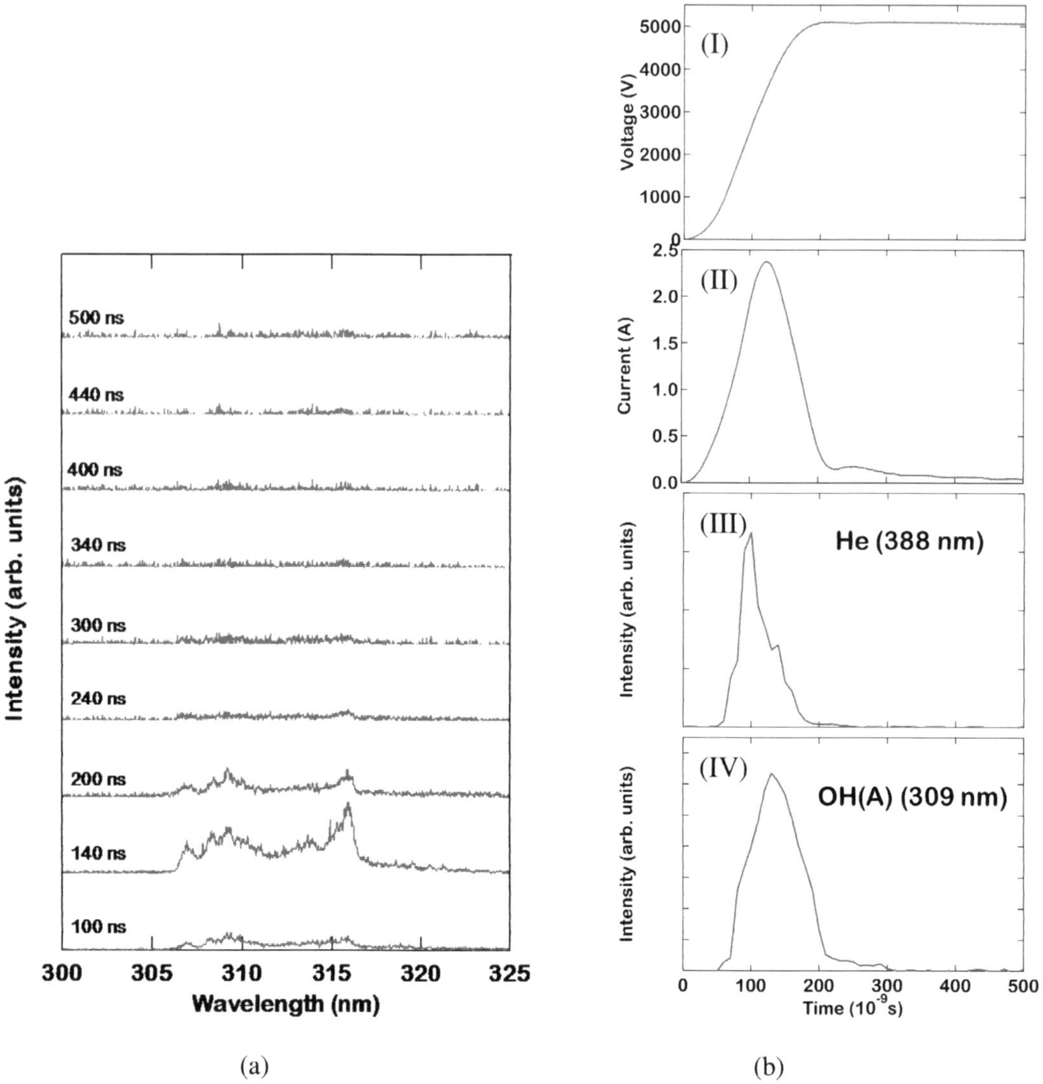

Figure 3. (a)Time-resolved OES for the wavelength region of 300-325 nm, which corresponds to OH emission band ($A^2\Sigma^+ \rightarrow X^2\Pi$). (b) Temporal variation of (I) voltage (II) current and optical emission intensity of (III) He (388 nm), (IV) OH (A) (309 nm).

In both cases of Ar/H_2O and He/H_2O, the OH(A) peak observed during 0 to 200 ns is considered to be caused mainly by the reaction (1) because energetic electrons exist during this period, which can be confirmed from the fact that the peak discharge current is also observed at the same time.

Regarding the time after 200 ns, energetic electrons does not exist, but low energy electrons and ions are considered to exist because weak current and Ar or He emission is observed. In this period, contribution of electron impact dissociation (1) is regarded as a minor reaction path to produce OH(A). Instead of reaction (1), following reaction may be expected [26, 27].

$$H_2O + e(> 5.1\ eV) \rightarrow OH(X) + H(n=1) + e \qquad (3)$$

And metastable Ar_m assisted H_2O dissociation (2) may be dominant.

On the other hand, in the case of He/H_2O mixture, contribution of electron impact dissociation (1) and (3) is similar in Ar/H_2O mixture. But, unlike Ar/H_2O mixture, metastable He_m assisted H_2O dissociation is not expected. Instead, following penning ionization reaction is expected because energy of He_m (approximately 20 eV) is sufficient for ionizing H_2O molecules [28];

$$H_2O + He_m \rightarrow H_2O^+ + He \tag{4}$$

Thus, one of the cause of relatively lower OH(A) intensity after 200 ns in the case of He/H_2O than in the case of Ar/H_2O is considered to be due to the difference in the effects of metastable excited He and Ar. Although dissociative recombination reactions of H_2O^+ and/or H_3O^+ ions with electrons to produce OH(A) are also known to occur [4, 24, 29, 30], especially when electron energy is low as in the case of the period after 200 ns in this study, this will be discussed elsewhere using numerical simulation.

4. Conclusion

Time-averaged and time-resolved OES have been performed on the 3D IMSP in He/H_2O in order to compare to the OES results on Ar/H_2O. From the result of wide-range OES on 3D IMSP in He/H_2O mixture, optical emission of Hα is stronger than that in Ar/H_2O mixture. This is due to difference in contribution of metastable Ar_m (11.5 eV) and He_m (19.8 eV). Because of this difference, He_m contributes to generate Hα line, while Ar_m does not.

The results of time-resolved OES on 3D IMSP in He/H_2O has also been different from those in Ar/H_2O. Duration of optical emission of OH(A) in the case of He/H_2O is shorter than that in the case of Ar/H_2O. Intensity of OH(A) emission in the case of He/H_2O is lower than that in the case of Ar/H_2O. The cause of these differences can be attributed to the difference in the reactions of Ar_m and He_m with H_2O, in which Ar_m can directly produce OH(A) while He_m can produce H_2O^+ instead of OH(A). These qualitative conclusion will be verified quantitatively by using numerical simulation of reactions including the other OH(A) production reaction paths.

Acknowledgement

This work has been partly supported by the Grant-in-Aid for Scientific Research on Innovative Areas "Frontier Science of Interaction between Plasmas and Nano-Interfaces" from the Ministry of Education, Culture, Sports, Science and Technology, Japan.

References

[1] Parvulescu V I, Magureanu M and Lukes (Eds.) P 2012 *Plasma Chemistry and Catalysis in Gases and Liquids* (Wiley-VCH, Weinheim, Germany)
[2] Mededovic S and Locke B R 2007 *Ind. Eng. Chem.* Res. **46** 2702
[3] Locke B R, Sato M, Sunka P, Hoffmann M R and Chang J S 2006 *Ind. Eng. Chem. Res.* **45** 882
[4] Bruggemen P and Leys C 2009 *J. Phys. D: Appl. Phys.* **42** 053001
[5] Malik M A 2010 *Plasma Chem. Plasma Process.* **30** 21
[6] Bruggeman P, Cunge G and Sadeghi N 2012 *Plasma Sources Sci. Technol.* **21** 035019
[7] Hibert C, Gaurand I, Motret O and Pouvesle J M 1999 *J. Appl. Phys.* **85** 7070
[8] Liu Z W, Yang X F, Zhu A M, Zhao G L and Xu Y 2008 *Eur. Phys. J. D.* **48** 365
[9] Dilecce G, Ambrico P F, Simek M and De Benedictis S 2012 *Chem. Phys.* **398** 142
[10] Srivastava N and Wang C 2011 *J. Appl. Phys.* **110** 053304
[11] Maehara T, Toyota H, Kuramoto M, Iwamae A, Tadokoro A, Mukasa S, Yamashita H , Kawashima A and Nomura S 2006 *Jpn. J. Appl. Phys.* **45** 8864
[12] Ishijima T, Hotta H, Sugai H and Sato M 2007 *Appl. Phys. Lett.* **91** 121501
[13] Ishijima T, Sugiura H, Saito R, Toyoda H and Sugai H 2010 *Plasma Sources Sci. Technol.* **19** 015010

Journal of Physics: Conference Series **518** (2014) 012021 doi:10.1088/1742-6596/518/1/012021

[14] Tachibana K 2006 *IEEJ Trans. Electr. Electron. Eng.* **1** 145

[15] Shirafuji T, Saito N, Takai O 2012: *12th European Plasma Conf. 24-29 June 2012, Bologna, Italy*

[16] Shirafuji T, Ueda J, Nakamura A, Cho S-P , Saito N and Takai O 2013 *Jpn. J. Appl. Phys.* accepted.

[17] Shirafuji T, Himeno Y, Saito N and Takai O 2013 *J. Photopolym. Sci. Technol.* **26** 507

[18] Shirafuji T and Himeno Y 2013 *Jpn. J. Appl. Phys.* **52** 11NE03

[19] Hieda J, Saito N and Takai O 2008 *J. Vac. Sci. Technol.* A **26** 854

[20] Hieda J, Shirafuji T, Noguchi Y, Saito N and Takai O 2009 *J. Jpn. Inst. Metals.* **73** 938 [in Japanese]

[21] Saito N, Hieda J and Takai O 2009 *Thin Solid Films* **518** 912

[22] Shirafuji T, Ogura Y and Himeno Y 2013 *Jpn. J. Appl. Phys.* to be publishied.

[23] Matsui Y, Takeuchi N, Sasaki K, Hayashi R and Yasuoka K 2011 *Plasma Sources Sci. Technol.* **20** 034015

[24] Leblond J B, Collier F, Hoffbeck F and Cottin P 1981 *J. Chem. Phys.* **74** 6242

[25] Hibert C, Gaurand I, Motret O and Pouvesle J M 1999 *J. Appl. Phys.* **85** 7070

[26] Beenakker C I M, de Heer F J, Krop H B and Mohlmann G R 1974 *Chem. Phys.* **6** 445

[27] Becker K, Stumpf B, Schulz G 1980 *Chem. Phys. Lett.* **73** 102

[28] Arfa M B, Coz G L, Sinou G, Nadan A L, Tuffin F and Tannous C 1994 *J. Phys. B: At. Mol. Opt. Phys* **27** 2541

[29] Mul P M, McGowan J W, Defrance P and Mitchell J B A 1983 *J. Phys. B: At. Mol. Phys.* **16** 3099.

[30] Sonnenfroh D M and Caledonia G E 1993 *J. Chem. Phys.* **98** 2872.

Effects of the electron irradiation energy on synthesis of gold nanoparticles using gas-liquid interfacial plasma

S Takahashi and T Kaneko

Department of Electronic Engineering, Tohoku University, 6-6-05 Aoba, Aramaki, Aoba-ku, Sendai 980-8579, Japan

E-mail: takahashi11@ecei.tohoku.ac.jp

Abstract. Structure controlled gold nanoparticles (AuNPs) are synthesized at the surface of the ionic liquid by the plasma irradiation under the strong magnetic field. The AuNPs are easily synthesized in the region where the edge of the plasma is irradiated, while hardly synthesized in the region where the core plasma is irradiated. In the plasma edge region, high energy electrons collide with neutral gas and the diffused low energy electrons generate the AuNPs by reducing Au ions of the gold chloride. On the other hand, in the core plasma region, the irradiated high energy electrons can dissociate the ionic liquid and generate the new compounds with oxidation effect, which inhibit the synthesis of the AuNPs.

1. Introduction

Recently, highly-ordered periodic structures of metal nanoparticles have attracted much attention due to their high catalytic activity [1], unique photosensitive reactivity [2], and so on. Especially, the synthesis of various kinds of nanoparticles using the plasma-liquid interfaces [3-6] is advantageous in that a reducing agent is the plasma itself, and then, the synthesis is continuous during the plasma irradiation. To realize the periodic structure of the nanoparticles, we adopt a novel plasma technique combined with introduction of ionic liquids [7-9] under strong magnetic fields up to tesla. Since the plasma generated under the strong magnetic field keeps its structure due to confinement along the magnetic field lines, the plasma structure can be transcribed to the liquid surface, resulting in the synthesis of the structured nanoparticles at the gas-liquid interface when the plasma reduces the metal chlorides in the liquid.

When the mesh anode is used to generate the plasma, the shadow region is made which have the shape of the wire of the mesh. Surprisingly, the nanoparticles are synthesized selectively in the shadow region [10]. However, the mechanism of the nanoparticle synthesis in the shadow region has not clarified.

In this paper, we focus on the effects of the electron irradiation energy to clarify the synthesis mechanism of the structure controlled nanoparticles. For that purpose, the difference of the gold nanoparticles (AuNPs) synthesis is measured by changing the discharge voltage (V_{DC}) which can control the electron irradiation energy. In addition, the relationship between the electron irradiation energy and the dissociation of the ionic liquid to generate the oxidizing agent is discussed.

2. Experimental setup

Figure 1 shows the schematic of an experimental setup for the discharge plasma in contact with the ionic liquid [11] containing a gold chloride ($HAuCl_4$), where the nitrogen plasma is generated between a disk cathode (ϕ15 mm in diameter) and a mesh anode (ϕ20 mm in diameter), and is irradiated to the ionic liquid on the teflon plate (30 mm × 30 mm). Figure 2 shows the picture of the teflon plate for explaining the positional relationships between the plasma and the ionic liquid. The ionic liquid is placed in small holes (1) - (5) formed on the teflon plate at a distance of 45 mm from the anode electrode. When the strong magnetic field is applied along the machine axis, the generated plasma is strongly magnetized. The AuNPs are synthesized in the ionic liquid by the plasma reduction of $HAuCl_4$.

The experimental conditions are as follow: gas pressure P_{gas}=30 Pa, gas flow is 1.6 sccm, magnetic field B=0.1 T.

A kind of the ionic liquid is N, N, N – Trimethyl – N – propylammonium Bis (trifluoromethanesulfonyl) imide, and compositional formula is $C_8H_{16}F_6N_2O_4S_2$. The concentration of the $HAuCl_4$ is 20 mg/ml.

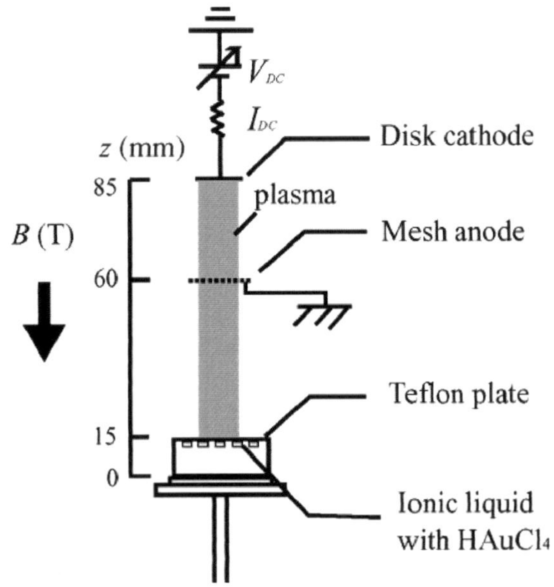

Figure 1. Schematic of the experimental setup.

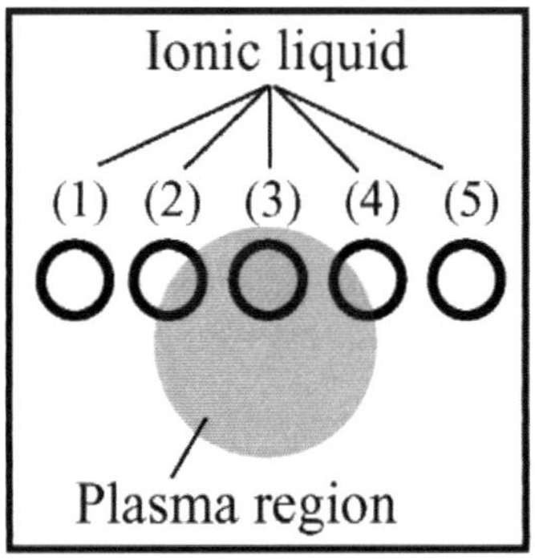

Figure 2. The picture of the teflon plate for explaining the positional relationships between the plasma and the ionic liquid.

3. Results and Discussion

Figure 3 shows the pictures of the AuNPs which are formed at the gas-liquid interface under the condition that the discharge voltage and irradiation time are (a) V_{DC} = 0.9 kV, t_i = 20 s and (b) V_{DC} = 0.9 kV, t_i = 40 s, respectively, where the small holes on the teflon plate are not prepared. The AuNPs are easily synthesized in the region where the edge of the plasma is irradiated. However, the synthesized AuNPs quickly diffuse radially, and therefore the exact synthesized region cannot be identified. To investigate the relationship between the region where the AuNPs are easily synthesized and the region where the edge of the plasma is irradiated in detail, the ionic liquid is separately placed in each hole as shown in Figure 2.

Figure 4 shows the pictures of the ionic liquid, to which the plasma is irradiated under the condition that the discharge voltage and the irradiation time are (a) V_{DC} = 0.8 kV, t_i = 120 s and (b) V_{DC} = 1.1 kV, t_i = 120 s. A tiny amount of the ionic liquid is sputtered and dissociated by the plasma irradiation. The region where the blown materials are deposited on the teflon plate indicates the plasma irradiation region. In both cases, the AuNPs are synthesized in the holes (2), (4) where the sharp density and potential gradient are generated. However, at the hole (3), synthesis of the AuNPs depends on V_{DC}, namely, the AuNPs are synthesized for V_{DC} = 0.8 kV, while the AuNPs are not synthesized for V_{DC} = 1.1 kV.

The difference in the AuNPs synthesis at the hole (3) is attributed to the difference in the electron irradiation energy. For V_{DC} = 0.8 kV, the low energy electrons are irradiated to the ionic liquid, and generate the AuNPs by reducing Au^{3+} of the gold chloride. For V_{DC} = 1.1 kV, on the other hand, the electron irradiation energy is higher than that for V_{DC} = 0.8 kV. The high energy electrons can dissociate the ionic liquid and generate the new compounds which are considered to affect the synthesis of the AuNPs. Here, optical emission spectra in the plasma region shows the H_α peak derived from H atom, which is dissociated from the ionic liquid.

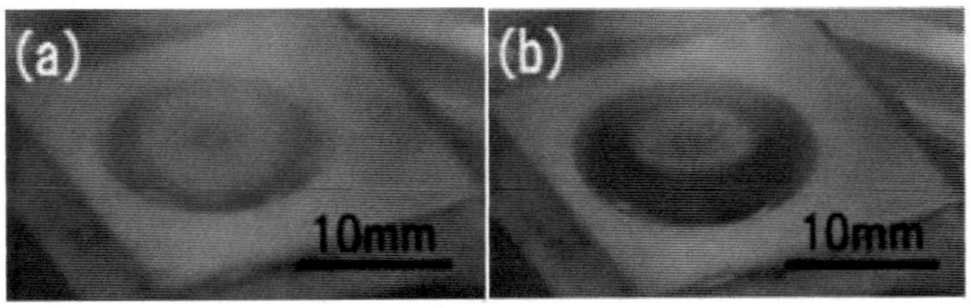

Figure 3. The pictures of the AuNPs synthesis which are formed at the gas-liquid interface under the condition (a) V_{DC} = 0.9 kV, t_i = 20 s and (b) V_{DC} = 0.9 kV, t_i = 40 s.

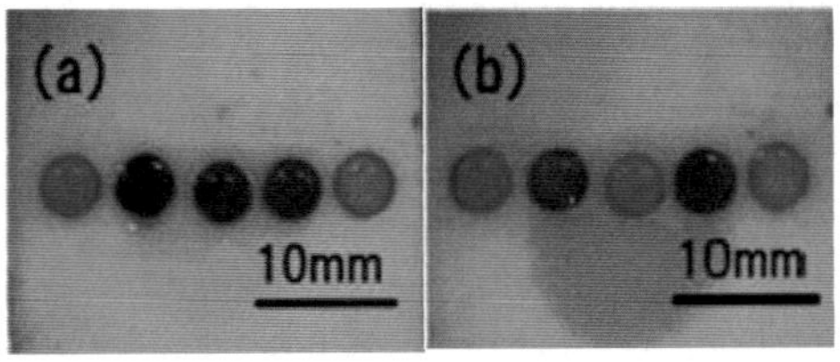

Figure 4. The pictures of the ionic liquid, to which the plasma is irradiated, (a) V_{DC} = 0.8 kV, t_i = 120 s and (b) V_{DC} = 1.1 kV, t_i = 120 s.

Figure 5 shows the Fourier transform infrared spectrophotometer (FT-IR) spectra of the new compounds in the ionic liquid for V_{DC}=0.7 -1.1 kV and t_i=120 s. The samples are made by the plasma irradiation to the ionic liquid on the glass plate as shown in Figure 3. The peaks around 3250 and 3350 cm^{-1} gradually increase by increasing V_{DC}. The peaks are identified to be N-H bond, and therefore, the candidate of the new compounds consisting of N-H bond and derived from the used ionic liquid is determined to be Trifluoromethanesulfonamide (TFMS).

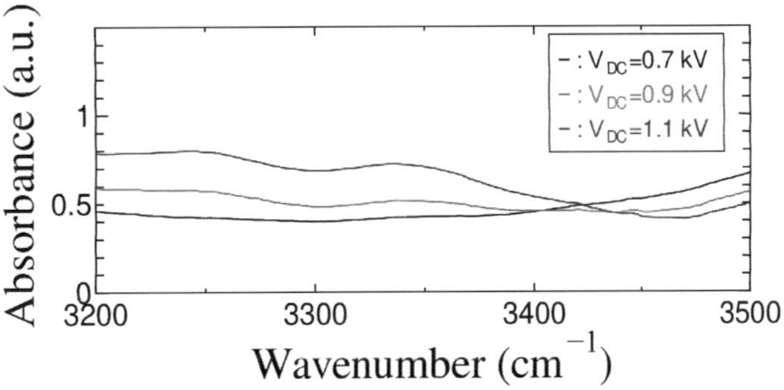

Figure 5. The FT-IR spectra of the new compounds.

Figure 6 shows the relative production rate of AuNPs and TFMS as a function of the discharge voltage V_{DC} which corresponds to the electron irradiation energy. Here, the production rates of AuNPs and TFMS are evaluated by the peaks of UV-Vis spectra (550 nm) and FT-IR spectra (3350 cm^{-1}), respectively. As the electron irradiation energy becomes higher, the production rate of TFMS monotonically increases. By contrast, the production rate of AuNPs is decreased when the production rate of TFMS is over a threshold. Therefore, the production of TFMS caused by the high energy electron irradiation affects the AuNPs synthesis.

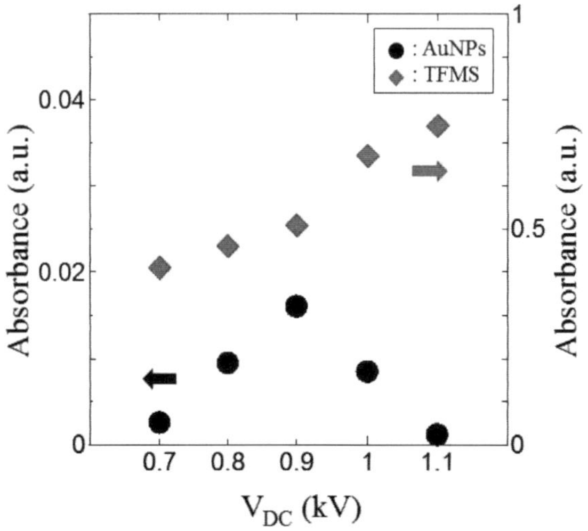

Figure 6. The relative production rate of AuNPs and TFMS as a function of the discharge voltage.

Figure 7 shows the structures of (a) the ionic liquid used in this experiment and (b) TFMS. TFMS could be generated by dissociating the N-S bond of the ionic liquid and combining with H, and the acidity of TFMS is reported to be very strong. Hence, the TFMS shows the oxidative effect and inhibits the AuNPs synthesis when the electron irradiation energy is high. In the edge region, on the other hand, even high energy electrons collide with neutral gas and become low energy electron. Therefore, the AuNPs can be synthesized in the plasma edge region even for V_{DC}=1.1 kV.

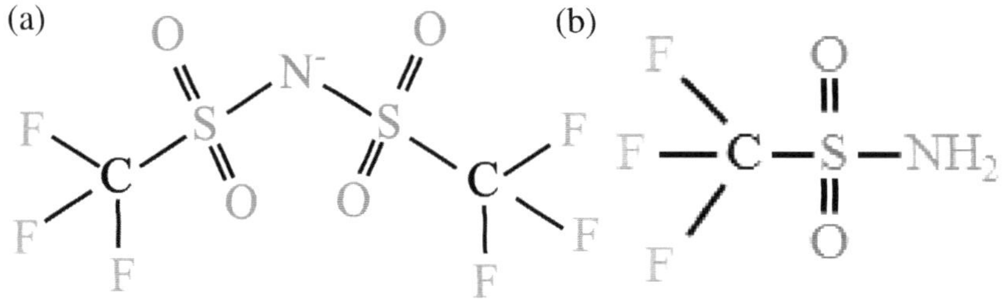

Figure 7. The structures of (a) the ionic liquid and (b) TFMS.

Figure 8 shows the plasma-irradiated ionic liquid (V_{DC}=0.8 kV) (a) without additional TFMS and (b) with additional TFMS. The concentration of additional TFMS is TFMS/IL=50 mg/ml and the plasma irradiation time is 120s. In the case of the ionic liquid with additional TFMS, it is found that AuNPs are not synthesized. It indicates that TFMS has the inhibitory effect on the AuNPs synthesis.

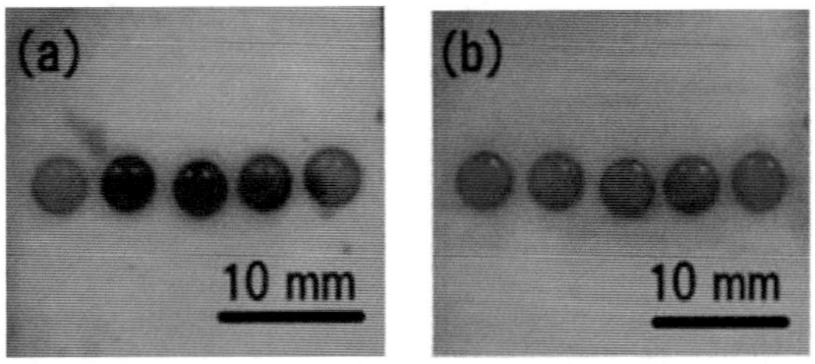

Figure 8. The plasma-irradiated ionic liquid (V_{DC}=0.8 kV) (a) without additional TFMS and (b) with additional TFMS.

4. Conclusion

We focus on the effects of the electron irradiation energy to clarify the synthesis mechanism of the structure controlled nanoparticles. We clarify the low energy electrons generate the AuNPs by reducing Au^{3+} of the gold chloride, however, the high energy electrons can dissociate the ionic liquid and generate the new compounds which are considered to affect the synthesis of the AuNPs. The new compounds are determined to be TFMS, whose acidity is reported to be very strong. Hence, the TFMS shows the oxidative effect and inhibits the AuNPs synthesis when the electron irradiation energy is high.

These results are the first step of elucidating the mechanism of AuNPs synthesis. Furthermore, using this mechanism, we can freely form the patterned AuNP structure, which is expected to be applied to the biosensors and the bioimaging devices using a principle of the surface plasmon resonance of the AuNPs.

References

[1] G. G. Wildgoose, C. E. Banks and R. G. Compton 2006 *Small* **2** 182
[2] J. R. Kreen 2003 *Nature Mater.* **2** 210
[3] T. Kaneko, K. Baba, and R. Hatakeyama 2009 *J. Appl. Phys.* **105** 103306
[4] T. Kaneko, Q. Chen, T. Harada, and R. Hatakeyama 2011 *Plasma Sources Sci. Technol.* **20** 034014
[5] J. Hieda, N. Saito and O. Takai 2008 *J. Vac. Sci. Technol. A* **26** 854
[6] I. G. Koo, M. S. Lee, J. H. Shim, J. H. Ahn and W. M. Lee 2005 *J. Mater. Chem.* **15** 4125
[7] K. BaBa, T. Kaneko and R. Hatakeyama 2007 *Appl. Phys. Lett.* **90** 201501
[8] T. Kaneko, K. BaBa, T. Harada and R. Hatakeyama 2009 *Plasma Process. Polym.* **6** 713
[9] K. BaBa, T. Kaneko and R. Hatakeyama 2009 *Appl. Phys. Express* **2** 035006
[10] T. Kaneko, S. Takahashi, and R. Hatakeyama 2012 *Plasma Phys. Control. Fusion* **54** 124027
[11] K.R. Seddon 2003 *Nature Mater.* **2** 363

Preparation of Si nanoparticles by laser ablation in liquid and their application as photovoltaic material in quantum dot sensitized solar cell

P Chewchinda[1], K Hayashi[1], D Ichida[2], H Seo[2], G Uchida[2], M Shiratani[2], O Odawara[1] and H Wada[1]

[1] Interdisciplinary Graduate School of Science and Engineering, Tokyo Institute of Technology, 4259 Nagatsuta, Midori-ku, Yokohama 226-8502 Japan
[2] Department of Electronics, Faculty of Information Science and Electrical Engineering, Kyushu University, 744 Motooka, Nishi-ku, Fukuoka 819-0395 Japan

E-mail: chewchinda.p.aa@m.titech.ac.jp

Abstract. Quantum dot sensitized solar cell was fabricated using silicon nanoparticles prepared by laser ablation in liquid. The obtained particles are spherical and have an average size of 6 nm. These particles were utilized as photosensitizing material in solar cell with polysulfide as an electrolyte. The conversion efficiency of solar cell with silicon nanoparticles was 5.3 times higher than the one with only TiO_2 particles. Electrical impedance spectroscopy also revealed the decrease in resistance in the former case led to higher current density than the latter one.

1. Introduction

Solar cell has attracted much attention from researchers for decades and several models had been proposed. For conventional solar cell, single material was used to convert the entire range of solar spectrum. As a result, there exists Shockley-Queisser limit, which was calculated based on the assumption that each photon resulted in only one electron-hole pair. Based on this assumption, the maximum conversion efficiency for silicon solar cell is limited to only 30% [1]. To overcome this limitation, different concepts have been proposed including intermediate band solar cell, dye-sensitized solar cell (DSSC), multi-junction solar cell, and hot carriers solar cell [2-8].

After the discovery of multiple exciton generation process (MEG) in PbSe quantum dot, the concept of quantum dot sensitized solar cell (QDSSC) has become another possible prospect for improving conversion efficiency [9-10]. This process starts with the absorption of photon with energy at least twice of the energy gap, generating an exciton. Then the excess energy, normally lost into heat, will be utilized to produce other excitons. Thus, one-photon absorption results in two or more electron-hole pairs. If these additional charge carriers are harvested before they recombine, it is possible to enhance the efficiency of photovoltaic device [11]. Later, MEG was proved to occur in several kinds of material including silicon [12-13].

Silicon nanoparticles can be synthesized by diverse techniques such as microemulsion route, chemical vapour deposition, and aerosol synthesis. Unfortunately, these fabrication processes are complicated and sometimes it is difficult to control the purity of the product [14-16]. Recently, laser ablation in liquid technique was proposed. This process is very simple and the purity of the obtained

Journal of Physics: Conference Series **518** (2014) 012023 doi:10.1088/1742-6596/518/1/012023

product is outstanding [17-19]. It also reported that with this technique, silicon nanoparticles could be prepared without oxidation [20].

In this study, silicon nanoparticles were prepared by laser ablation in liquid. After that, these particles are utilized further to fabricate QDSSC. The device characteristics were analyzed and studied.

2. Experiment

2.1. Silicon nanoparticles preparation
P-type <100> silicon wafer (t = 600 μm) was used as a target material immersed in 40 mL of 99.5% ethanol. This experiment was conducted in the closed system to prevent the evaporation of solvent. Magnetic stirrer was also used. Nd:YAG laser at second harmonic generation (λ = 532 nm) with pulse duration of 13 ns and repetition rate of 10 Hz was employed, and silicon target was irradiated for 30 minutes. Laser power was 4.18 W/pulse, which corresponded to energy density of 0.44 J/pulse·cm^2. After irradiation, the colloidal solution was dropped onto carbon-coated copper grid and dried. Finally, the size and morphology of prepared silicon nanoparticles were observed by Transmission Electron Microscope (H-8100 TEM) with acceleration voltage of 200 kV.

2.2. Fabrication of quantum dot solar cell
Ethanol in the colloidal solution was evaporated in order to obtain silicon nanoparticles powder. FTO glasses were used as photo and counter electrodes. For photo electrode, FTO glass was coated with TiO_2 paste and sintered at 450 °C for 30 minutes. This TiO_2-coated electrode was again coated with a mixture of silicon nanoparticles and TiO_2 paste, and sintered for another 30 minutes. For counter electrode, FTO glass was sputtered with gold. Finally, both photo and counter electrodes were assembled together and the gap between two electrodes was filled with polysulfide electrolyte. Another device without silicon nanoparticles was also fabricated as a reference. Their properties were analyzed by solar simulator and electrochemical impedance spectroscopy (EIS).

3. Results and discussion

3.1. Size and morphology of silicon nanoparticles
Both size and morphology of silicon nanoparticles were observed by TEM. figure 1 shows that silicon nanoparticles prepared by laser ablation in ethanol possess spherical shape. Although some agglomeration was observed, most particles well dispersed. From TEM images, the histogram of nanoparticles size was constructed by measuring the diameter of 100 nanoparticles. The result was shown in figure 2 The average size obtained was around 6 nm with standard deviation of 2.68 nm. This obtained particle size was smaller than Bohr diameter of silicon (8 nm) [21-23]. Therefore, quantum confinement effect and also MEG can occur.

Figure 1. TEM of Si nanoparticles.

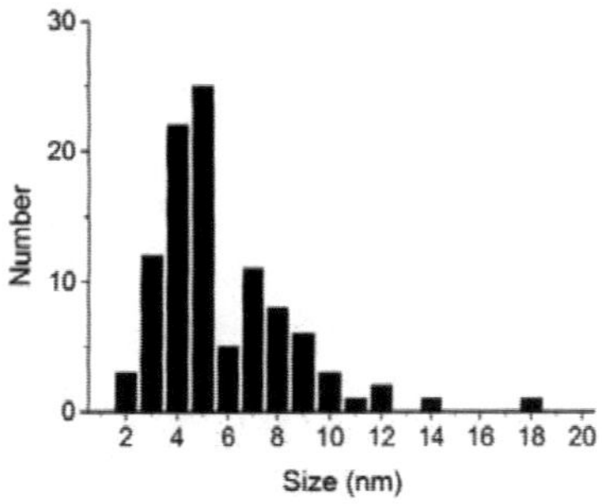

Figure 2. Size distribution obtained from TEM image.

3.2. Photovoltaic properties

Figure 3 shows J-V characteristic of quantum dot sensitized solar cell with silicon nanoparticles along with the one without them. The conversion efficiency of solar cell with silicon nanoparticles was improved up to 0.016% compared to that of the cell with only TiO_2 particles (0.003%). The current density and open circuit voltage of the cell with silicon nanoparticles were also found to be two times higher than that of the device without them. From these results, an improvement of the device performance by the addition of silicon nanoparticles was confirmed. Other electrical parameters are demonstrated in Table 1.

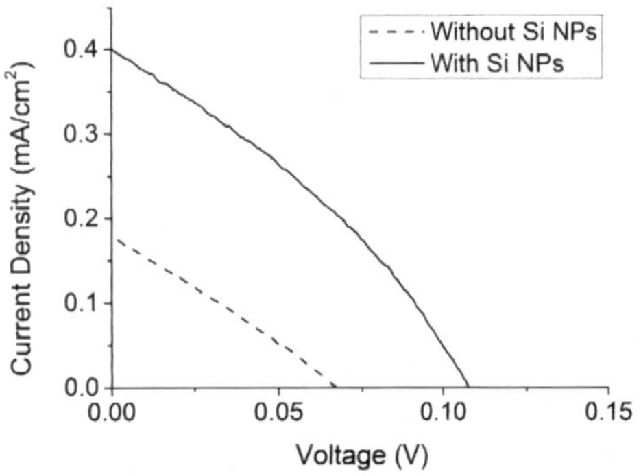

Figure 3. J-V characteristic of solar cell with (solid line) and without silicon nanoparticles (dashed line).

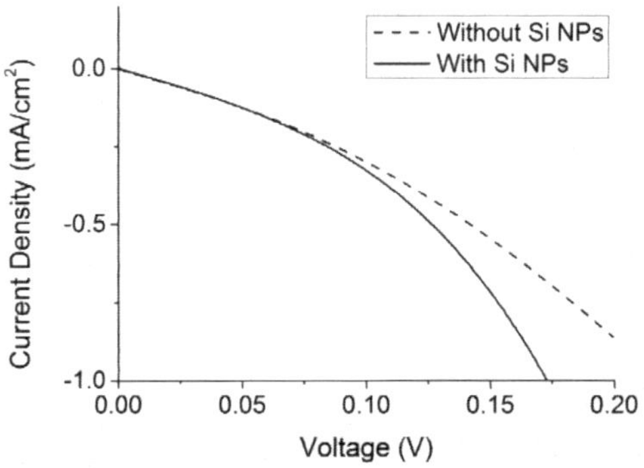

Figure 4. J-V characteristic of dark current with (solid line) and without silicon nanoparticles (dashed line).

Table 1. Electrical parameters of fabricated solar cell.

Sample	V_{oc} (V)	I_{sc} (mA)	J_{sc} (mA/cm^2)	P_{max}	FF	Eff. (%)
Without Si NPs	0.067	0.044	0.178	0.001	0.274	0.003
With Si NPs	0.114	0.104	0.418	0.004	0.330	0.016

Dark current is another important parameter for solar cell. It represents the electron recombination, which in turn, affects the device performance. Ideally, this dark current should be as small as possible for the device to achieve the highest current density. As shown in figure 4, the onset of dark current shifted to lower bias for the cell with silicon nanoparticles. This implies an increase in dark current density [24]. Although this unwanted current was enhanced in this case, the total current density were still higher than that of the other one. Since the total current density composed of photo-generated current and dark current, the increase of dark current density might be so small that the total current was dominated by the photo-generated term. As a consequence, the addition of silicon nanoparticles can still increase the total current density of this device.

Electrical impedance spectroscopy (EIS) was used to further study on the device performance. The result is shown in figure 5. The intercept at high frequency region corresponds to series resistance including that of electrolyte and conducting electrode. The semi-circle at high frequency range represents charge transfer resistance at counter electrode/electrolyte interface, while the one at low frequency range corresponds to charge transfer resistance by ions in electrolyte [24-27]. Since the electrolyte and counter electrode are the same, the values of series resistance (R_s) in both devices were almost unchanged [24, 27]. However, the resistances at low and middle frequency were decreased in the cell with silicon nanoparticles. This reduction led to higher current density in this device. EIS parameters are listed in Table 2 as additional information.

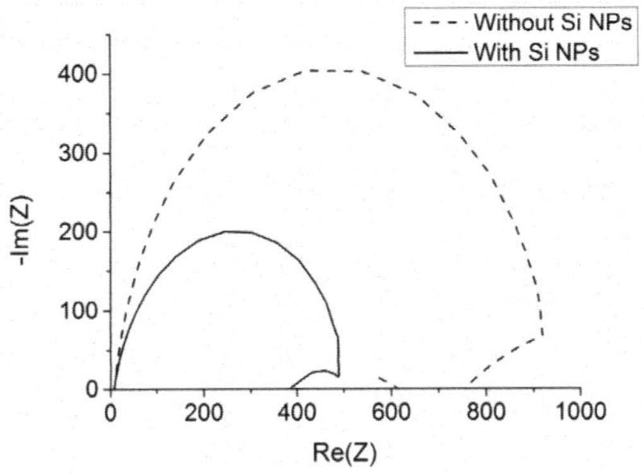

Figure 5. Impedance of solar cell with (solid line) and without silicon nanoparticles (dashed line).

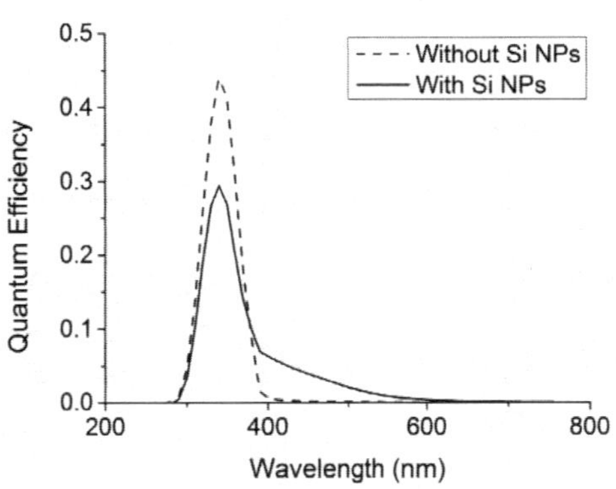

Figure 6. IPCE of solar cell with (solid line) and without silicon nanoparticles (dashed line).

Table 2. EIS parameters of fabricated solar cell.

Sample	R_s (Ω)	Z_1 (Ω)	Z_2 (Ω)
Without Si NPs	8.224	1113	927.8
With Si NPs	7.240	382	487.2

Incident photon-to-current efficiency (IPCE) indicates the ratio of the number of photons incident on photovoltaic device to the number of generated charge carriers. From figure 6, IPCE of solar cell with silicon nanoparticles increases in the range of 400-600 nm compared with that of the device without these particles. It seems that the addition of silicon nanoparticles contributes to light absorption at this region. However, no significant change was observed above 600 nm. This result emphasizes that the current density of solar cell with silicon nanoparticles was enhanced leading to an improvement of the device efficiency.

4. Conclusions

Silicon nanoparticles are one of the promising materials in the photovoltaic application. Although several techniques can be employed to prepared silicon nanoparticles, laser ablation in liquid was preferred in this work due to its simplicity and purity of product. The average size of prepared silicon nanoparticles was small enough so that MEG can be expected to occur. The conversion efficiency of QDSSC fabricated by these nanoparticles was shown to be five-fold higher than that of the device without them. This result can be explained by the reduction in electrical impedance and the increase in photon-to-current conversion efficiency from 400 to 600 nm.

Acknowledgements

We would like to thank Prof. K. Nakamura (Nd:YAG laser) and Mr. T. Fuchigami (TEM) from Tokyo Institute of Technology for their contribution to this research. This work was supported by the Materials and Structure Laboratory (Tokyo Tech. collaborative research) and JSPS KAKENHI Grant Number 23119506.

References

[1] Shockley W and Queisser H J, 1961 *J. Appl. Phys.* **32** 510
[2] Castan H, Perez E, Garcia H, Duenas S, Bailon L, Olea J, Pastor D, Garcia-Hemme E, Irigoyen M and Gonzalez-Diaz G, 2012 *AIP Conf. Proc.* **1496** 189
[3] Hu W G, Igarashi M, Lee M Y, Li Y M and Samukawa S, 2013 *Nanotechnology* **24** 265401
[4] Lin L Y, Chen C Y, Yeh M H, Tsai K W, Lee C P, Vittal R, Wu C G and Ho K C, 2013 *J.*

Power Sources **243** 535
[5] Lou Y Y, Yuan S, Zhao Y, Hu P F, Wang Z Y, Zhang M H, Shi L Y and Li D D, 2013 *Chem. Eng. J.* **229** 190
[6] Steinmann V, Umbach T E, Schadel M, Krumrain J, Lenze M R, Burckstummer H, Wurthner F and Meerholz K, 2013 *Org. Electron.* **14** 2029
[7] Koenig D, Casalenuovo K, Takeda Y, Conibeer G, Guillemoles J F, Patterson R, Huang L M and Green M A, 2010 *Physica E: Low-Dimentional Systems & Nanostructures* **42** 2862
[8] Solanki C S and Beaucarne G, 2007 *Energy for Sustainable Development* **11** 17
[9] Schaller R and Klimov V, 2004 *Phys. Rev. Lett.* **92** 1
[10] Nozik A J, 2002 *Physica E: Low-dimensional Systems and Nanostructures* **14** 115
[11] Semonin O E, Luther J M, Choi S G, Chen H Y, Gao J B, Nozik A J and Beard M C, 2011 *Science* **334** 1530
[12] Nozik A J, 2008 *Chem. Phys. Lett.* **457** 3
[13] Beard M C, Knutsen K P, Yu P R, Luther J M, Song Q, Metzger W K, Ellingson R J and Nozik A J, 2007 *Nano Lett.* **7** 2506
[14] Liong W L, Sreekantan S and Hutagalung S D, 2010 *Proc. of SPIE* **7743** 774306–1
[15] Heitmann J, Muller F, Zacharias M and Gosele U, 2005 *Adv. Mater.* **17** 795
[16] Wu X L, Siu G G, Tong S, Liu X N, Yan F, Jiang S S, Zhang X K, and Feng, 1996 *Appl. Phys. Lett.* **69** 523
[17] Liu P, Cui H, Wang C X and Yang G W, 2010 *Phys. Chem. Chem. Phys.* **12** 3942
[18] Perminov P A, Dzhun I O, Ezhov A A, Zabotnov S V, Golovan L A, Ivlev G D, Gatskevich E I, Malevich V L and Kashkarov P K, 2011 *Laser Phys.* **21** 801
[19] Sajti C L, Sattari R, Chichkov B N and Barcikowski S, 2010 *J. Phys. Chem.* C **114** 2421
[20] Yang S, Cai W, Zhang H, Xu X and Zeng H, 2009 *J. Phys. Chem.* C **113** 19091
[21] Warner J H, Rubinsztein-Dunlop H and Tilley R D, 2005 *J. Phys. Chem.* B **109** 19064
[22] Lu W B, Bian Y L, Liu H F, Han L, Yu W and Fu G S, 2010 *Mat. Lett.* **64** 1073
[23] Lin S W and Chen D H, 2009 *Small (Weinheim an der Bergstrasse, Germany)* **5** 72
[24] Li X M, Qiu Y, Wang S S, Lu S, Gruar R I, Zhang X H, Darr J A and He T, 2013 *Phys. Chem. Chem. Phys.* **15** 14729
[25] Seo S J, Hinsch A, Veurman W, Brandt H, Kang M S, Shin S H and Moon S H, 2013 *J. Appl. Polym. Sci.* doi: 10.1002/APP.39739
[26] Chen H, Zhu L, Liu H and Li W, 2013 *J. Phys. Chem* C **117** 3739
[27] Seo H W, Wang Y T, Uchida G, Kamataki K, Itagaki N, Koga K, Shiratani M, 2013 *Electrochimica Acta* **95** 43

Journal of Physics: Conference Series **518** (2014) 012024 doi:10.1088/1742-6596/518/1/012024

Surface modification of dye-sensitized solid-state solar cells by plasma jet

K Shimizu[1], S Kaneta[2], M Blajan[1], K Ogi[2] and A Konno[3]

[1] Organization for Innovation and Social Collaboration, Shizuoka University, 3-5-1 Johoku, Naka-ku, Hamamatsu, Shizuoka 432-8561, Japan
[2] Graduate School of Engineering, Shizuoka University, 3-5-1 Johoku, Naka-ku, Hamamatsu, Shizuoka 432-8561, Japan
[3] Graduate School of Science and Technology, Shizuoka University, Shizuoka University, 3-5-1 Johoku, Naka-ku, Hamamatsu, Shizuoka 432-8561, Japan

shimizu@cjr.shizuoka.ac.jp

Abstract. Surface treatment of dye-sensitized solid-state solar cells (DSSSCs) by atmospheric pressure plasma jet was carried out to improve its characteristics using Ar/O_2 as process gas. It is a dry process and a low energy consumption method for surface modification of DSSSCs due to its generation at atmospheric pressure. Active species and radicals were generated by the atmospheric pressure plasma jet. Because these were flown by the gas, the TiO_2 surface was modified. The important parameters of DSSSCs, such as, the short circuit photocurrent densities (J_{sc}) and the open circuit photovoltage (V_{oc}) were measured by the solar cell evaluation system. Surface condition of DSSSCs was analyzed by an X-ray photoelectron spectroscopy (XPS), Ultraviolet Visible Absorption Spectroscopy (UV-VIS) and Scanning Electron Microscope (SEM). As a result, solar conversion efficiency of DSSSCs increased by 22.5% after 15min of dry treatment compare with before treatment.

1. Introduction

The dye-sensitized solar cells (DSSCs) which were invented by Grätzel are well known as clean and low cost photovoltaic devices due to their simple structure and use of low cost materials [1-5]. DSSCs have the disadvantage of leaks and deterioration of electrolytic solution. The dye-sensitized solid-state solar cells (DSSSCs) could have the potential to be alternatives to the existent DSSCs [6-8]. DSSSCs use CuI as solid electrolyte, which does not leak from the cell and does not deteriorate. However, solar conversion efficiency of DSSSCs is lower than that of a conventional DSSCs efficiency [9].

One method for surface treatments of DSSCs is using chemicals [10-12]; wet process, which makes the process expensive and potentially harmful to the human body. TiO_2 surface modification by low pressure plasma treatment has been carried out, and an increase of the solar conversion efficiency has been reported using the RF-plasma [13-16] and vacuum deposition [17-21]. Modification methods in atmospheric condition, dielectric barrier discharge (DBD) [22, 23], corona discharge [24], and plasma jet [25-27], have been also reported. By using atmospheric pressure plasma, there is the advantage that does not require a vacuum device. It is necessary to apply about 30 kV for DBD and corona discharge [23, 24]. In this study, Surface treatment of DSSSCs by atmospheric pressure plasma jet was studied to improve its efficiency. The discharge gap of atmospheric plasma jet was set at 2.1 mm. Thus only a relatively low voltage is required to energize the discharge which furthermore leads to a reduced electricity consumption. Effect on the DSSSCs by plasma exposure was investigated by using the solar cell evaluation system, an X-ray photoelectron spectroscopy (XPS), Ultraviolet Visible Absorption Spectroscopy (UV-VIS) and Scanning Electron Microscope (SEM).

2. Experimental setup for TiO_2 surface modification

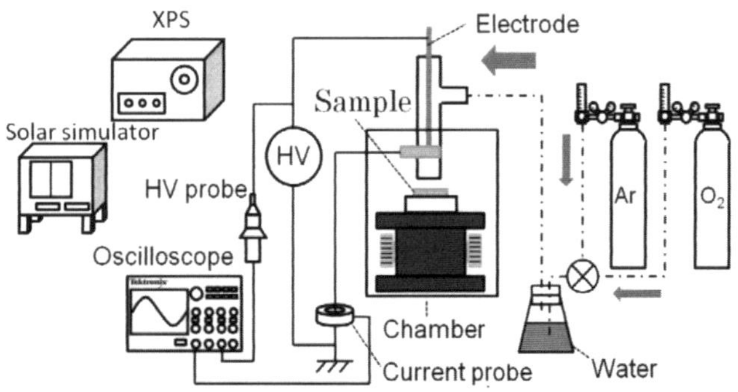

Figure 1. Experimental setup for TiO_2 surface modification by atmospheric plasma jet. High frequency neon transformer was utilized to energize the plasma jet. Ar and O_2 gas were used as process gases. X-ray photoelectron spectroscopy (XPS) Solar simulator and Scanning Electron Microscope (SEM) were used for evaluation.

Figure 1 shows the experimental setup for DSSSC sample surface modification by atmospheric pressure plasma jet. Atmospheric pressure plasma was generated between two electrodes. The high voltage electrode was made of tungsten wire (0.8 mm in diameter) and was fixed to the center of the glass tube, while the grounded electrode was an Aluminum-tape wrapped around a glass tube (5.0 mm outer diameter, 3.2 mm inner diameter). The plasma jet electrode was energized by an AC high voltage power supply. The process gas was fed through the glass tube from gas cylinders.

DSSSC sample and the plasma jet electrode were placed into a chamber to control the ambient gas. An Ar/O_2 mixture (5.00 + 0.25 L/min.) was used as the process gas. Additionally, the process gas can be humidified (relative humidity 65%) to confirm the water vapor effect on modification of DSSSC. The frequency of AC high voltage was set at 16.1 kHz. The distance between electrode and sample was fixed to 13 mm. The size of the TiO_2 sample was 0.5 cm^2. After the plasma exposure treatment, the sample TiO_2 sample was soaked to dye again for about 18 hours (N719 Ruthenium-based dye). I-V characteristics and various parameters of the sample were measured under simulated solar light (Air Mass 1.5) using a solar cell evaluation system (JASCO, CEP-25BX).The UV-VIS (UV-1600PC, SHIMADZU) was used to measure the dye absorption quantity. The XPS (ESCA-3400, SHIMADZU) was used to analyze chemical composition of the TiO_2 surface. TiO_2 surface before and after treatment was observed by the SEM (JEOL, JSM-7001F).

3. Result and discussion

3.1. Photovoltaic measurements

Figure 2 and Table 1 shows a comparison of photocurrent voltage (I-V) characteristics between untreated DSSSCs and treated DSSSCs. Treatment time of sample surface was 5, 10 and 15 min. The target sample was attached in the chamber and relative humidity in chamber was 0%. Three samples were treated simultaneously each time.

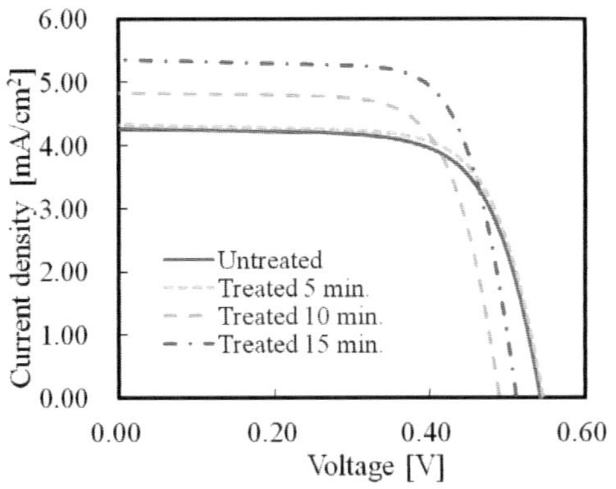

Figure 2. I-V characteristics before and after various treatment times (5, 10 and 15 min.). Applied voltage was set at 5.0 kV. After plasma treatment, current density and conversion efficiency were increased depending on treatment time.

Table 1. Characteristics of untreated and treated DSSCs.

Treated time [min.]	J_{SC} [mA/cm^2]	V_{OC} [V]	FF	η_C [%]
Untreated	4.25	0.54	0.69	1.60
5	4.32	0.55	0.70	1.65
10	4.82	0.49	0.71	1.67
15	5.35	0.51	0.72	1.96

As shown in Figure 2 and Table 1, the short circuit photocurrent density (Jsc) and the solar conversion efficiency (η_C) of untreated TiO_2 sample were 4.25 mA/cm^2 and 1.60%, respectively. J_{SC} and η_C measured for the DSSSCs sample treated for 15 min. were increased to 5.35 mA/cm^2 and 1.96% respectively, which corresponds to an increase with 25% and 22.5%. However, the open circuit photovoltage (V_{OC}) decreased. Fill factor (FF) was not changed because J_{SC} was increased and V_{OC} was decreased. Conversion efficiency was improved with increasing treatment time.

3.2. Effect of humidity on the surface modification of DSSSCs

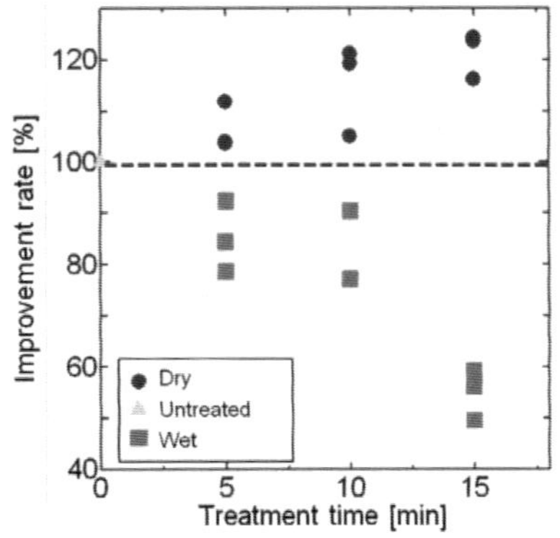

Figure 3. Improvement rate versus treatment time for 3 types of DSSSCs samples treatment. Applied voltage was set at 5.0 kV. Average conversion efficiency of untreated samples was 100%.

The dry treatment (dry process gas with relative humidity 0%), and the wet treatment (humid process gas with relative humidity 65%) were carried out to compare the effect of humidity. Figure 3 shows a comparison of the conversion efficiency between the dry treatment and the wet treatment. The dry treatment improved the conversion efficiency with about 20% after 15 min treatment time. The wet treatment decreased the conversion efficiency with about 45% after 15 min treatment time.

3.3. Dye adsorption quantity

Dye adsorption quantity of TiO_2 surface versus treatment time are shown in Figure 4. After Dry treatment, dye adsorption quantity increased with about 5%. On the other hand, after Wet treatment, dye adsorption amount decreased with about 20%. Hydrophilic group modification could be occurred both treatments (result will be next session). However, humidity in process gas inhibited the dye adsorption and decreased the conversion efficiency.

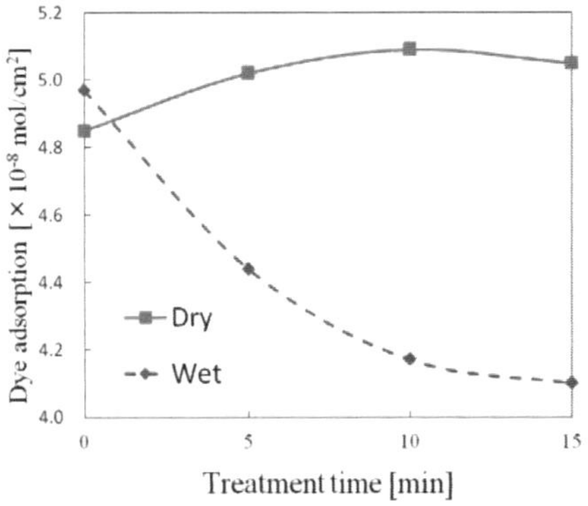

Figure 4. Dye adsorption versus treatment time for dry and wet treatment. Applied voltage was set at 5.0 kV. The analysis of dye adsorption amount using UV-VIS. Solid line corresponds to the dye adsorption amount after the dry treatment. The broken line corresponds to the dye adsorption after the wet treatment.

3.4. XPS analysis

Ti 2p and O 1s XPS spectra of TiO_2 sample surface treated for 15 min with AC high voltage are shown in Figure 5. Table 2 shows the stoichiometry value of the atomic content of Ti 2p and O1s peak.

Figure 5 shows the XPS spectra of Ti 2p on the TiO_2 surface after the dry treatment (15 min.). Ti 2p spectra has been analyzed for four peaks, i.e. Ti^{4+} $2p_{1/2}$, Ti^{3+} $2p_{1/2}$, Ti^{4+} $2p_{3/2}$, and Ti^{3+} $2p_{3/2}$. The surface stoichiometry was determined by calculating the relative peaks area as ratio of the total Ti 2p and O 1s in XPS spectra. The Ti^{3+} surface state in Ti 2p XPS spectra increased from 3.27% to 3.9% after 15 min of Ar/O_2 plasma treatment. O 1s XPS spectra was measured as well as Ti 2p to investigate the surface reaction. Figure 6 shows the XPS spectra of O1s. The Ti_2O_3 peak in O 1s was increased from 27.01% to 32.19% after 15 min of Dry treatment. $Ti^{4+} \rightarrow Ti^{3+}$ reduced reaction could be occurred at TiO_2 sample surface [28-30]. Part of TiO_2 surfaces were changed to Ti_2O_3 as shown in Figure 6. This surface reaction determined the increase of the electric conductivity of the DSSSCs sample surface and consequently, solar conversion efficiency was improved.

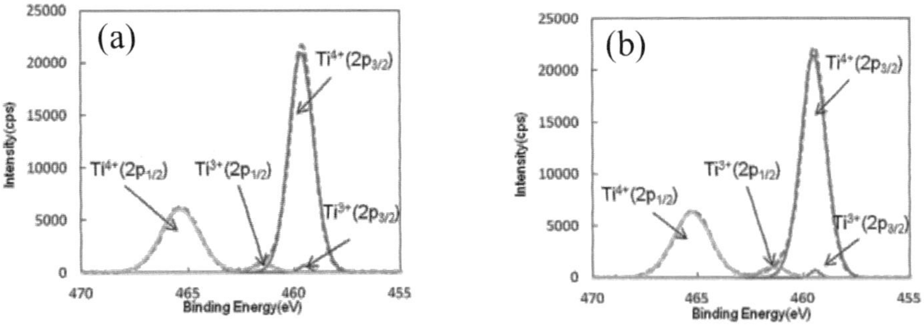

Figure 5. (a) XPS spectra of Ti 2p before treatment. (b) XPS spectra of Ti 2p after 15 min treatment. Solid line shows analysis of four peaks, dotted line shows original peak.

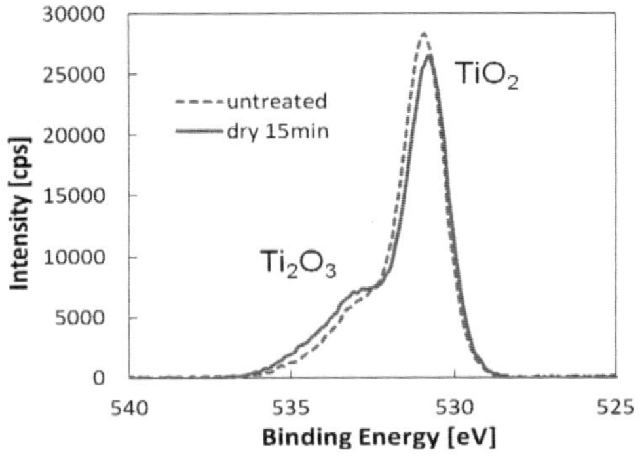

Figure 6. XPS spectra of TiO_2 and Ti_2O_3 before and after 15 min treatment. Applied voltage was set at 5.0 kV. After 15min Dry treatment, Ti_2O_3 peak was increased with 5% after treatment. Solid line corresponds to untreated sample. The broken line corresponds to Dry treated sample.

Table 2. XPS relative peak areas of Ti2p and O1s before and after treatment.

	Ti^{4+} $2p_{1/2}$	Ti^{4+} $2p_{3/2}$	Ti^{3+} $2p_{1/2}$	Ti^{3+} $2p_{3/2}$	TiO_2	Ti_2O_3
Untreated [%]	29.52	67.21	2.39	0.88	72.99	27.01
15min [%]	29.61	66.48	2.93	0.97	67.81	32.19

3.5. SEM analysis

As shown in Figure 7, the photos of TiO_2 surface before and after treatment were taken by a SEM. Treated sample was treated by Dry treatment. Treatment time was 15 min.

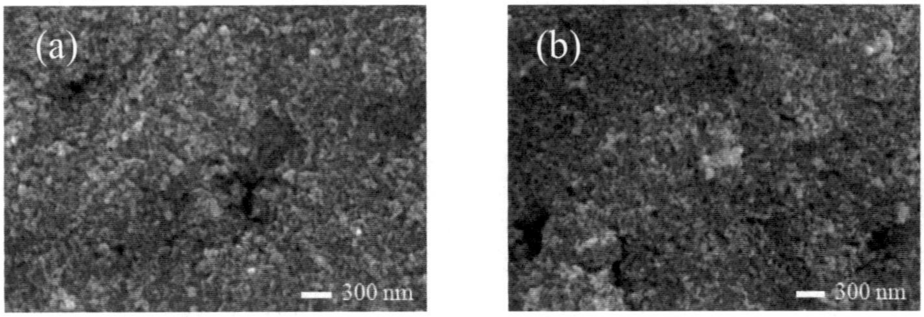

Figure 7. (a) FESEM image of Untreated TiO_2 surface. (b) FESEM image of Dry treated TiO_2 surface. Magnification was 30000.

No significant change on the TiO_2 surface was observed. After plasma treatment by atmospheric plasma jet, TiO_2 surface was changed not as physical appearance but only as chemical composition.

4. Conclusions

Surface modification of dye-sensitized solid-state solar cell using plasma jet was carried out and the following conclusions were obtained:

1. The current density was increased and conversion efficiency was improved with about 20% after 15 min of the dry treatment. It could be due to the increase of dye adsorption quantity and the electron conductivity of the DSSSCs samples.
2. Conversion efficiency and dye adsorption quantity were decreased after the wet treatment. Inhibition of dye adsorption due to humidity occurred.
3. XPS analysis showed the Ti_2O_3 peak increase in TiO_2 surface. Electron conductivity and conversion efficiency increased due to the reduction action of Ti_2O_3 from TiO_2.

4. SEM analysis showed no significant change before and after the plasma treatment. TiO_2 surface was changed but only as chemical composition not as physical appearance.

References

[1] O'Regan B and Grätzel M 1991 *Nature* **353** 737
[2] Grätzel M 2009 *Accounts of Chemical Research* **42** 1788
[3] Hagfeldt A, Boschloo G, Sun L,Kloo L and Pettersson H 2010 *Chem. Rev.* **110**, 6595
[4] Jena A, Mohanty SP, Kumar P, Naduvath J, Gondane V, Lekha P, Das J, Narula HK, Mallick S and Bhargava P 2012 *Trans. Ind. Ceram. Soc.* **71** 1
[5] Narayan MR 2012 *Renewable and Sustainable Energy Reviews* **16**, 208
[6] Chung I, Lee B, He J, Chang RPH and Kanatzidis MG 2012 *Nature* **485** 486
[7] Bouclé J and Ackermann 2012 *J Polymer International* **61** 355
[8] Yum JH, Chen P, Grätzel M and Nazeeruddin MK 2008 *ChemSusChem* **1** 699
[9] Konno A and Premalal EVA 2010 *J. Photopolym. Sci. Technol.* **23** 279
[10] Ito S, Murakami TN, Comte P, Liska P, Grätzel C, Nazeeruddin MK and Grätzel M 2008 *Thin Solid Films* **516** 4613
[11] Heil PE, Kang H, Choi H, Kim K 2013 *Applied Physics* A **112** 371
[12] Lin YH, Wu YC and Lai BY 2012 *Int. J. Electrochem. Sci.* **7** 9478
[13] Parveza MK, Yooa GM, Kima JH, Kob MJ and Kima SR 2010 Chemical Physics Letters **495**, 69
[14] Park KH and Dhayal M 2009 *Electrochemistry Communications* **11** 75
[15] Kim HJ, Kim J and Honga B 2013 *Applied Surface Science* **274** 171
[16] Lee B, Buchholz DB, Guo P, Hwang DK and Chang RPH 2011 *J. Phys. Chem. C* **115** 9787
[17] Dang BHQ, MacElroya D and Dowlinga DP 2013 *Applied Surface Science*. **275** 289
[18] Wu S, Han H, Tai Q, Zhang J, Chen BL, Xu S, Zhou C, Yang Y, Hu H and Zhao XZ 2008 *Appl. Phys. Lett.* **92** 122106
[19] Wang M, Huang C, Cao Y, Yu Q, Guo W, Liu Q, Liang J and Hong M 2009 *Nanotechnology* **20** 285311
[20] Shanmugam M, Baroughi MF and Galipeau D 2010 Thin Solid Films **518** 2678
[21] Kim JT and Kim SH 2011 *Solar Energy Materials & Solar Cells* **95** 336
[22] Zen S, Teramoto Y, Hanawa K, Kobayashi S, Ono R and Oda T *Proc. 2012 Joint Electrostatics Conference* 1
[23] Zen S, Teramoto Y, Ono R and Oda T 2012 *Jpn. J. Appl. Phys.* **51** 056201
[24] Teramoto Y, Shiono T, Ono R and Oda T 2010 *IEEJ Trans. FM* **130** 525
[25] Yuji T, Akatsuka H, Mungkung N, Park BW, Sung YM 2009 *Vacuum* **83** 124
[26] Chang H, Yang YJ, Li HC, Hsu CC, Cheng IC and Chen JZ 2013 *J. Power Sources* **234** 16
[27] Seo HK, Elliott CM, and Shin HS 2010 *ACS Appl. Mater. Interfaces* **2** 3397
[28] Liu R, Yang WD and Qiang LS 2012 *J. Power Sources* **199** 418
[29] Kim Y, Yoon CH, Kim KJ and Lee Y 2007 *J. Vac. Sci. Technol. A* **25** 1219
[30] Kim Y, Yoo BJ, Vittal R, Lee Y, Park NG and Kim KJ 2008 *J. Power Sources* **175** 914

Journal of Physics: Conference Series **518** (2014) 012025 doi:10.1088/1742-6596/518/1/012025

Formation mechanism of metal embedded amorphous silicate nanoparticles by induction thermal plasmas

M Tanaka[1], J Noda[1], T Watanabe[2], J Matsuno[3], and A Tsuchiyama[3]

[1] Dept. Environmental Chemistry and Engineering, Tokyo Institute of Technology, Yokohama, Japan
[2] Dept. Chemical Engineering, Kyushu University, Fukuoka, Japan
[3] Dept. Geology and Mineralogy, Kyoto University, Kyoto, Japan

watanabe@chem-eng.kyushu-u.ac.jp

Abstract. The purpose of this paper was to synthesize metal embedded amorphous silicate nanoparticles by induction thermal plasmas. Powder of Fe, Ni, Sn, and Zn were used as precursor as well as SiO_2 and MgO. In the case of Fe, Ni, and Sn, metal embedded nanoparticles were successfully synthesized. While most of Zn nanoparticles were attached on the surface of the amorphous silicate nanoparticles. Internal rate of each metal in the synthesized nanoparticles and nucleation temperature were evaluated to understand the formation mechanism of the metal embedded amorphous silicate nanoparticles. Results indicated that the relationship between the nucleation temperature of silicon and the melting temperature of each metal has important role on the metal embedded nanoparticles formation.

1. Introduction

Induction thermal plasmas has many unique advantages including high enthalpy, high chemical reactivity, large plasma volume, long residence time, and selective oxidation or reduction atmosphere according to the required chemical reactions. The high temperature of thermal plasma enables the evaporation of a large amount of raw materials even with high melting and boiling temperatures [1-3]. The formation of nanoparticles in supersaturated state by homogeneous nucleation and heterogeneous condensation can be accomplished due to the rapid quenching rate in the tail flame [4-7]. Furthermore, it is available to synthesize nanoparticles with high purity by RF thermal plasma, because the thermal plasma is generated in the torch without an internal electrode. In the present work, synthesis of metal embedded amorphous silicate nanoparticles by RF thermal plasmas was tried based on following motivations.

First, crystalline silicate in circumstellar regions and in primitive materials of the solar system is believed to be crystallized from amorphous silicates. GEMS (glass with embedded metal and sulfide) is a major component of cometary dust. According to recent work [8], GEMS is composed of amorphous silicate including iron metal in the size of few tens of nanometer. The alteration experiments with GEMS analogue materials are required to understand the evolution of amorphous silicate in the early solar system.

Second, the metal nanoparticles are expected to be applied in magnetic storage media, nanosensors and devices, drug delivery system and so on. However, the metal nanoparticles are difficult and dangerous to handle due to their instability and high reactivity. Therefore, surface coated metal nanoparticles have attracted to solve this issue. In particular, amorphous silicate as shell material could be useful because of its high thermal resistance and corrosive properties.

The objective of the present work is to synthesize the metal embedded amorphous silicate nanoparticles by induction thermal plasma. Another purpose is to investigate the formation mechanism of metal embedded amorphous silicate nanoparticles. The different metals are tried to synthesize the objective nanoparticles.

2. Experimental procedure

2.1. Experimental setup

Figure 1 shows experimental setup for metal embedded amorphous nanoparticles preparation, which is composed of the RF torch and the synthesis chamber. Precursors of metal, SiO_2, and MgO powders with particle size of several micrometers are injected from the powder feeder by Ar carrier gas. Different metals including Fe, Sn, Ni, and Zn were compared to investigate the formation mechanism. The composition ratio of metal to Si was 0.56, which was adjusted based on the GEMS mean composition. The powder feed rate was fixed at 250 mg/min. The input power of RF thermal plasma was controlled around 30 kW. Argon was used for the carrier gas of the raw powder, and argon and helium were used for plasma forming gases to evaporate the raw materials efficiently. Table 1 shows operating conditions for the synthesis system. In these experimental conditions, complete evaporation of the raw materials was confirmed by SEM observation.

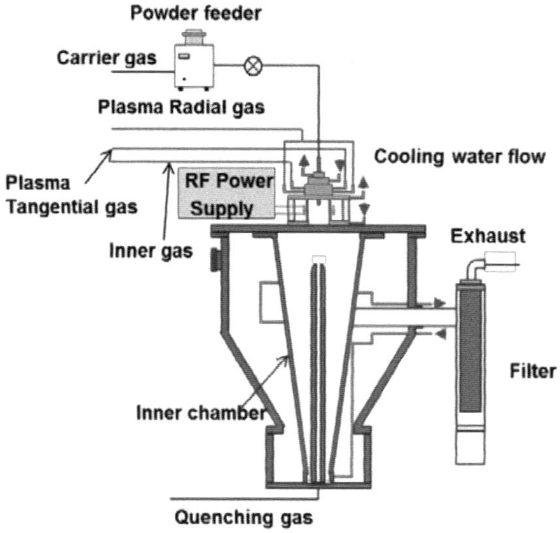

Figure 1. Experimental setup for the synthesis of metal embedded nanoparticles

2.2. Analyses

The phase identification of the prepared nanoparticles were determined by X-ray diffractometry (XRD, Miniflex, Rigaku Co.) operated with Cu Kα source (λ = 0.1541 nm). The diffraction data was collected using a continuous scan mode with speed of 4 degree/min in the region 10–90degree with a step size of 0.04degree. The accelerating voltage and applied current was 40 kV and 50 mA, respectively.

The particle morphology and size distribution of synthesized nanoparticles were observed using a TEM (S4700, Hitachi Ltd.) operating at an accelerating voltage of 5 kV.

Surface composition was measured by X-ray photoelectron spectroscopy (XPS, 1700R ESCA system, ULVAC-PHI Inc.), while inductively coupled plasma atomic emission spectroscopy (ICP-AES, ICPS-8100, Shimadzu Co.) was used to measure the composition of each element in the synthesized nanoparticles.

Table 1 Operating conditions for synthesis of metal embedded nanoparticles

Plasma power plate	30 kW
Frequency	4 MHz
Pressure	101.3 kPa
Sheath gas 1	Ar 60 L/min
Sheath gas 2	He 5 L/min
Inner gas	Ar 5 L/min
Carrier gas	Ar 6.5 L/min
Raw Material	Me, MgO, SiO_2
Me/Si	0.56
Me	Fe, Ni, Sn, or Zn
Feeding rate	250 mg/min

3. Results and discussion

XRD spectra of synthesized nanoparticles in different systems are presented in Figure 2. The diffraction peaks of Si, Mg, and their oxides were not observed. In contrast, the diffraction peaks of each metal, Fe, Sn, Ni, and Zn, were found in different systems. These results indicate that SiO_2-MgO silicate and metal co-exist in the synthesized nanoparticles. In the cases of Ni and Zn, the diffraction peaks of metal oxide were also observed. This will be discussed in following section.

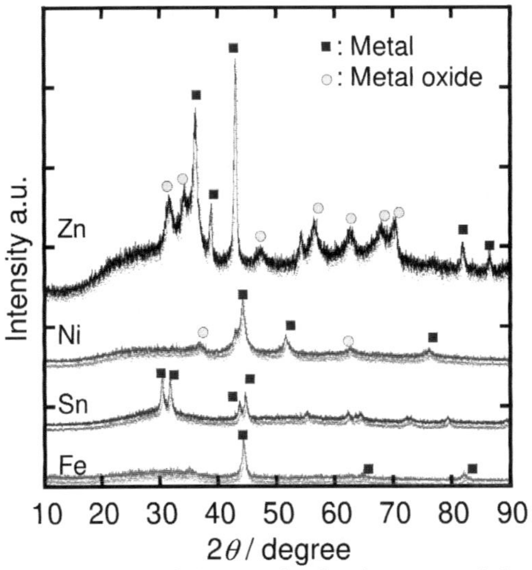

Figure 2. XRD spectra of the synthesized nanoparticles from different metal precursors.

TEM images of the synthesized nanoparticles in Fe-SiO_2-MgO system are presented in Figure 3. Core/shell structured nanoparticles were observed as shown in Figure 3 (a). Additional energy dispersive spectrometer (EDS) analysis revealed that the core contains Fe, while the shell part are composed of Si, Mg, and O. Figure 4 represents TEM images of the synthesized nanoparticles in different systems. In the case of Sn and Ni, core/shell structured nanoparticles were also observed as well as Fe case. While, Zn nanoparticles were not found in core of synthesized nanoparticles. The particle mean diameters of the main amorphous silicates were in the range from 29 nm to 34 nm in different metal-SiO_2-MgO systems. The effect of additional metals on the diameter of the amorphous silicate nanoparticles was not confirmed.

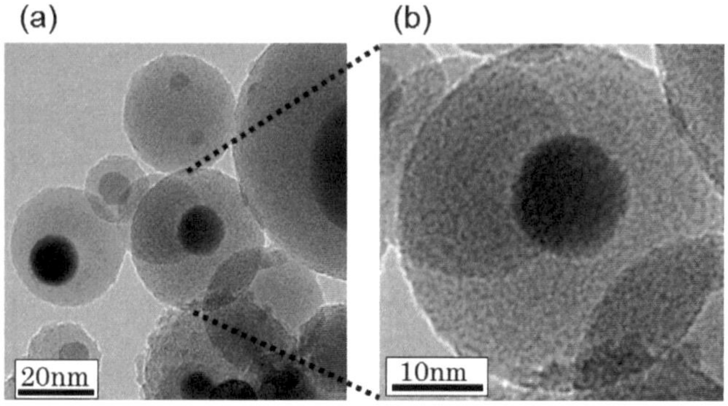

Figure 3. TEM images of Fe embedded nanoparticles synthesized by RF thermal plasmas.

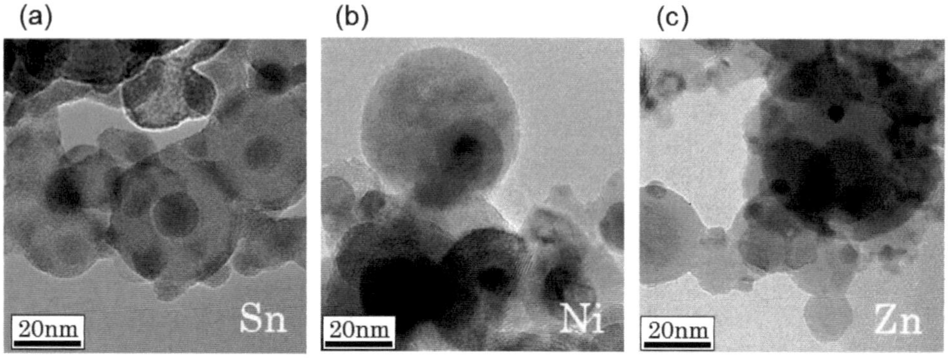

Figure 4. TEM images of synthesized nanoparticles from different precursors: (a) Sn-SiO_2-MgO, (b) Ni-SiO_2-MgO, (c) Zn-SiO_2-MgO.

Figure 5 shows the mole fraction of each metal in the synthesized nanoparticles measured by ICP-AES and that in the surface region of the synthesized nanoparticles measured by XPS. Since the mole fractions in the synthesized nanoparticles in different systems were determined by the initial composition of the raw material, these values were almost same among the used metals. In contrast, the mole fractions of each metal in the surface region of the synthesized nanoparticles were varied in different systems. The mole fraction of Fe in the surface region was lowest than other metals. This result suggests that the Fe nanoparticles were successfully coated by the shell of the amorphous silicate. In the case of Zn, the mole fraction of Zn in the surface region of the nanoparticles was higher than that in the whole nanoparticles. This result indicates that Zn nanoparticles were not coated by the amorphous silicate. This result also can be used to explain the reason of the existence of zinc oxide, as presented in the XRD spectrum. Zn nanoparticles were oxidized by the surrounding air because Zn nanoparticles were not coated by the amorphous silicate.

Internal rate of each metal was defined to evaluate the amount of the fully coated metal quantitatively. Internal rate is the ratio of coated metal to total metal in the synthesized nanoparticles. It was calculated based on the volume ratio between the metals in the surface of the particles and that in the whole particles. Figure 6 shows the internal rate of each metal in different systems. Internal rate of Fe was highest, while that of Zn was lowest. This trend has good agreement with the results of other analyses mentioned above. To discuss the reason of different internal rates among used metals in the present work and the formation mechanism of the metal embedded amorphous silicate nanoparticles, the nucleation temperature was evaluated based on the homogeneous nucleation model.

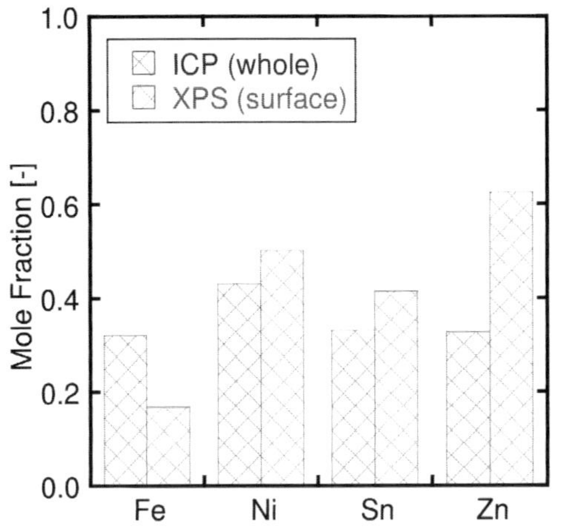

Figure 5. Mole fraction of each metal in surface region and whole region of the synthesized nanoparticles.

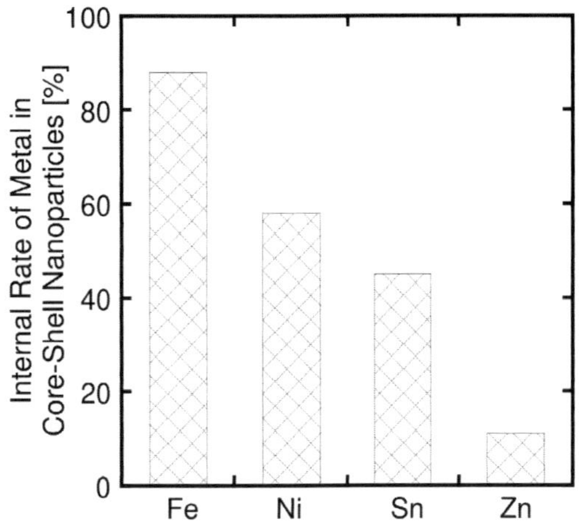

Figure 6. Internal rate of each metal in the synthesized nanoparticles from different precursors.

Girshick et al. derived the homogeneous nucleation models as an extension of the kinetic nucleation theory [9]. The proposed model which is shown in Equation (1) was used for the estimation of the critical saturation ratio.

$$J = \frac{\beta_{ij} n_s^2 S}{12}\sqrt{\frac{\Theta}{2\pi}}\exp\left(\Theta - \frac{4\Theta^3}{27(\ln S)^2}\right) \tag{1}$$

where n_s is the equilibrium saturation monomer concentration at temperature T, β is the collision frequency function between monomers, and S is the saturation ratio. The dimensionless surface tension Θ is given by the following Equation:

$$\Theta = \frac{\sigma s_1}{kT} \tag{2}$$

where σ is the surface tension and s_1 is the monomer surface area. The collision frequency function can be estimated using Equation (3) as follows:

$$\beta_{ij} = \left(\frac{3v_1}{4\pi}\right)^{1/6}\sqrt{\frac{6kT}{\rho_P}\left(\frac{1}{i}+\frac{1}{j}\right)} \times \left(i^{1/3} + j^{1/3}\right)^2 \tag{3}$$

where ρ_p is the particle mass density and v_1 is the monomer volume. Since the particle is nucleated by the monomer collision in this model, i and j are equal to 1. The nucleation rate is strongly dependent on the surface tension and the saturation ratio. Particle formation is observed experimentally when the nucleation rate is over 1.0 $cm^{-3}s^{-1}$. Hence, the corresponding value of the saturation ratio is defined as the critical saturation ratio.

The nucleation temperature at the critical saturation ratio is presented in Figure 7 as well as melting and boiling points. Because of the unknown properties of the silicon oxide, only melting and boiling points of SiO and SiO_2 are plotted without nucleation temperature. As shown in Figure 7, nucleation temperature of Si was highest among that of the considered species. Furthermore, the melting point of SiO and SiO_2 were higher than the nucleation temperature of considered other metals of Fe, Ni, Sn, and Zn. These results indicate that the silicon or its oxide firstly nucleated and other species except Zn then co-condensed on the nuclei. Finally, the metals are considered to start solidification in the molten silicate in the quenching process to form the metal embedded amorphous silicate nanoparticles. Zn has the lowest boiling point, which is lower than the melting point of silicate. Therefore, Zn vapour cannot condense on the silicate particles. These results can explain the obtained experimental results.

However, the difference among Fe, Ni, and Sn has not been understood yet and is under investigation. In this study, the relationship between the nucleation temperature and melting and boiling point was considered to discuss the formation mechanism. In the future study, the effect of the temperature decay rate, temperature history, on the formation mechanism will be investigated.

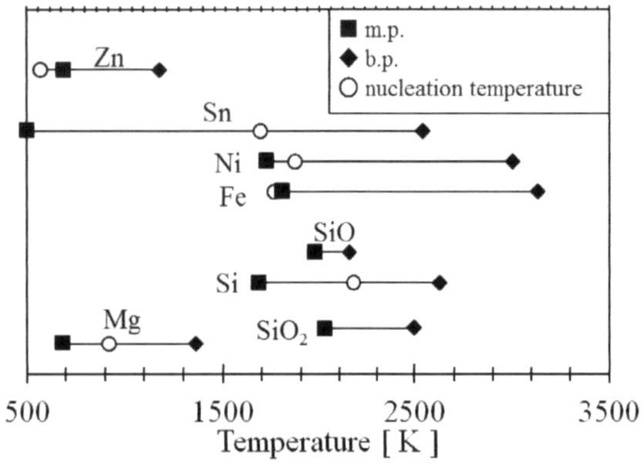

Figure 7. Nucleation temperature, melting and boiling points of considered species.

4. Conclusion

The metal embedded amorphous silicate nanoparticles as an analogue material of GEMS were successfully synthesized in RF thermal plasma process. Internal rate and nucleation temperature of each metal were evaluated to discuss the formation mechanism of the metal embedded nanoparticles. In the case of Fe, the highest internal rate of 90% was obtained, while the internal rate of Zn was less than 20%. This difference can be explained by the relationship between the nucleation temperature of metal and that of silicon or silicon oxide. The synthesized metal embedded nanoparticles as an analogue materials of GEMS would help to reveal the evolution of amorphous silicate in the early solar system. Furthermore, synthesized nanoparticles as surface coated metal nanoparticles would be applied in nanomagnetic field.

References

[1] Watanabe T, Nezu A, Abe Y, Ishii Y and Adachi K 2003 *Thin Solid Films* **435** 27
[2] Watanabe T and Okumiya H 2004 *Sci. Technol. Adv. Mater.* **5** 639
[3] Szepvolgyi J, Markovic Z, Scheier P and Feil S 2006 *Plasma Chem. Plasma Process.* **26** 597
[4] Shigeta M and Watanabe T 2005 *J. Mater. Res.***20** 2801
[5] Desilets M, Bilodeau JF and Proulx P 1997 *J. Phys. D: Appl. Phys.* **30** 1951
[6] Gonzalez NYM, Morsli ME and Proulx P 2008 *J. Therm. Spray Technol.* **17** 533
[7] Shigeta M and Watanabe T 2010 *J. Appl. Phys.* **108** 043306
[8] Kemper F, Vriend WJ and Tielens AGGM 2004 *Astrophys. J.* **609** 826
[9] Girshick SL, Chiu CO and McMurry PH 1990 *Aerosol Sci. Technol.* **13** 465

Journal of Physics: Conference Series **518** (2014) 012026 doi:10.1088/1742-6596/518/1/012026

Synthesis of Ni_2B nanoparticles by RF thermal plasma for fuel cell catalyst

Y Cheng[1], M Tanaka[1], T Watanabe[1,2], S Y Choi[3], M S Shin[3] and K H Lee[3]

[1] Dept. Environmental Chemistry and Engineering, Tokyo Institute of Technology, 4259 Nagatsuta-cho, Midori-ku, Yokohama 226-8502, Japan
[2] Dept. Chemical Engineering, Kyushu Univ., 744 Motooka, Nishi-ku, Fukuoka 819-0395, Japan
[3] Plasma Nano Division, Cheorwon Plasma Research Institute, 4620 Hoguk-ro, Galmal-eup, Cheorwon-gun, Gangwon-do, 269-802, Korea

E-mail: watanabe@chemenv.titech.ac.jp

Abstract. The catalyst of Ni_2B nanoparticles was successfully prepared using nickel and boron as precursors with the quenching gas in radio frequency thermal plasmas. The generating of Ni_2B needs adequate reaction temperature and boron content in precursors. The quenching gas is beneficial for the synthesis of Ni_2B in RF thermal plasma. The effect of quenching rate, powder feed rate and boron content in feeding powders on the synthesis of nickel boride nanoparticles was studied in this research. The high mass fraction of 28 % of Ni_2B nanoparticles can be generated at the fixed initial composition of Ni:B = 2:3. Quenching gas is necessary in the synthesis of Ni_2B nanoaprticles. In addition, the mass fraction of Ni_2B increases with the increase of quenching gas flow rate and powder feed rate.

1. Introduction

The development of novel materials is a fundamental point of chemical research, which has been promoted by progress in nanotechnology. Because of their refractory nature, resistances to sulfur poisoning, and desulphurization ability in organic synthesis and catalytic properties [1], metal borides have been considered as potentially desirable catalysts. Among various metal borides, nickel boride, Ni_2B, is regarded as a potential industrial catalyst for hydrogenation reactions [2]. Previous investigations indicated that nickel boride nanoparticle is useful in catalysts for electrodes of fuel cell [2, 3]. The remarkable catalytic activity of nickel boride has been attributed to the ability of boron to donate an electron to nickel [4]. Several routes have been proposed to synthesize Ni_2B, i.e., sodium borohydride, boronizng of pure nickel by the powder-pack method [5] and ball milling of element components [6]. To the best of our knowledge, there are hardly any reports on Ni_2B nanoparticles prepared by Radio frequency (RF) thermal plasma route, which is a great challenge.

RF thermal plasmas has many unique merits including high enthalpy, high chemical reactivity, large plasma volume, long residence time, and selective oxidation or reduction atmosphere according to the required chemical reactions. The high temperature of thermal plasma enables the evaporation of a large amount of raw materials even with high melting and boiling temperatures [7,8]. The formation of nanoparticles in supersaturated state by homogeneous nucleation and heterogeneous condensation can be accomplished due to the rapid quenching rate in the tail flame [9-11]. Furthermore, it is available to synthesize nanoparticles with high purity by RF thermal plasma, because the thermal plasma is generated in the torch without an internal electrode.

Few investigations for the synthesis nickel boride Ni_2B by RF thermal plasma have been reported so far. The purpose of this study is to examine the effect of experimental parameters on the synthesis of nickel boride nanoparticles in RF thermal plasma. The experiments were carried out with controlling the powder feed rate, boron content in feeding powders and quenching gas flow rate to produce expected Ni_2B nanoparticles. Adequate phase composition and temperature play the important role in the synthesis of Ni_2B nanoparticles. The phase composition of product and mass fraction of nickel boride were measured by X-ray diffractometry (XRD).

2. Experimental procedure

The schematic apparatus of RF thermal plasma system for the production of boride nanoparticles is shown in figure 1. The system consists of a plasma torch, a reaction chamber, a particle collection filter, and a power supply. The plasma torch is composed of a water-cooled quartz tube and a water-cooled induction coil (3 turns), coupling its electromagnetic energy to the plasma at a frequency of 4 MHz. The feeding powders were nickel (particle size: 45 μm, purity: 99.9%, Kojundo Chemical Laboratory Co. Ltd., Japan) and crystalline boron (particle size: 45 μm, purity: 99%, Kojundo Chemical Laboratory Co. Ltd., Japan). Nickel and boron particles were mixed at the fixed initial molar composition before they were injected into the RF thermal plasma through the powder feeder.

After the precursors were introduced into the plasma region with argon carrier gas, they were instantaneously evaporated due to the very high enthalpy of the RF thermal plasma. Therefore, the nickel boride nanoparticles are synthesized from the gas phase.

The operating conditions are summarized in Table 1. Argon was introduced as the carrier gas of 3 L/min, the inner gas of 5 L/min, the sheath gas of 60 L/min and the quenching gas. Helium of 5 L/min was also used as the sheath gas. The sheath gases were injected from the outer slots of the plasma torch located between the injection tube and the quartz tube to protect the inner surface of the quartz tube and to stabilize the plasma. The powder feed rate was 0.1, 0.2 and 0.5 g/min and the molar content of boron in the feeding powder was controlled from 33.3 to 75.0 at. %. The quenching gas flow rate was from 5 to 20 L/min. The RF plasma was operated at the fixed input power of 30.0 kW under atmospheric pressure.

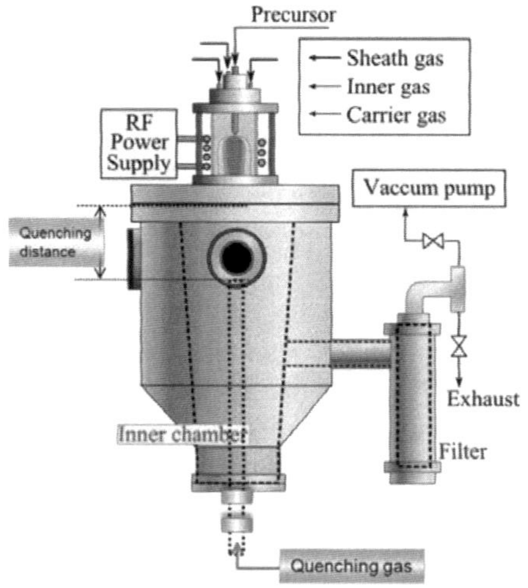

Figure 1. Experimental set-up of RF thermal plasma for the synthesis of metal boride nanoparticles.

The synthesized nanoparticles were characterized for phase identification by XRD (MXP3TA, Mac Science). The crystalline size was calculated from the full widths at the half maximum (FWHM) of the most intensive diffractions according to the Scherrer's equation. Quantitative phase analysis by using XRD data was carried out based on the adiabatic method [12]. This method evaluates the mass fraction of each phase from the relative intensity of the diffraction peak of product. The mass fraction of *X* phase can be calculated by following equation (1)

$$W_X = \frac{I_X}{K_A^X \sum_{X=A}^{N} \frac{I_X}{K_A^X}} = \frac{I_X}{K_{Ni}^X \cdot \left(\frac{I_{Ni}}{Ki} + \frac{I_{NiB}}{K_{Ni}^{NiB}} + \frac{I_{Ni_4B_3}}{K_{Ni}^{Ni_4B_3}} + \frac{I_{Ni_2B}}{K_{Ni}^{Ni_2B}} + \frac{I_{Ni_3B}}{K_{Ni}^{Ni_3B}}\right)} \quad (1)$$

where, *X* (= Ni, NiB, Ni_4B_3, Ni_2B, Ni_3B) denotes each phase in the product. I_X presents the intensity of *X* phase in the product from the XRD spectrum. K_A^X is the ratio of the reference intensity ratio (*RIR*) values of *X* phase to that of the reference phase *A*, *i.e.*, $K_A^X = RIR_X/RIR_A$. In the experimental measurement of Ni-B system, peaks for Ni, NiB, Ni_4B_3, Ni_2B and Ni_3B were identified in XRD spectrum, and Ni was selected as the reference materials. Based on the powder diffraction file cards, *RIR* values for Nb, NiB, Ni_4B_3, Ni_2B and Ni_3B are determined to be 7.41, 2.88, 1.43, 5.68 and 1.45, respectively.

Table 1. Operating condition for the synthesis of niobium boride nanoparticles.

Process parameters	Value
Sheath gas and flow rate	Ar-He (60:5) 65 L/min
Inner gas and flow rate	Ar 5 L/min
Carrier gas and flow rate	Ar 3 L/min
Quenching gas flow rate	Ar 5, 10 and 20 L/min
Plasma power	30.0 kW
Reactor pressure	101.3 kPa
Frequency	4 MHz
Powder Feed rate	0.1-0.5 g/min
Boron molar content in feeding powder	33.3-75.0 at.%

3. Results and discussion

XRD spectra of as-prepared product with or without quenching gas at powder feed rate of 0.2 g/min are displayed in figure 2. Figure 3 displays the magnification of figure 2 from 40° to 50°. The peaks of Ni_2B and Ni_4B_3 can be identified with quenching gas. The peaks of Ni_3B are observed without quenching gas. Generating Ni_2B needs some special temperature and phase composition according to the phase diagram [13]. Quenching gas can stop the reaction at certain temperature. Therefore, Ni_2B was prepared with adequate quenching gas and phase composition.

Figure 4 shows the XRD spectra of product at the fixed initial composition with different quenching gas flow rates. From the XRD measurement, Ni, Ni_4B_3 and Ni_2B were identified. Gibbs free energy of Ni_2B is negative, indicating Ni_2B can be generated spontaneously and easily. The highest intensity of the peak was unreacted Ni, while the dominant product is Ni_4B_3. This XRD result indicates that the synthesis of Ni_2B nanoparticle is difficult.

Figure 5 indicates the relationship between the mass fraction of each phase in the products and quenching gas flow rate. The mass fraction of Ni_2B in product increases with increasing quenching gas flow rate. Adequate quenching gas can stop the reaction at certain temperature, which can improve the

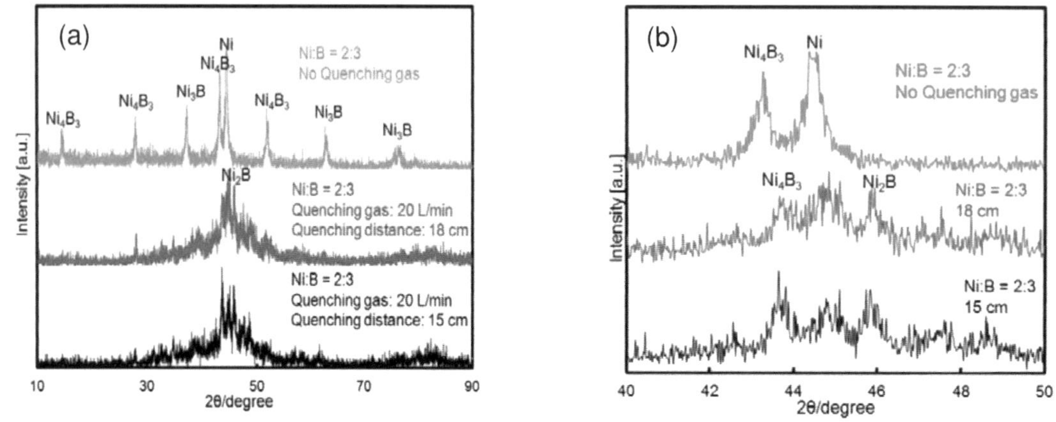

Figure 3. XRD patterns of product with different quenching distance at Ni:B = 2:3 (a) and magnified patterns (b).

generating of Ni_2B nanoparticles. The unreacted boron, which was not reacted with nickel but evaporated, increases with increasing the quenching gas flow due to shorter residence time, resulting from the higher temperature gradient. The enthalpy of the complete evaporation was about several hundred J/s when the input power was adjusted at 30 kW. The energy efficiency for the evaporation was around 2%. Complete evaporation of the raw materials was confirmed by SEM observation of the collected powders.

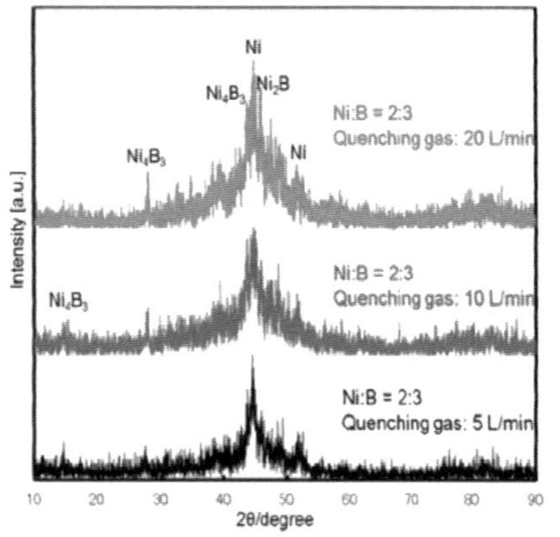

Figure 4. XRD patterns of product with different quenching gas flow rate at Ni:B = 2:3

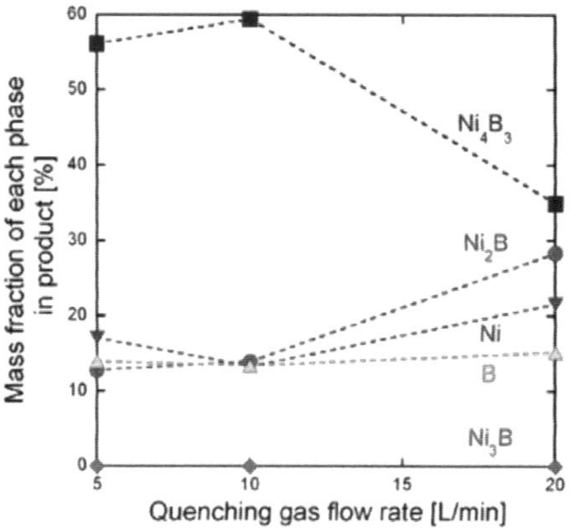

Figure 5. Mass fraction of each phase in product with different quenching gas flow rate at Ni:B = 2:3

Figure 6 displays the XRD patterns of product with different boron molar content in feeding powders at fixed quenching gas flow rate. In boron poor condition, the dominant product is Ni_3B. The peak of unreacted Ni has the highest intensity in boron rich condition, while Ni_3B are not observed. The highest intensity of Ni_2B in product is generated with the phase composition of Ni:B = 2:3.

The effect of boron molar content in feeding powders on the mass fraction of each phase in product is shown in figure 7. The highest mass fraction Ni_2B in product is observed with the initial composition of Ni:B = 2:3. In the boron poor condition, Ni_2B cannot be generated, while the mass fraction of Ni_2B decreases with increasing boron content in feeding powders in boron rich condition. We can get the high mass fraction of Ni_4B_3 with the initial composition of Ni:B = 1:2. Adequate temperature and boron content improve the synthesis of Ni_2B and Ni_4B_3. The synthesis of Ni_2B is the most difficult compared with that of Ni_4B_3 and Ni_3B due to the narrow composition region in phase diagram. The Ni_3B were generated in the boron poor condition. In addition, the mass fraction of unreacted boron increases with the increase of boron content in feeding powders because of larger amount boron in feeding powders.

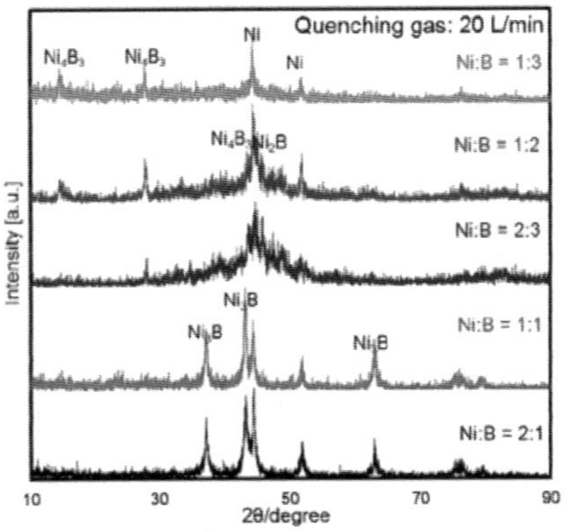

Figure 6. XRD patterns of product with different boron molar content in feeding powders.

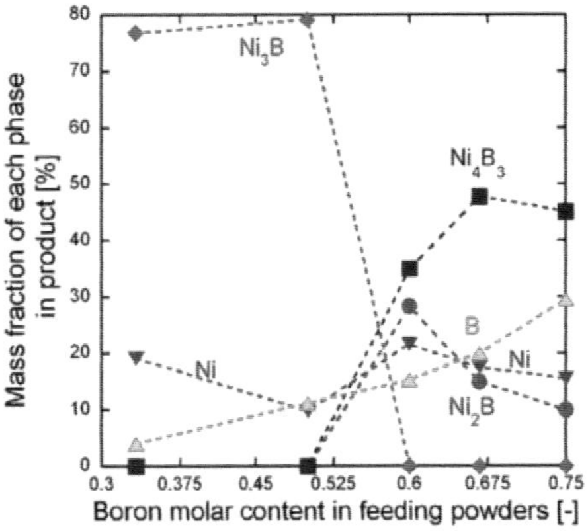

Figure 7. Mass fraction of each phase in product with different boron molar content in feeding powders.

Figure 8 displays the XRD spectra of product with different powder feed rate at the fixed initial composition of Ni:B = 2:3. With the low powder feed rate, Ni_3B can be identified. With the high powder feed rate, only the peaks of Ni and Ni_2B are observed.

The relationship between the powder feed rate and the mass fraction of each phase in product is observed in figure 9. The high mass fraction of Ni_2B and Ni_4B_3 in product is prepared at the powder feed rate of 0.2 g/min.

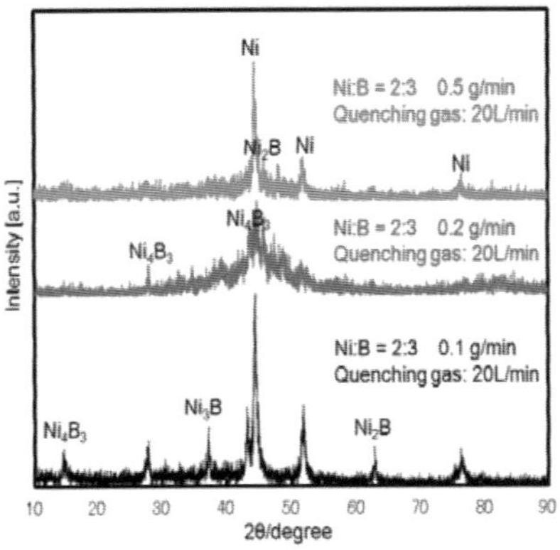

Figure 8. XRD patterns of product with different powder feed rate at Ni:B = 2:3

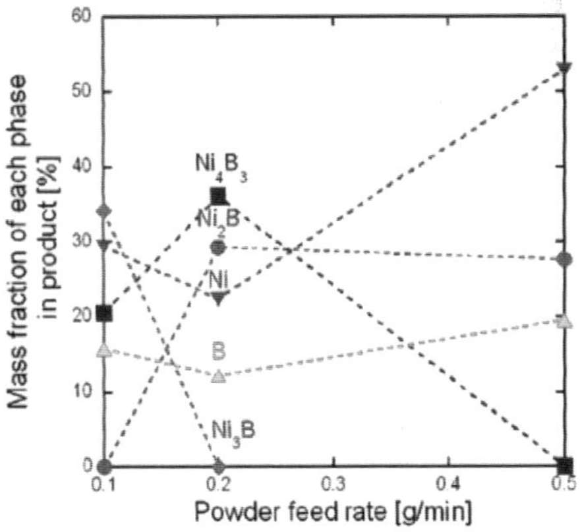

Figure 9. Mass fraction of each phase in product with different powder feed rate at Ni:B = 2:3

5. Conclusions

Experimental study were carried out to prepare nickel boride nanoparticles and to investigate the effect of powder feed rate, quenching gas flow rate and the boron molar content in feeding powders on the phase composition of product in RF thermal plasma. In this work, nickel boride Ni_2B nanoparticles were successfully synthesized in RF thermal plasma. The quenching gas and boron content in feeding powders play an important role in the controlled synthesis of nickel boride nanoparticles in RF thermal plasma. The highest mass fraction Ni_2B in product is observed in the initial composition of Ni:B = 2:3 with quenching gas flow rate of 20 L/min. The high mass fraction of Ni_4B_3 in the initial composition of Ni:B = 2:3 with quenching gas flow rate of 10 L/min. The Ni_3B were generated only in the boron poor condition.

References

[1] Ganem B, et al. 1986 *Chem. Rev.* **86** 763
[2] Brown HC, et al. 1963 *J. Amer. Chem. Soc.* **85** 1005
[3] Jasinski R 1963 Symposium on Fuel Cells, Division of Fuel Chemistry, 145th National Meeting of the American Chemical Society, N.Y., Sept. 8-13,
[4] Okamoto Y, et al. 1979 *J. Chem. Soc.* **75** 2027
[5] Ueda N, Mizukoshi T, Demizu K, Sone T, Ikenaga A and Kawamoto M 2000 *Surf. Coat. Tech.* **126** 25
[6] Corrias A, Ennas G, Marongiu G, Musinu A, Paschina G and Zedda D 1995 *Mat. Sci. Eng. A.* **204** 211
[7] Watanabe T, Nezu A, Abe Y, Ishii Y and Adachi K 2003 *Thin Solid Films* **435** 27
[8] Szepvolgyi J, Markovic Z, Scheier P and Feil S 2006 *Plasma Chem. Plasma Process.* **26** 597
[9] Shigeta M and Watanabe T 2005 *J. Mater. Res.* **20** 2801
[10] Desilets M, Bilodeau. JF and Proulx P 1997 *J. Phys. D: Appl. Phys.* **30** 1951
[11] Gonzalez NYM, Morsli ME and Proulx P 2008 *J. Therm. Spray Technol.* **17** 533
[12] Cheng YY and Watanabe T 2011 *J. Chem. Eng.Jpn.* **44** 583
[13] Passerone A, Muoloa ML, Valenza F and Kaufman L 2010 *Comput. Coupl. Phase Diag. Thermochem.* **34** 6

Effect of arc behaviour on the temperature fluctuation of carbon electrode in DC arc discharge

F Liang[1], M Tanaka[1], S Choi[1] and T Watanabe[1, 2]

[1] Dept. Environmental Chemistry and Engineering, Tokyo Institute of Technology, Yokohama, Japan
[2] Dept. Chemical Engineering, Kyushu University, Fukuoka, Japan

watanabe@chem-eng.kyushu-u.ac.jp

Abstract. Diffuse and multiple arc-anode attachment modes were observed in a DC arc discharge with a carbon electrode. During the arc discharge, the surface temperature of the electrode was successfully measured by two-colour pyrometry combined with a high-speed camera which employs appropriate band-pass filters. The relationship between the arc-anode attachment mode and the temperature fluctuation of electrode surface was investigated. The diffuse arc-anode attachment mode leads to relatively large temperature fluctuation on anode surface due to the rotation of the arc spot. In the case of diffuse mode, the purity of synthesized multi-wall carbon nanotube was deteriorated with temperature fluctuation.

1. Introduction

The arc discharge is an attractive method to prepare nanomaterials with several advantages [1]. Firstly, it is an environmentally benign technique without generating any toxic by-products and hazardous gases. In addition, the high temperature of arc plasma provides the products with fewer defects. Finally, high productivity in the arc discharge method reduces the manufacturing cost.

The arc discharge method has also been employed to prepare lots of carbon nanomaterials successfully. Fullerene was scraped from the chamber wall in helium arc discharge at low pressure [2]; multi-wall carbon nanotube (MWNT) was collected from cathode deposit at low pressure argon arc [3]; single-wall carbon nanotube was found in a “collaret” around the cathode deposit, the web-like structure suspended between cathode and chamber wall, and cloth-like soot suspended in the chamber wall by using the catalyst in helium arc [4]; and graphene flakes was produced in the inner wall of the chamber under hydrogen atmosphere [5].

In spite of the considerable efforts on the arc discharge method, however, the formation mechanism of carbon nanomaterial has not been understood yet. Understanding the characteristics of thermal plasma is necessary to explain the formation mechanism of carbon nanomaterial. The anode provides the carbon species for the production of carbon nanomaterial, and the anode ablation rate determines the productivity. Therefore, further analysis of the anode region and anode phenomena is required to improve the production of carbon nanomaterial by arc discharge method. For the anode boundary, there are three typical arc-anode attachment modes, diffuse attachment, multiple attachments, and constricted attachment [6]. Heberlein *et al* investigated the formation mechanism and the characteristics of arc-anode attachment modes with a metal anode [7-9]. However, the arc-anode attachment in a carbon electrode and the effect of arc-anode attachment mode on the electrode temperature fluctuation have not been reported.

The temperature of carbon electrode is an important parameter in determining the characteristics of carbon nanomaterial. Therefore, an accurate temperature measurement is required to understand the growth of carbon nanomaterial in dynamic interaction between the electrode and arc plasma. The measurement of electrode temperature, however, is difficult with conventional methods because of the strong radiation and dynamic behavior of the arc column.

In the present work, the two-colour pyrometry combined with a high-speed camera which employed appropriate two band-pass filters was applied to measure the surface temperature of carbon electrode during the arc discharge. The objective of this work is to investigate the effect of arc-anode attachment mode on the electrode temperature fluctuation. In order to achieve the purpose, the effect of arc current on the arc behavior and the electrode temperature was investigated.

2. Experimental

2.1. Experimental setup

Figure 1 indicates the schematic illustration of an arc discharge system for the preparation of carbon nanomaterials and a high-speed camera system for temperature measurement. A graphite anode rod of 30 mm in diameter (99.99%, Toyo Tanso Co., Ltd.) with an inclined top plane was put on a water-cooling copper plate. A graphite rod of 6 mm in diameter (99.99%, Toyo Tanso Co., Ltd.) was placed at an oblique angle from the anode. The arc was generated for 5 min in helium environment at atmospheric pressure. The arc current was controlled at 100, 150, and 200 A and the electrode gap

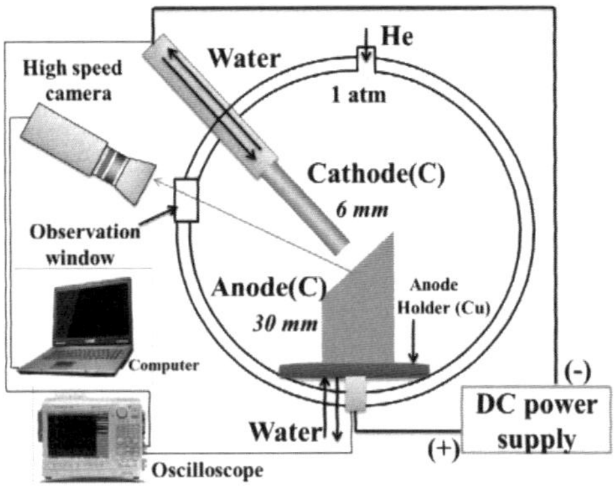

Figure 1. Schematic diagram of arc generation system.

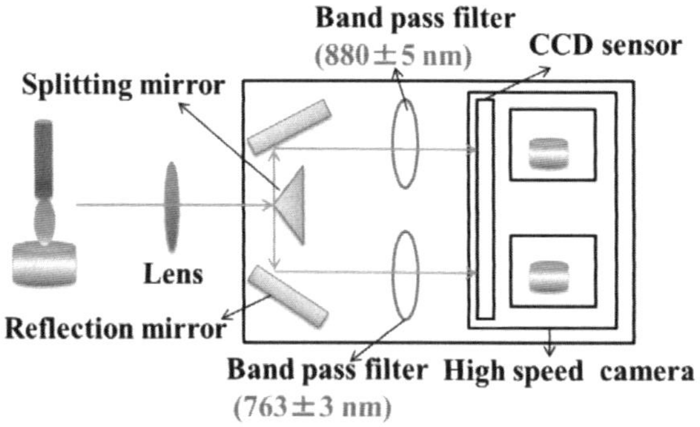

Figure 2. Schematic illustration of temperature measurement system by two-color pyrometry.

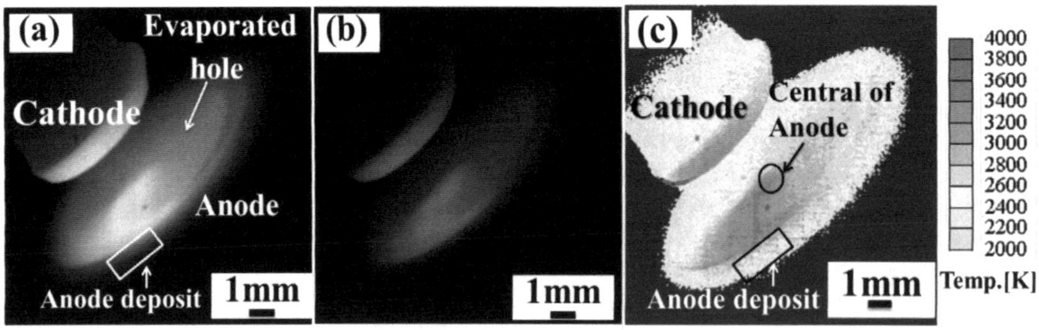

Figure 3. Images of the electrode at the wavelength of (a) 880±5 nm and (b) 763±3 nm recorded by the high-speed camera, and (c) a corresponding temperature distribution map calculated by Plank's law based on radiation intensity from images (a) and (b).

distance was fixed at 3 mm. The evaporated carbon species were deposited on the peripheral of the evaporated hole of the anode to form the anode deposit, where grows the carbon nanomaterial.

2.2. Temperature measurement
Figure 2 shows the schematic illustration of two-colour pyrometry for the temperature measurement of electrode surface during the arc discharge. The radiation emitted from the electrode surface was divided by a splitting mirror, and then it passed through two band-pass filters. In order to eliminate the influence of line emission from the arc, two band-pass filters of 763±3 and 880±5 nm were employed. According to optical emission spectrum, band-pass filters were selected with no line emission in the bandwidth of them. As shown in Figures 3 (a) and (b), images of the electrode in different wavelengths at 880±5 and 763±3 nm were simultaneously recorded by the high-speed camera (FASTCAM-SA WTI, Photron). The frame rate was fixed at 10,000 fps (frame per second) and the shutter speed was fixed at 20μs.

Plank's law was used to evaluate 2-dimensional temperature distribution of the electrode surface in two-colour pyrometry. Since the emissivity ε is unknown for many applications, the grey body assumption $\varepsilon(\lambda_1) = \varepsilon(\lambda_2)$ is used where it is hypothesized that the surface emissivity is independent of wavelength. Therefore, surface temperature is expressed as follows:

$$T = \frac{C_b(\lambda_1 - \lambda_2)}{\lambda_1 \lambda_2} \frac{1}{ln(I_1 \lambda_1^5 / I_2 \lambda_2^5)} \tag{1}$$

where, λ_1 and λ_2 are selected wavelengths (nm) of two band-pass filters; C_b equals to 0.014388 mK; and I_1 and I_2 correspond to radiation intensities ($W \cdot m^{-3} \cdot sr^{-1}$) at the wavelengths of λ_1 and λ_2, respectively. Considering bandwidths of band-pass filters and the sensitivity of high-speed camera, the two-colour pyrometry system was calibrated by a tungsten lamp which was used as a standard light source. As shown in Figure 3 (c), the temperature distribution map was obtained by the above equation with the measured intensity ratio of selected radiations. Since the sensitivity of the high-speed camera was poor at a low temperature region, only high temperature region where the temperature is over than 2,000 K was calculated in the temperature distribution map. In Figure 3, the anode deposit region locating at the peripheral area of the evaporated hole was measured to investigate the temperature fluctuation effect on the characteristics of synthesized carbon nonmaterial.

2.3. Voltage and arc behavior measurement
The voltage of arc discharge was recorded by an oscilloscope (Scope Corder DL 850, Yokogawa) at 10 MHz. For the observation of dynamic arc behavior, the same high-speed camera was used at the frame rate of 10,000 fps.

3. Results and discussion
Figure 4 shows the snapshots of high-speed camera at different arc currents of 100, 150 and 200 A. A band-pass filter which has the wavelength of 763±3 nm was used. Several small bright regions appear on the anode surface randomly without periodicity at the lowest arc current of 100 A. These bright regions represent high temperature areas on the anode surface, which correspond to the anode spots. It means that the multiple arc-anode attachment mode was formed [7], which was confirmed by high-speed camera observation at a shorter shutter speed of 0.37 μs. The multiple arc-anode attachment mode was also observed at 150 A. In contrast, a bright region with a relatively large size appears on the anode surface when the arc current reaches to 200 A. In addition, the bright region rotates on the anode surface periodically. A single arc spot is explained by the formation of diffuse attachment mode [7]. However, one bright region is separated into several small regions at 4 and 5 ms in Figure 4 like as the multiple attachment mode. Therefore, the arc-anode attachment mode is the mixture of diffuse attachment and multiple attachment modes and is dominated by diffuse attachment mode in the case of 200 A.

As shown in Figure 4, the rotation frequency of bright area is about 125 Hz when the arc current is 200 A. Although the reason of this periodical arc rotation is unclear, the Lorenz force induced by the arc itself is related to the periodical arc behavior [10]. The formation of different arc-anode attachment modes is explained by the competition of the cathode jet and the anode jet. The strong cathode jet leads to the transition of arc-anode attachment mode from diffuse attachment mode into multiple attachment mode with increasing the arc current.

In order to examine the characteristics of dynamic arc discharge behavior, the change of arc voltage

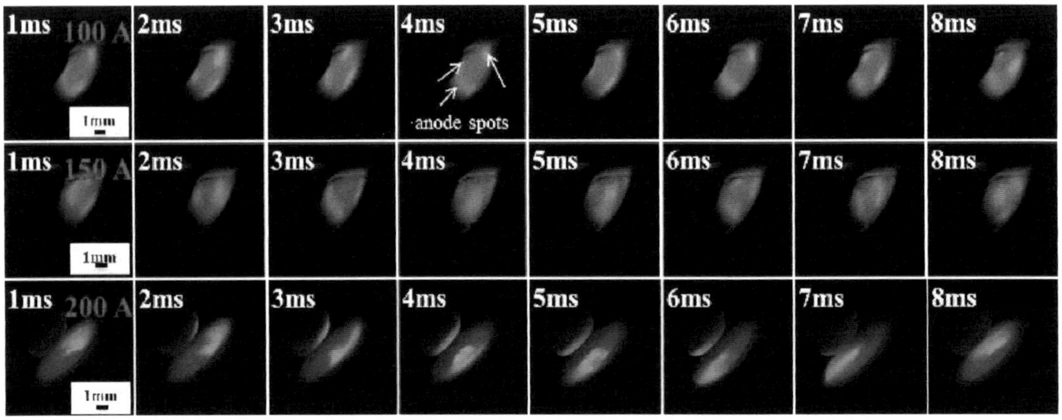

Figure 4. Representative snapshots of high-speed camera at different arc currents.

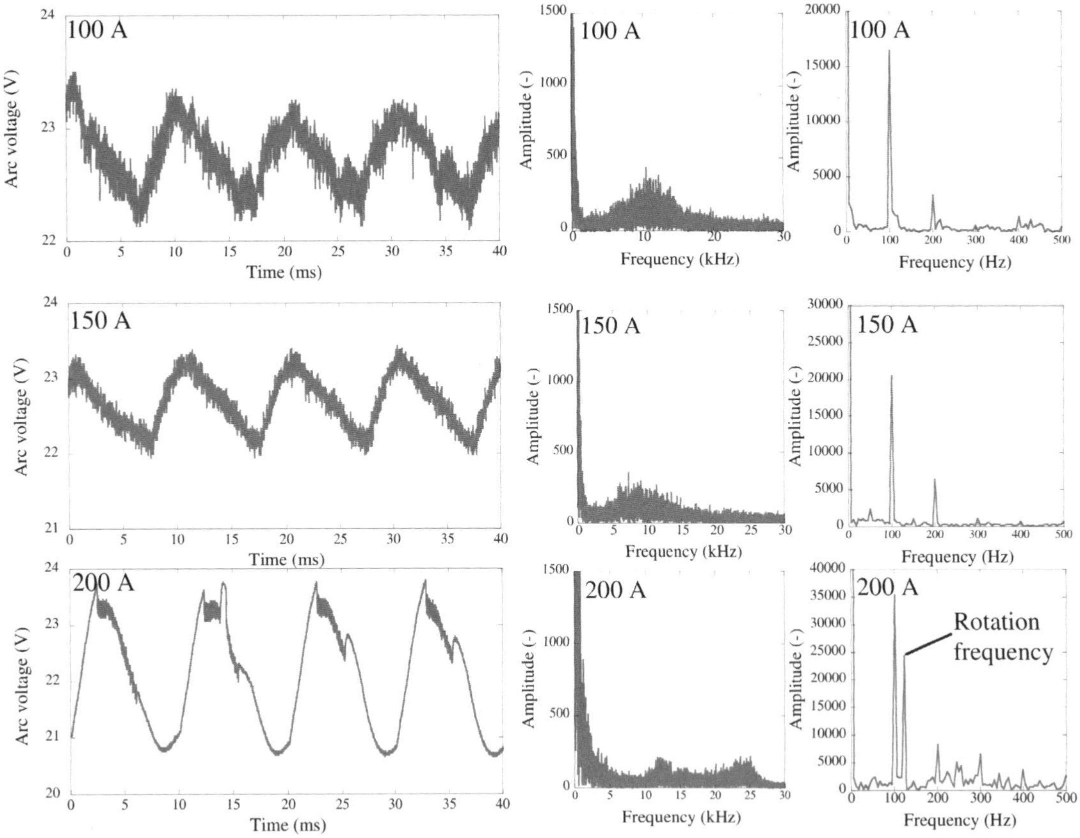

Figure 5. Voltage waveform at different arc currents: 100, 150, and 200 A. The corresponding FFT spectra of voltage waveform at different frequency ranges.

according to time was traced. Figure 5 shows the waveforms of the arc voltage and the corresponding Fast Fourier Transform (FFT) spectra at different arc currents. These voltage waveforms exhibit similar shapes derived from the current ripple of the power supply. As a result the main frequency is 100 Hz, which comes from the power supply. According to the FFT spectrum of the voltage waveform in a wide range, the arc voltage fluctuated with high frequency about 10 kHz in multiple attachment mode. In contrast, the relatively small amplitude of peak is observed in the FFT spectrum of voltage waveform with the highest arc current of 200 A due to the diffuse attachment mode. In addition to the common peak at 100 Hz and its harmonics, the peak at 125 Hz is found in the case of 200 A. This characteristic frequency corresponds to the arc rotation which was measured by the high-speed camera in Figure 4.

Figure 6 shows the snapshots of electrode temperature distribution according to time at different arc currents. The high temperature regions on anode surface correspond to the bright area in the snapshots of high-speed camera in Figure 4. In the arc current of 200 A, it was evaluated that the rotation frequency of high temperature region is about 125 Hz which is the same frequency with the arc spot rotation.

Figure 7 indicates the FFT spectra of anode temperature at different arc currents. No significant peak can be observed in the range of 0 to 500 Hz in multiple attachment mode when the arc current are 100 and 150 A. However, a dominant peak around 125 Hz is appeared in the diffuse attachment mode when the arc current reaches to 200 A. This fluctuation frequency of anode temperature comes from the rotation of arc spot.

Figure 8 shows the time variation of the temperature of anode deposit during 30 ms. The anode deposit temperature fluctuated with the frequency of 125 Hz in the diffuse attachment along with the arc rotation. Although the average temperatures of anode surface in different attachment modes were similar as 2,700 K for the multiple mode and 2,690 K for the diffuse mode, the standard deviations of

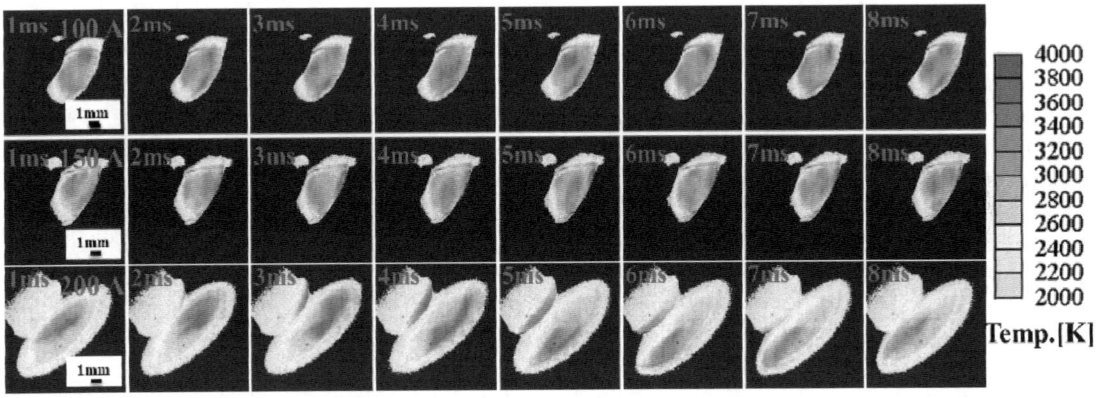

Figure 6. Snapshots of temperature distribution according to time at different arc currents.

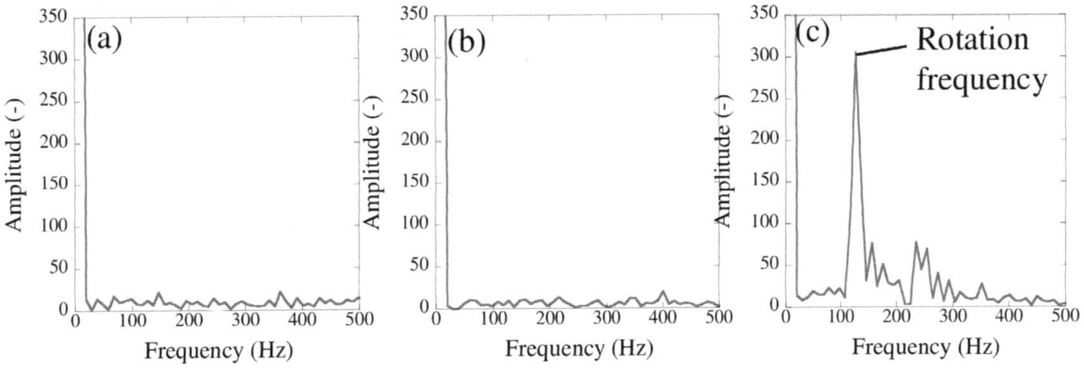

Figure 7. The FFT spectra of anode temperature at different arc currents, (a), (b), and (c) corresponding to the arc current of 100,150, and 200 A, respectively.

measured temperature were quite different as 181 K for the multiple mode and 260 K for the diffuse attachment mode, respectively. According to temperature distribution of the anode deposit, the MWNT with higher purity was obtained in multiple attachment mode than diffuse attachment mode, due to the large temperature fluctuation in diffuse attachment mode. Since different carbon nanomaterials can be obtained in different temperature ranges, the temperature with large fluctuation amplitude and small fluctuation frequency results into the low purity of MWNT on the anode deposit.

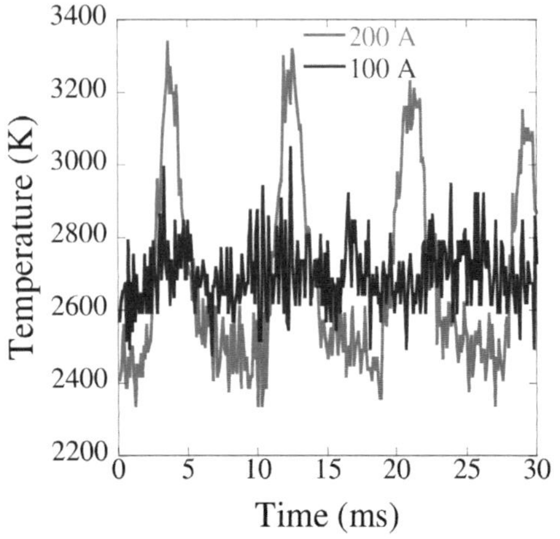

Figure 8. Temperature variation of anode deposit region with the time at different arc currents.

4. Conclusion

The multiple arc-anode attachment mode and the diffuse arc-anode attachment mode were observed by changing the arc current in the DC arc discharge with carbon electrode. The arc-anode attachment mode was transformed from the multiple mode into the diffuse mode when the arc current changed from 100 to 200 A. The FFT spectrum of voltage waveform indicates the frequency around 10 kHz can be observed in the multiple attachment mode. In contrast, the arc spot moved with a constant frequency of 125 Hz in the diffuse attachment mode. The temperature of carbon electrode was successfully measured by using two-colour pyrometry combined with a high-speed camera during the arc discharge. The anode deposit temperature fluctuated at the frequency of 125 Hz in diffuse attachment at 200 A due to the rotation of arc spot. Enhanced temperature fluctuation results in the degradation of purity for the final product from the anode deposit.

References

[1] Keidar M, Shashurin A, Li J, Volotskova O, Kundrapu M and Zhuan T 2011 *J. Phys. D: Appl. Phys.* **44** 174006
[2] Krätschmer W, Lamb L, Fostiropoulos K and Huffman D R 1990 *Nature* **347** 354
[3] Iijima S 1991 *Nature* **354** 56
[4] Keidar M, Levchenko I, Arbel T, Alexander M, Waas A M and Ostrikov K 2008 *Appl Phys Lett.* **92** 043129
[5] Subrahmanyam K S, Panchakarla L S, Govindaraj A and Rao C N R 2009 *J. Phys. Chem. C.* **113** 4257
[6] Sanders N, Etemadi, Hsu K C and Pfender E 1982 *J. Appl. Phys.* **53** 4136
[7] Heberlein J V, Mental J and Pfender E 2010 *J. Phys. D: Appl. Phys.* **43** 023001
[8] Yang G, Heberlein J V and Pfender E 2006 *J. Phys. D: Appl. Phys.* **39** 2764
[9] Yang G and Heberlein J V 2007 *Plasma Sources Sci.Technol.* **16** 529
[10] Collares M P and Pfender E 1997 *IEEE Trans. Plasma Sci*, **25** 864

Metal plating removal from insulator substrate using pulsed arc discharge

K Imasaka[1], S Gnapowski[2] and H Akiyama[2]

[1] Department of Electrical Engineering and Information Technology, Kyushu Sangyo University, 2-3-1 Matsukadai, Higashi-ku Fukuoka, 813-8503, Japan
[2] Institute of Pulsed Power Science, Kumamoto University, 2-39-1 Kurokami, Chuo-ku, Kumamoto, 860-8555, Japan

E-mail: imasaka@te.kyusan-u.ac.jp

Abstract. Removal technique of metal materials from a metal plating insulator substrate using a pulsed arc discharge was proposed and its fundamental characteristics were investigated. The metal plating substrate with three metal-layers structure (cupper, nickel and gold layers) is used as the sample substrate. Repetitive pulsed arc discharge plasma is generated using three types of electrode systems. Effects of the electrode systems on the metal plating removal from the insulator substrate were investigated. The metal plating was removed by the pulsed arc discharge between the electrode and substrate surface. A part of the gold layer, which is the topmost metal layer on the insulator substrate is vaporized and removed by the repetitive pulsed arc discharges.

1. Introduction

Environmental destruction and depletion of resources has become a problem due to the economic growth of recent years on a global scale. Therefore, the conversion to sustainable and recycling-oriented society utilizing 3R (Reduce, Reuse and Recycle) from the traditional mass disposal type society is demanded. There is recycle of printed circuit board for mounting electronic circuit components with respect to the 3R. In order to recycle the printed circuit board, it is necessary to remove and separate the metal plating materials from the insulator substrate such as the plastics.

Printed circuit boards composed of the multilayer metal plating and the insulator substrate are widely used. In general, cupper (Cu), nickel (Ni), gold (Au) and chromium (Cr) are utilized as the multilayer metal plating materials. As you know, Ni and Cr are rare metals and also Au is one of the high additional value metals. Therefore, it is important and useful to remove and separate these metal plating materials from the printed circuit boards concerning to 3R. Pulsed power technology seems to be one of promising methods for it. Early works related to the recycle of the printed circuit board using pulsed power technology showed that it was possible to separate electronic circuit components such as resistor, capacitor and soldered metallic wires from the board by its fragmentation into small pieces using the shock waves which was generated by the pulsed discharge [1-3]. However, these studies could not remove the metal plating materials from the board.

The final purpose of this study is to develop the recycle technology to remove and separate the metal plating material from the insulator substrate using the pulsed arc discharge. In this work, basic phenomena of the metal plating removal from the substrate by the proposed method were investigated.

2. Experimental setup

Schematic diagram of the metal plating insulator substrate used in this experiment is shown in figure 1. The substrate has a structure in which three metal plating layers were formed on an insulator substrate made of a fiber-glass reinforced plastic (FRP). Materials of the three metal-layers are composed of Cu, Ni and Au, respectively. This metal plating substrate is widely used for the printed circuit boards.

On the other hand, it is reported that a thin metal wire or rod electrode covered with an insulator such as a ceramic tube is effective electrode system as an electrode system for the removal of metal plating by the pulsed arc discharge [4-6]. Therefore, the similar electrode system was also used in this study. In the previous studies [4-6], however, only one type of electrode system was used and the metal plating substrate was composed of Cu, Ni and Cr in electroplating on an acrylonitrile butadiene styrene (ABS) used as the plastic substrate, which was different one from this study. In this study, three types of electrode system (Type-A, Type-B and Type-C) were used to investigate effects of the relative positional relationship between the electrode system and the surface of metal plating layers on the removal of metal plating materials. These electrode systems were shown in figure 2.

<Type-A> This electrode system was same as one used in the previous studies [4-6]. Rod electrodes made of tungsten inserted into the ceramic tubes were used as the high voltage and ground electrode, respectively. Both electrodes were set up above the metal plating substrate at the angle of 45 degrees. In this case, we defined the separation between two electrodes and the distance from the substrate as *d* and *g*, respectively.

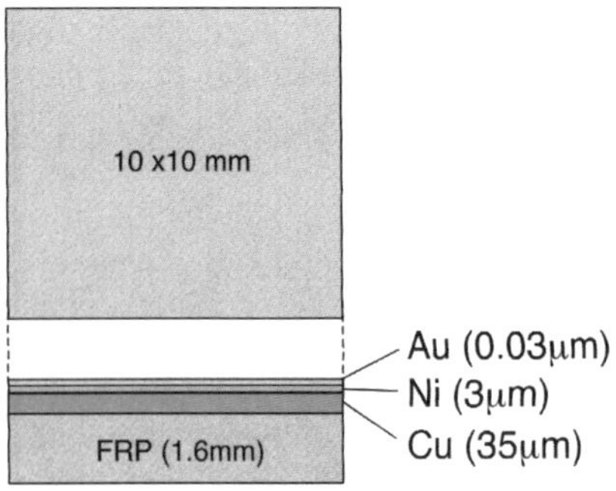

Figure 1. Metal plating substrate.

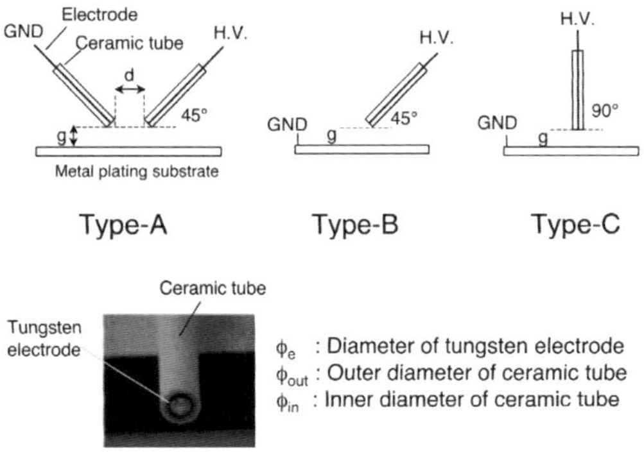

Figure 2. Configuration of electrode systems.

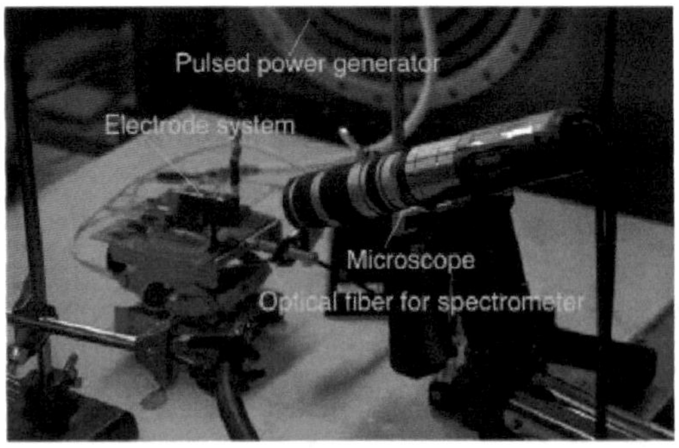

Figure 3. Experimental setup.

<Type-B> The tungsten rod electrode inserted into the ceramic tube and the metal plated substrate were used as the high voltage and ground electrode, respectively. The rod electrode was set up at the distance of g from the surface of metal plating substrate, which was used as the ground electrode. Angle of the rod and substrate was 45 degrees.

<Type-C> Type-C electrode system was similar that of Type-B, however, Angle of the rod and substrate in the Type-C was 90 degrees.

Several diameters of the tungsten wire or rod and the ceramic tube were paired and used. The diameters of tungsten, ϕ_e were 0.1, 0.3, 0.7 and 3.0 mm. Outer and inner diameters of the ceramic tube, ϕ_{out} and ϕ_{in} , respectively were ϕ_{out}/ϕ_{in} = 2.0/1.0, 2.5/1.3, 6.0/3.5 mm.

Experimental apparatus was shown in figure 3. A pulsed power generator was used to generate the pulsed arc discharge on the metal plating substrate. The specifications of the generator were maximum output voltage of 120 kV, discharge repetition rate of 40 pps and the energy of 40 J/shot. A feature of this generator is higher repetition rate of discharge in comparison with the previous studies [4-6] in which a generator has the repetition rate of 6 pps. Output voltage and current were measured with a high voltage probe (Pulse electronics co. ltd, EP-100K) and Rogowski coil (Pearson electronics, Inc, Model101). A high-speed microscope (Keyence co., VW-6000) was used to observe the surface condition of the metal plating substrate and behavior of the repetitive pulsed arc discharges. An electron probe microanalyser (Shimazu co, EPMA-1720) was employed in order to analyze the material elements of the metal plating layers.

3. Results and discussion

3.1. Effects of electrode systems on the removal of metal plating

Figure 4 shows the surface condition of the metal plating substrate after the pulsed arc discharge generated in the three types of electrode systems. Settings of the output voltage, repetition rate and the number of shots of the pulsed power generator were 100 %, 40 pps and 200 shots, respectively. Details of the typical results for each electrode system are as follows.

<Type-A> Pulsed arc discharge was generated between electrodes through the metal plating surface by applying the pulsed high voltage between electrodes. After the discharge of 200 shots, two removed regions were formed on the metal plating layer just below the high voltage and ground electrode. Size of the removed region was several mm in diameter. The removed region tended to become larger as increasing the number of shots and output voltage setting value. However, the removal regions were formed at only two positions below the each electrode and it was difficult to

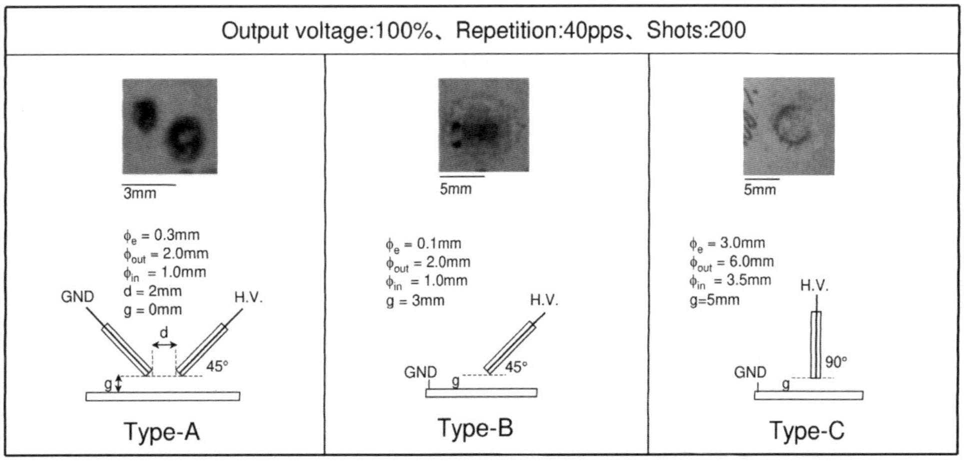

Figure 4. Metal removal using pulsed arc discharge on different type of electrode systems.

remove the metal plating from the substrate efficiently between electrodes as a whole.

<Type-B> When the repetitive pulsed high voltage was applied to the tungsten electrode, pulsed arc discharge was generated between electrodes and the metal plating was removed. The removed region was formed on the metal plating below the tungsten electrode. Size of the removed region was increased to approximately 10 mm in diameter in comparison with that of the Type-A.

<Type-C> The metal plating on the substrate below the tungsten electrode was removed by the repetitive pulsed arc discharge. Size of the removed region was approximately 10 mm in diameter, which was almost same size in case of the Type-B electrode system. However, ring-shaped arc discharge spots were formed in the removed region. It was different from the result of the Type -A and Type-B electrode systems.

Figure 5 shows a relationship between the removed area of the metal plating and the 3-types of electrode systems in which the experiments were performed by changing the combination of parameters such as ϕ_e, ϕ_{out}, ϕ_{in}, d and g. These parameters for each electrode system are shown in Table 1. The combination of ϕ_e, ϕ_{out}, and ϕ_{in}, is all the same for the three electrode systems. The d and g was changed in 3.0 mm and 2.0 mm ore less, respectively. Settings of the output voltage, repetition rate and the number of shots of the pulsed power generator were 100 %, 40 pps and 200 shots, respectively. All of the removed area generated by the pulsed arc discharge under these experimental conditions for each electrode system was summarized in figure 5. It should be noticed that figure 5 shows only the ranges of diameters without statistical viewpoint because the diameters were obtained under many different experimental conditions. The removed area was evaluated by its diameter. Diameter of the removed area formed in the Type-A electrode system was widely varied from about 0.5 to 6.0 mm. Meanwhile, Diameter of the removed area formed by using the Type-B or Type-C electrode system was increased to about 2 times compare to that of the Type-A. From figure 5, it was also found that the Type-C electrode system was more effective to the removal of the metal plating in this experiment. In particular, the characteristic ring-shaped discharge region was formed in the Type-C electrode system as shown in figure 4. Therefore, then we investigated relationship between the removed region and the discharge phenomena by using the Type-C electrode system.

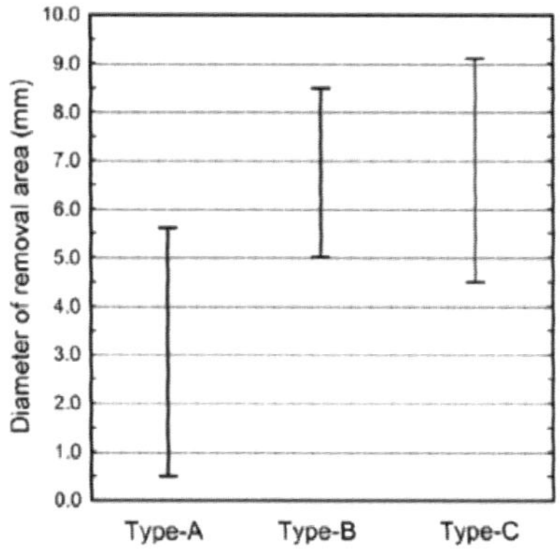

Figure 5. Dependency of diameter of removal area on 3-types of electrode systems.

Table 1. The parameters ϕ_e, ϕ_{out}, ϕ_{in}, d and g for each electrode system.

	ϕ_e (mm)	ϕ_{out} (mm)	ϕ_{in} (mm)	g (mm)	d (mm)
Type-A	0.1, 0.3	2.0	1.0	0, 1.0, 3.0	0, 1.0, 2.0
	0.7	2.5	1.3	0, 1.0, 3.0	0, 1.0, 2.0
	3.0	6.0	3.5	0, 1.0, 3.0	0, 1.0, 2.0
Type-B and Type-C	0.1, 0.3	2.0	1.0	1.0, 3.0	-
	0.7	2.5	1.3	1.0, 3.0	-
	3.0	6.0	3.5	1.0, 3.0	-

3.2. Relationship between removed area of metal plating and discharge phenomena

Figure 6 shows the metal removal region and the pulsed arc discharge phenomena by changing the output voltage setting value from 10% to 100% in case of the Type-C electrode system. Repetition rate of the discharge and the number of shots were 40 pps and 400 shots, respectively. It was found that the removed region of the metal plating by the discharge tended to become larger as increasing the voltage setting value and the ring-shaped discharge region was clearly formed at the output voltage setting value of 50%. The high-speed microscope observation revealed that the formation of the ring-shaped discharge region was due to the behavior of the repetitive pulsed arc discharge. The repetitive discharge plasma channels were produced between electrodes shot to shot and these plasma channels were moved in the shape of the ring between the tungsten rod and the inner space of ceramic tube.

Output voltage and current waveforms were also shown in figure 6. The peak output voltage and current value for 1 shot discharge at the voltage setting value of 10% was 6kV and 400A, respectively. When the voltage setting value was increased to 100%, the peak output voltage was almost same because the gap distance between electrodes was same as 5mm in both cases but the current was increased to 800A. As increasing the number of discharge shots up to 400 the peak current value was decreased to 700A. The decrease of current was thought to be due the increase of impedance between electrodes because of the removal of metal plating from the substrate.

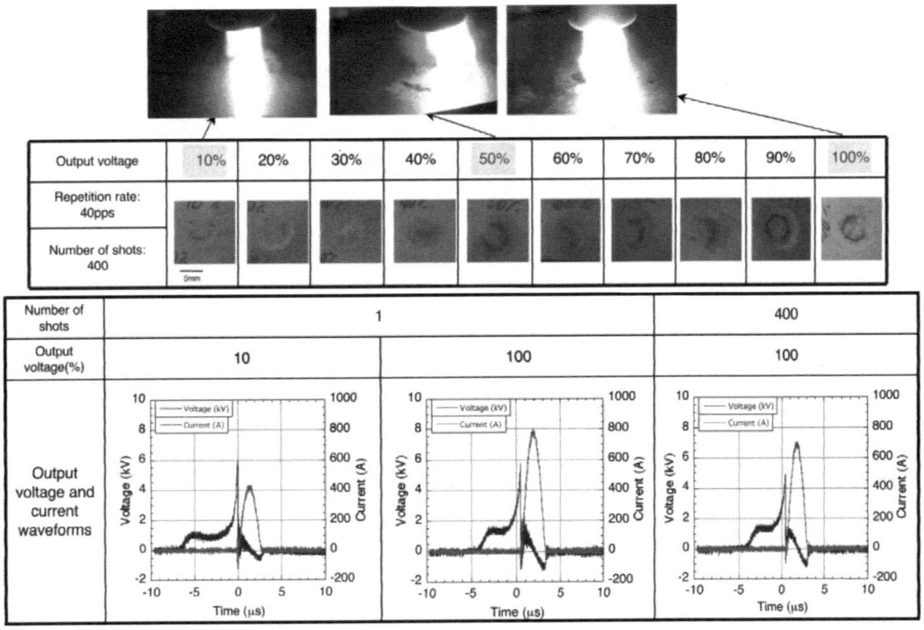

Figure 6. Effects of pulsed arc discharge on metal removable (Top: photos of pulsed arc discharge, middle: surface condition after discharge, bottom: output voltage and current waveforms).

3.3. Surface analysis of the metal-plated substrate after the discharge

Figure 7 shows the material element analysis of the metal plating layers after the repetitive pulsed arc discharge. The metal plating layers were composed of Cu, Ni and Au. Material of the topmost surface of the layers was Au and the middle layer material was Ni. It was found from figure 7 that the Au layer was removed by the repetitive pulsed arc discharge. However, the Ni layer could not be removed by the discharge. This result suggest that the Au layer was evaporated and removed by the pulsed arc discharge but the Ni and Cu layers could not be removed from the substrate. Although the reason for this result was under consideration, one of the possible reasons for it was thought that not only

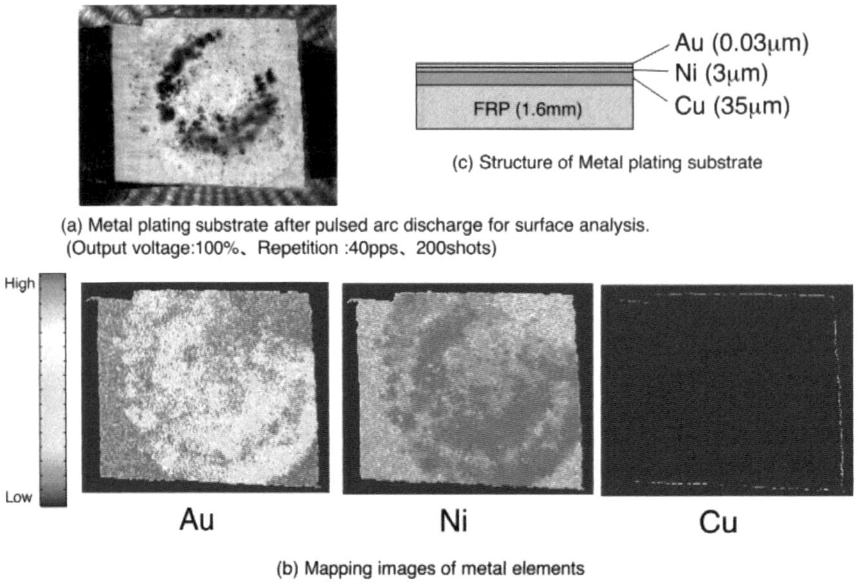

Figure 7. Metal elemental analysis of metal plating substrate after pulsed arc discharge.

experimental conditions such as the generator and electrode systems but also a chemical process of the metal plating on the FRP substrate might be influenced to the discharge phenomena.

4. Conclusions

The removal method of the metal plating layers from the insulator substrate by the repetitive pulsed arc discharge was proposed and the characteristics were basically investigated. Obtained results were summarized below.

(1) The rod electrode inserted into ceramic tube to plane electrode system (Type-C) was effective to the removal of metal plating.

(2) The metal plating was removed by the repetitive pulsed arc discharge, which formed the ring-shaped discharge region.

(3) Au layer, which was the topmost material of the metal plating layers could be evaporated and removed by the pulsed arc discharge.

In the future experiment, we will try to investigate the relationship between the discharge phenomena and the mechanism of metal plating removal in more detail by using an optical emission spectrometer and other metal plating substrates.

References

[1] Nakashi H, Hirooka T, Katsuki S and Akiyama H 2003 *IEEJ Trans. on Fundamentals and Materials* **123**, 531-36

[2] Bluhm H, Frey W, Giese H, Hoppe P, Schulthe C and Straner R 2000 *IEEE Tranc. Electrical Insulation* **7** 625-36

[3] Weise Th H G G, Hofman J and Loffler M J 1995 *Digest of Technical Papers-10th IEEE Int. Pulsed Power Conf.* **2** 1194-99

[4] Nagashima T, Akiyama H, Namihira T and Katsuki S 2007 *Digest of Technical Papers-IEEE Int.Pulsed Power Conf.* 1437-40

[5] Akiyama H, Nagashima T, Namihira T, Kato Y, Shimomura N. Katsuki S and Hisazumi T 2005 *IEEJ Trans. on Fundamentals and Materials* **125**, 1006-10

[6] Nagashima T, Akiyama H, Namihira T and Katsuki S 2009 *J. Plasma Fusion Res. SERIES* **8** 1471-76

AUTHOR INDEX